CD files for this title can now be found by entering the ISBN

9780471763581

on booksupport.wiley.com

HISTOLOGICAL ANALYSIS OF ENDOCRINE DISRUPTIVE EFFECTS IN SMALL LABORATORY FISH

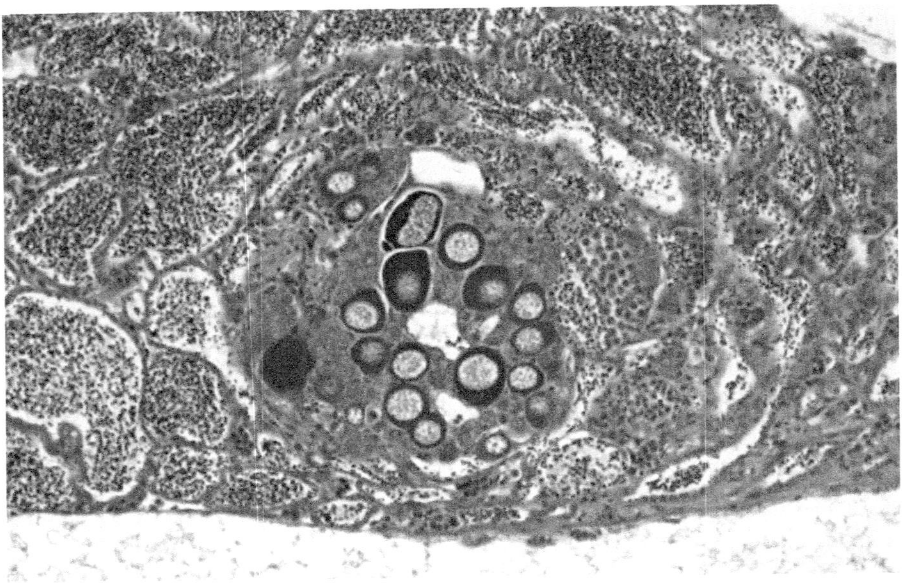

Testis–ova in the testis of an adult untreated medaka (*Oryzias latipes*). (Courtesy of John W. Fournie, U.S. EPA, Gulf Breeze, Florida.)

HISTOLOGICAL ANALYSIS OF ENDOCRINE DISRUPTIVE EFFECTS IN SMALL LABORATORY FISH

Daniel R. Dietrich
Heiko O. Krieger

WILEY
A JOHN WILEY & SONS, INC., PUBLICATION

Published by John Wiley & Sons, Inc., Hoboken, New Jersey
Published simultaneously in Canada

Library of Congress Cataloging-in-Publication Data:

Histological analysis of endocrine disruptive effects in small laboratory fish / Daniel R. Dietrich, Heiko O. Krieger.
p. ; cm.
Includes bibliographical references and index.
ISBN 978-0-471-76358-1 (cloth)
1. Endocrine disrupting chemicals. 2. Histology, Pathological. 3. Fishes–Endocrinology. 4. Fishes–Effect of chemicals on. I. Dietrich, Daniel R. II. Krieger, Heiko O., 1972-
[DNLM: 1. Endocrine Disruptors–analysis. 2. Fishes–anatomy & histology. 3. Animals, Laboratory–anatomy & histology. 4. Histological Techniques–veterinary. 5. Xenobiotics–adverse effects. SH 171 H673 2009]
RA1224.2.H57 2009
616.4′ 06–dc22

2008033273

10 9 8 7 6 5 4 3 2 1

CONTENTS

PREFACE

In view of the ever-expanding literature on xenobiotic [endocrine-disrupting compounds (EDCs)]-induced endocrine disruption in fish populations one is:

1. Easily overwhelmed and thus prone to losing focus.
2. Often unsure whether the parameters and changes reported in the literature are really important or even relevant for the task(s) defined at the outset.
3. More and more left with the impression that many of the deleterious population effects reported are not as ubiquitous as reported but rather, represent a very restricted local issue.

In view of these concerns latter and the fact that most studies employ fish gonadal histopathology in the assessment of compound-induced endocrine-disruptive effects, the European Chemical Industry Council–Long Range Initiative (CEFIC-LRI) sponsored a laboratory fish gonad histopathology workshop. This workshop was organized by the Department of Human and Environmental Toxicology, University of Konstanz, Germany, and the company IPMC-TMC GmbH in Switzerland, and was carried out in Hannover, Germany in February 2004 with the participation of the leading fish histopathologists in Europe and the United States: Helmut Segner (University of Bern, Switzerland), Leo van der Ven (RIVM, The Netherlands), Jeff Wolf (EPL Inc., Sterling, Virginia), John Fournie (U.S. Environmental Protection Agency), and Daniel Dietrich (University of Konstanz, Germany), as well as with statistical support by Nelly van der Hoeven, EcoΣtat, Leiden, Netherlands.

The workshop clearly identified several issues that appeared to hamper understanding of experimental results with EDCs in fish:

- Limited understanding of normal physiology and reproduction of the fish species used
- Variant terminology in describing effects
- Insecurities as to the relevance of the parameters determined for evaluating EDC-mediated effects
- Limited understanding of the natural variation of the parameters determined
- Flaws in the actual technical determination of parameters (variant limits of detection or quantification)

- Flaws in the experimental design (i.e., number of parameters determined, statistics used, replication and/or number of animals used per replicate, etc.)

In view of these issues and in the spirit of a successful workshop, the experts who were assembled at the Hanover workshop with Daniel Dietrich in the lead wanted to develop a book on fish gonadal histopathology that addresses just these points. Meanwhile, over 1200 publications and reports were reviewed, multiple book chapters and tables generated, and a book assembled that hopefully intuitively leads the reader through basic fish biology to fish sexual reproduction, female and male gonad development and histopathology, EDC-mediated effects, methods for histopathological processing and quantification of effects, experimental design issues, and finally, a short conclusion chapter reiterating the most important findings. The book aims to provide a thorough overview and, where possible, more in-depth information to students, researchers, and experts who seek information on the state-of-the-art of research in the field of endocrine disruption in fish.

Power point files of the illustrations that appear in this book are provided in the attached CD-ROM.

Konstanz, Germany
March 2009

DANIEL R. DIETRICH

ACKNOWLEDGMENTS

This book and all the research, reviewing, and editing work could not have been carried out without the support, understanding, patience, and love of our families. In view of this, it is especially difficult when one of our loved ones passes away prematurely. This book is therefore dedicated to Maike Wester, whose death has made us realize how precious time is when it is dedicated to research in lieu of spending it with loved ones.

We would also like to acknowledge Nelly van der Hoeven, EcoΣtat, Leiden, the Netherlands, for her support in the preparation of Chapter 9, as well as the numerous researchers who made tissue slides and artwork available. Last but not least, Karin Rieder, secretary at the Human and Environmental Toxicology Department at the University of Konstanz, and the team at John Wiley & Sons, Inc., are thanked for their patience and support in making this book happen.

We have organized the book to enlighten scientists interested in endocrine disruption and gonadal histopathology in small fish. The information presented represents the combined knowledge of the contributing authors and data from published information, and as such is part of a rapidly developing field. In this sense:

> The minute one utters a certainty, the opposite comes to mind.
> —May Sarton, *Mrs. Stevens Hears the Mermaids Singing*, 1965

CONTRIBUTING AUTHORS

Prof. Daniel Dietrich (Editor), Human and Environmental Toxicology, University of Konstanz, D-78457 Konstanz, Germany

Dr. John (Jack) Fournie, U.S. Environmental Protection Agency, National Health and Environmental Effects Research Laboratory, Gulf Ecology Division, Gulf Breeze, Florida 32561

Dr. Sylvia Gimeno, Procter & Gamble, Central Product Safety, Strombeek-Bever B-1853, Belgium

Heiko Krieger, M.Sc. (Co-Editor), Human and Environmental Toxicology, University of Konstanz, D-78457 Konstanz, Germany

Dr. Silke Rumpf, IPMC-TMC GmbH, CH-8566 Neuwilen, Switzerland

Prof. Helmut Segner, Centre for Fish and Wildlife Health, University of Bern, CH-3001 Bern, Switzerland

Dr. Leo van der Ven, Rijksinstituut voor Volksgezondheid en Milieu (RIVM), Bilthoven NL-3720BA, The Netherlands

Dr. Piet Wester, Rijksinstituut voor Volksgezondheid en Milieu (RIVM), Bilthoven NL-3720BA, The Netherlands

Dr. Jeff Wolf, EPL Inc., Sterling, Virginia 20166

CHAPTER 1

Introduction

Histopathology is a tool employed routinely in disease diagnosis for the detection and assessment of xenobiotically mediated adverse effects in laboratory animals [1–19]. Recently, histopathological analysis has become a method applied more frequently for the assessment of potential effects of endocrine-active compounds (capable of endocrine modulation, often referred to as EDCs (*endocrine-disrupting chemicals* [20]) in fish. EDCs have been defined by the WHO-IPCS [21] as follows: "An endocrine disruptor is an exogenous substance or mixture that alters function(s) of the endocrine system and consequently causes adverse health effects in an intact organism, or its progeny, or (sub)populations" and "A potential endocrine disruptor is an exogenous substance that possesses properties that might be expected to lead to endocrine disruption in an intact organism, or its progeny, or (sub)populations."

EDCs can have an effect at both the cellular and whole-organ levels and have been demonstrated to induce histopathological changes within the gonads of exposed fish within the confines of routine aquaculture procedures or in an environmental risk assessment (ecotoxicology) setting (see Figure 1.1) [20,22–26]. Chemicals with endocrine-active properties and proven potential for adverse effects are considered chemicals of concern and are subject to authorization under recent European legislation.

Concurrent with the definition of EDCs, effects of EDCs on the level of the cell or the organ could then be derived. Exposure of small laboratory fish to estrogens [e.g., 17β-estradiol (E_2) and 17α-ethinylestradiol (EE_2)] has been reported to induce gonadal changes, such as atresia of vitellogenic oocytes [27–33] and inhibition of spermatogenesis (via various modes of action) [28,33–45]. Additional effects have been reported, such as pronounced interstitial fibrosis [34,46] and induced interstitial accumulation of protein, in some cases identified as vitellogenin [47,48], or Sertoli cell hypertrophy [49]. However, systematic evaluation of the effects of hormonally active compounds, or any other xenobiotic for that matter, on the gonads of small laboratory fish based on peer-reviewed publications has been hampered to some extent by procedural restrictions, interpretational problems (i.e., no clear nomenclature of observed structures and effects), and graphical issues (figures inadequate for proper assessment of the effect), thus limiting the use and extrapolation potential of the data

Histological Analysis of Endocrine-Disruptive Effects in Small Laboratory Fish
By Daniel R. Dietrich and Heiko O. Krieger

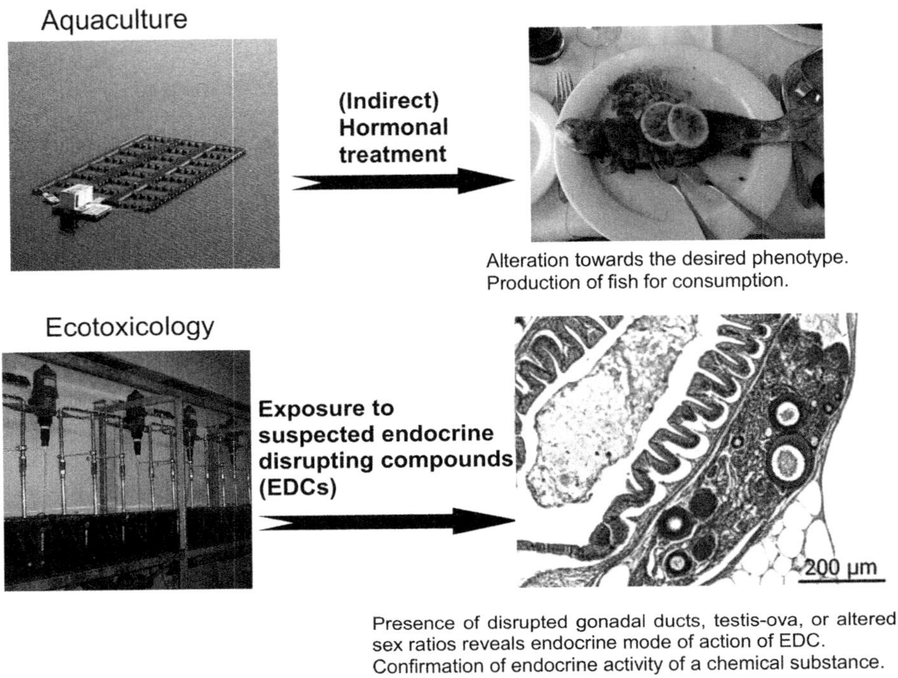

FIGURE 1.1 Control of sexual differentiation in fish: various applications. (Picture upper left courtesy of Kevin Frost, upper right courtesy of Susanne Beeck, pixelio).

provided. High stringency and standardization in the evaluation process of effects are prerequisites if the findings from different studies with endocrine-active xenobiotics in small fish are to be compared reliably [22].

Comprehensive, reproducible, and objective histopathological analysis will be greatly facilitated if the following basic standards of pathological reporting are included, while by the same token, lack of the following factors will restrict the quality and thus the value of the histopathological analysis:

1. A standardized and optimized histological gonad preparation method (fixation, embedding, sectioning, and staining techniques), which would allow clear distinction of individual gonadal cell types and interpretation of changes.
2. A standardized gonadal staging and evaluation system and a clear nomenclature system for the various cell types, which would allow a qualitative and/or quantitative assessment of the effects of endocrine-modulating compounds.
3. A standardized methodology to allow quantification of the occurrence and frequency of individual gonadal cell types for effect assessment.
4. Agreement on reporting pathological findings using representative photographs with adequate resolution and magnification and, wherever possible, the inclusion of a size bar.

In the Appendix we provide standardized protocols and procedures as well as guidance for possible quantification and reporting approaches for points 1 to 4 above, and in the main body of the book we focus on an understanding of the processes visualized via histopathological changes and the interpretation of these findings. Indeed, a thorough assessment of histopathological tissue changes can provide key diagnostic information and will form the basis for an understanding of the underlying mechanism(s) of endocrine active substances [50,51]. A more detailed morphological and mechanistic understanding of effects observed will allow enhanced reproducibility and comparability of the studies conducted. However, the description of chemically induced changes in gonadal histology can be of profound value only if these changes can be put into context with and interpreted in light of the normal reproductive biology of the particular fish. Consequently, a prerequisite for the histopathological analysis of gonadal changes in fish is an understanding of the reproductive biology and normal gonadal histology of the fish such that an induced (e.g., chemical-mediated) effect can be distinguished from naturally occurring gonadal changes in fish during development, spawning, gonadal regression or recrudescence, or during aging. The distinction of such effects will then also allow better prediction of the ramifications (type and magnitude of consequences) of the chemically mediated gonadal effects for fish reproductive capacity and thus probably also for the fish population. It must be noted, however, that beyond the obvious—the complete lack of gonadal development or the destruction of the reproductive tissue and thus reproductive capacity *in toto*—no agreement has been found to date as to what small or moderate gonadal changes, whether the occurrence of intersex (testis–ova/ovo–testis) or outright pathological changes, in an individual's gonad will translate into at the level of the population.

Our goal in this book is to provide a detailed understanding of the normal gonad physiology, development, anatomy, and histopathology as well as of changes induced by xenobiotics (e.g., endocrine-active substances) primarily in three small fish species: fathead minnow (*Pimephales promelas*), medaka (*Oryzias latipes*), and zebrafish (*Danio rerio*). These three species are employed predominantly in routine experiments used for environmental risk assessment, including hazard and consequently, risk assessment of chemicals (e.g., endocrine-active compounds) [52]. However, wherever possible and considered of additional value for understanding differences or exceptions to the general concept developed with the three species chosen, references to other species, notably carp, trout, bream, stickleback, guppy, platyfish, eelpout, and sheepshead minnow are included.

REFERENCES

1. Wester, P.W., J.H. Canton, and A. Bisschop, Histopathological study of *Poecilia reticulata* (guppy) after long-term β-hexachlorocyclohexane exposure. *Aquatic Toxicology*, 1985, **6**(4):271–296.
2. Wester, P.W. and J.H. Canton, Histopathological study of *Oryzias latipes* (medaka) after long-term β-hexachlorocyclohexane exposure. *Aquatic Toxicology*, 1986, **9** (1):21–45.

3. Wester, P.W. and J.H. Canton, Histopathological study of *Poecilia reticulata* (guppy) after long-term exposure to bis(tri-*n*-butyltin)oxide (TBTO) and di-*n*-butyltindichloride (DBTC). *Aquatic Toxicology*, 1987, **10**(2–3):143–165.
4. Wester, P.W., Histopathological study of *Poecilia reticulata* (guppy) and *Oryzias latipes* (medaka) following methyl bromide and sodium bromide exposure, in *Pathology*. University of Utrecht, Utrecht, The Netherlands, 1988, p. 208.
5. Wester, P.W., et al., The toxicity of bis(tri-*n*-butyltin)oxide (TBTO) and di-*n*-butyltin-dichloride (DBTC) in the small fish species *Oryzias latipes* (medaka) and *Poecilia reticulata* (guppy). *Aquatic Toxicology*, 1990, **16**(1):53–72.
6. Wester, P.W., J.H. Canton, and J.A.M.A. Dormans, Pathological effects in freshwater fish *Poecilia reticulata* (guppy) and *Oryzias latipes* (medaka) following methyl bromide and sodium bromide exposure. *Aquatic Toxicology*, 1988, **12**(4):323–344.
7. Wester, P.W. and H.H. Canton, Histopathological effects in *Poecilia reticulata* (guppy) exposed to methyl mercury chloride. *Toxicologic Pathology*, 1992, **20**(1):81–92.
8. Wester, P.W. and J.G. Vos, Toxicological pathology in laboratory fish: An evaluation with two species and various environmental contaminants. *Ecotoxicology*, 1994, **2**(1):21–44.
9. Wester, P.W., Histopathological effects of environmental pollutants β-HCH and methyl mercury on reproductive organs in freshwater fish. *Comparative Biochemistry and Physiology, Part C: Toxicology and Pharmacology*, 1991, **100C**(1–2):237–239.
10. Wester, P.W. and J.H. Canton, The usefulness of histopathology in aquatic toxicity studies. *Comparative Biochemistry and Physiology, Part C: Toxicology and Pharmacology*, 1991, **100**(1–2):115–117.
11. Fenske, M. and H. Segner, Aromatase modulation alters gonadal differentiation in developing zebrafish (*Danio rerio*). *Aquatic Toxicology*, 2004, **67**(2):105–126.
12. Hoeger, B., et al., Sex and low-level sampling stress modify the impacts of sewage effluent on the rainbow trout (*Oncorhynchus mykiss*) immune system. *Aquatic Toxicology*, 2005, **73**(1):79–90.
13. Hoeger, B., et al., Water-borne diclofenac affects kidney and gill integrity and selected immune parameters in brown trout (*Salmo trutta* f. *fario*). *Aquatic Toxicology*, 2005, **75**(1):53–64.
14. Hoeger, B., et al., Effects of treated sewage effluent on immune function in rainbow trout (*Oncorhynchus mykiss*). *Aquatic Toxicology*, 2004, **70**(4):345.
15. Schwaiger, J., et al., Toxic effects of the non-steroidal anti-inflammatory drug diclofenac: Part I. Histopathological alterations and bioaccumulation in rainbow trout. *Aquatic Toxicology*, 2004, **68**(2):141.
16. Schwaiger, J., et al., How estrogenic is nonylphenol? A transgenerational study using rainbow trout (*Oncorhynchus mykiss*) as a test organism. *Aquatic Toxicology*, 2002, **59**(3–4):177–189.
17. Schwaiger, J., et al., Chronic toxicity of nonylphenol and ethinylestradiol: Haematological and histopathological effects in juvenile common carp (*Cyprinus carpio*). *Aquatic Toxicology*, 2000, **51**(1):69.
18. Triebskorn, R., et al., Toxic effects of the non-steroidal anti-inflammatory drug diclofenac: Part II. Cytological effects in liver, kidney, gills and intestine of rainbow trout (*Oncorhynchus mykiss*). *Aquatic Toxicology*, 2004, **68**(2):151.
19. van der Ven, K., et al., Development and application of a brain-specific cDNA microarray for effect evaluation of neuro-active pharmaceuticals in zebrafish (*Danio rerio*).

Comparative Biochemistry and Physiology, Part B: Biochemistry and Molecular Biology, 2005, **141**(4):408.

20. Vos, J.G., et al., Health effects of endocrine-disrupting chemicals on wildlife, with special reference to the European situation. *Critical Reviews in Toxicology*, 2000, **30**(1): 71–133.
21. Damstra, T., et al., *Global Assessment of the State-of-the-Science of Endocrine Disruptors.* International Programme on Chemical Safety. WHO-IPCS, Geneva, Switzerland, 2002, p. 180.
22. Brown, R.P., et al., A critical review of the scientific literature on potential endocrine-mediated effects in fish and wildlife. *Ecotoxicology and Environmental Safety*, 2001, **49**(1):17–25.
23. McLachlan, J.A., Environmental signaling: What embryos and evolution teach us about endocrine disrupting chemicals. *Endocrine Reviews*, 2001, **22**:319–341.
24. Jobling, S., et al., Altered sexual maturation and gamete production in wild roach (*Rutilus rutilus*) living in rivers that receive treated sewage effluents. *Biology of Reproduction*, 2002, **66**:272–281.
25. Jobling, S., et al., Widespread sexual disruption in wild fish. *Environmental Science and Technology*, 1998, **32**:2498–2506.
26. van Aerle, R., et al., Sexual disruption in a second species of wild cyprinid fish (the gudgeon. *Gobio gobio)* in United Kingdom freshwaters. *Environmental Toxicology and Chemistry*, 2001, **20**(12):2841–2847.
27. Nagahama, Y., The functional morphology of teleost gonads, in *Fish Physiology*, W.S. Hoar, D.J. Randall, and E.M. Donaldson, Editors. Academic Press, San Diego, CA, 1983, pp. 223–275.
28. Papoulias, D.M., D.B. Noltie, and D.E. Tillitt, An in vivo model fish system to test chemical effects on sexual differentiation and development: Exposure to ethinyl estradiol. *Aquatic Toxicology*, 1999, **48**(1):37–50.
29. Gray, M.A., K.L. Teather, and C.D. Metcalfe, Reproductive success and behaviour of Japanese medaka (*Oryzias latipes*) exposed to 4-*tert*-octylphenol. *Environmental Toxicology and Chemistry*, 1999, **18**(11):2587–2594.
30. Olsson, P.-E., et al., Effects of maternal exposure to estrogen and PCB on different life stages of zebrafish (*Danio rerio*). *Ambio*, 1999, **28**(1):100–106.
31. Ankley, G.T., et al., Description and evaluation of a short-term reproduction test with the fathead minnow (*Pimephales promelas*). *Environmental Toxicology and Chemistry*, 2001, **20**(6):1276–1290.
32. Miles-Richardson, S.R., et al., Effects of waterborne exposure to 4-nonylphenol and nonylphenol ethoxylate on secondary sex characteristics and gonads of fathead minnows (*Pimephales promelas*). *Environmental Research, Section A*, 1999, **80**(2):S122–S137.
33. van den Belt, K., R.Verheyen, and H. Witters, Reproductive effects of ethynylestradiol and 4*t*-octylphenol on the zebrafish (*Danio rerio*). *Archives of Environmental Contamination and Toxicology*, 2001, **41**(4):458–467.
34. Kang, I.J., et al., Effect of 17β-estradiol on the reproduction of Japanese medaka (*Oryzias latipes*). *Chemosphere*, 2002, **47**(1):71–80.
35. Seki, M., et al., Effect of ethinylestradiol on the reproduction and induction of vitellogenin and testis–ova in medaka (*Oryzias latipes*). *Environmental Toxicology and Chemistry*, 2002, **21**(8):1692–1698.

36. van den Belt, K., et al., Effects of ethynylestradiol on the reproductive physiology in zebrafish (*Danio rerio*): Time dependency and reversibility. *Environmental Toxicology and Chemistry*, 2002, **21**(4):767–775.

37. Zillioux, E.J., et al., The sheepshead minnow as an in vivo model for endocrine disruption in marine teleosts: A partial life-cycle test with 17α-ethynylestradiol. *Environmental Toxicology and Chemistry*, 2001, **20**(9):1968–1978.

38. Laenge, R., et al., Effects of the synthetic estrogen 17α-ethynylestradiol on the life-cycle of the fathead minnow (*Pimephales promelas*). *Environmental Toxicology and Chemistry*, 2001, **20**(6):1216–1227.

39. Miles-Richardson, S.R., *Effects of Estrogenic Compounds on the Fathead Minnow (Pimephales promelas)*. PhD thesis, Michigan State University, East Lansing, MI, 1997.

40. Nimrod, A.C. and W.H. Benson, Reproduction and development of Japanese medaka following an early life stage exposure to estrogens. *Aquatic Toxicology*, 1998, **44**(1–2): 141–156.

41. Palace, V.P., et al., Induction of vitellogenin and histologic effects in wild fathead minnows from a lake experimentally treated with the synthetic estrogen, ethynylestradiol. *Water Quality Research Journal of Canada*, 2002, **37**(3):637–650.

42. Panter, G.H., R.S. Thompson, and J.P. Sumpter, Adverse reproductive effects in male fathead minnows (*Pimephales promelas*) exposed to environmental relevant con centrations of the natural oestrogens, oestradiol and oestrone. *Aquatic Toxicology*, 1998, **42**:243–253.

43. van Aerle, R., T.J. Runnalls, and C.R. Tyler, Gonadal sex differentiation in fathead minnow. *Pimephales promelas*. *Annual Conference of SETAC Europe*, Vienna, Austria, 2002.

44. Kinnberg, K., B. Korsgaard, and P. Bjerregaard, Concentration-dependent effects of nonylphenol on testis structure in adult platyfish *Xiphophorus maculatus*. *Marine Enviromental Research*, 2000, **50**(1–5):169–173.

45. Christiansen, T., B. Korsgaard, and A. Jespersen, Effects of nonylphenol and 17β–estradiol on vitellogenin synthesis, testicular structure and cytology in male eelpout (*Zoarces viviparus*). *Journal of Experimental Biology*, 1998, **201**(2):179–192.

46. Gray, M.A., A.J. Niimi, and C.D. Metcalfe, Factors affecting the development of testis–ova in medaka, *Oryzias latipes*, exposed to octylphenol. *Environmental Toxicology and Chemistry*, 1999, **18**(8):1835–1842.

47. Wolf, J.C., et al., Qualitative and quantitative histomorphologic assessment of fathead minnow *Pimephales promelas* gonads as an endpoint for evaluating endocrine-active compounds: a pilot methodology study. *Toxicologic, Pathology*, 2004, **32**(5):600–612.

48. van der Ven, L.T.M. and P.W. Wester. http://www.rivm.nl/fishtoxpat, 2003.

49. Kinnberg, K., et al., Effects of nonylphenol and 17β-estradiol on vitellogenin synthesis and testis morphology in male platyfish *Xiphophorus maculatus*. *Journal of Experimental Biology*, 2000, **203**(2):171–181.

50. van der Ven, L.T.M., P.W. Wester, and J.G. Vos, Histopathology as a tool for the evaluation of endocrine disruption in zebrafish (*Danio rerio*). *Environmental Toxicology and Chemistry*, 2003, **22**(4):908–913.

51. Wester, P.W., et al., Aquatic toxicology: Opportunities for enhancement through histopathology. *Environmental Toxicology and Pharmacology*, 2002, **11**(3–4):289–295.

52. Ankley, G.T. and R.D. Johnson, Small fish models for identifying and assessing the effects of endocrine-disrupting chemicals. *ILAR Journal*, 2004, **45**(4):469–483.

CHAPTER 2

Fish Species of Interest

The three species described below are OECD-recommended fish species for regulatory purposes (e.g., acute, chronic toxicity and embryotoxicity; see www.oecd.org for guidelines for ecotoxicity testing) and are considered to be suitable test species for multigenerational studies and endocrine-related research, including genomic endpoints (www.zfin.org, zebrafish genomics). Their small size at maturity reduces maintenance costs and makes them relatively easy to maintain and breed [1]. As a result of their importance in regulatory procedures for environmental risk assessment of xenobiotics, they are raised commercially and thus are readily available. The time to sexual maturity is relatively short, thus reducing the time needed for multigenerational tests. In addition, these species have long been used extensively in toxicity testing, thus contributing to a broad basis in genetics (www.zfin.org), physiology, development, reproduction, and histopathology (http://www.rivm.nl/fishtoxpat; http://aquaticpath.umd.edu/fhm/; http://wedcdb.nies.go.jp/SHf/index.html). All three species are asynchronous spawners; that is, spawning takes place repeatedly over an extended period. Further advantages of these fish species are the relatively large brood size (allowing statistical detection of effects, e.g., changes in fecundity, increased incidences of intersex, or changes in sex ratio), possible histological examination of the entire fish body, and their relative resistance to diseases. However, it must be emphasized that although all three small fish species discussed here—zebrafish, fathead minnow, and medaka—can be bred continuously in the laboratory, the question remains as to whether or not the findings obtained with these fish species can readily be extrapolated to the wild (i.e., environmental systems) [2]. Indeed, the zebrafish is a nonseasonal spawner, whereas wild medaka and fathead minnows are seasonal spawners. Seasonality of reproduction, whether induced by dry seasons (Asia) or temperature (North America), has a major influence on the reproductive status (cell-type presence, abundance, and distribution) of wild fish gonads. Consequently, wild fish may be affected by exogenous xenobiotics in a more diverse if not different fashion than

Histological Analysis of Endocrine-Disruptive Effects in Small Laboratory Fish
By Daniel R. Dietrich and Heiko O. Krieger

the gonads of their laboratory counterparts [3–8]. Furthermore, laboratory-bred and laboratory-maintained populations may tend to form inbred strains and thus may display different or even unique effect patterns, as experienced previously with trout [9], when exposed to xenobiotics or their mixtures [10]. Consequently, these factors, discussed in more detail later in the book, need to be considered carefully when interpreting mechanistic or endpoint data from laboratory studies or when extrapolating endpoints and xenobiotic effect concentrations from laboratory studies to the wild (i.e., environmental systems).

2.1 FATHEAD MINNOW (*Pimephales promelas*)

Fathead minnows (Figure 2.1) are North American representatives of the ecologically important and ubiquitous Cyprinidae family, ranging throughout the

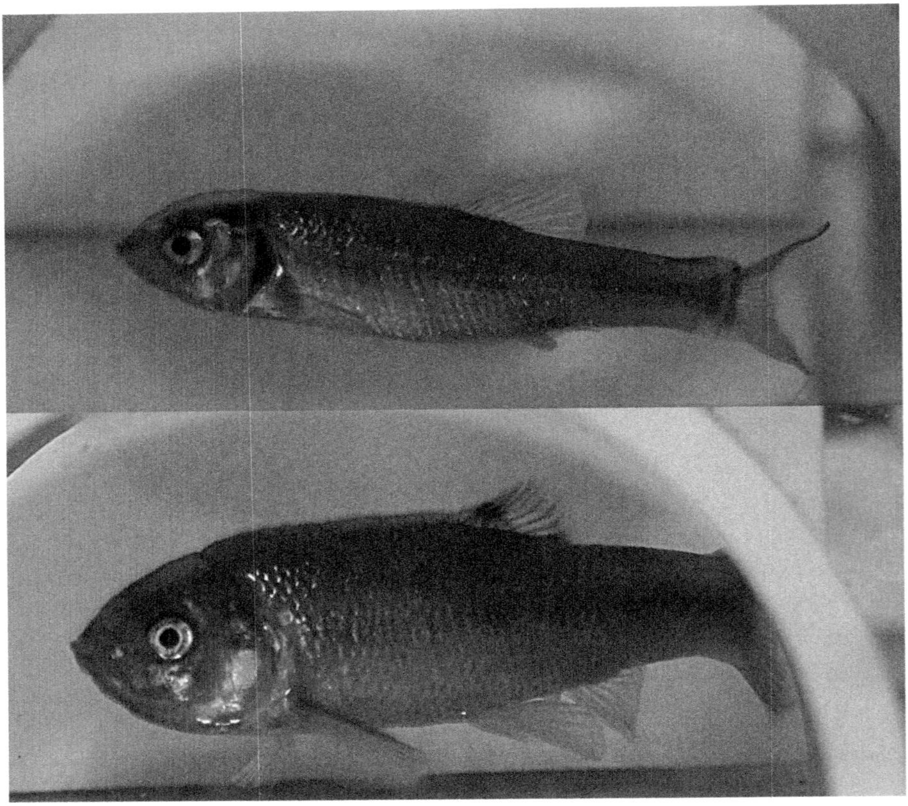

FIGURE 2.1 Fathead minnow, *Pimephales promelas*. Upper, female; lower, male. (From [11], with permission.) (See insert for color representation.)

Midwest and upper Mississippi River drainage, west to Utah, north to Canada, and east to Maine. Based on their ubiquitous occurrence, they have been used extensively in acute chronic toxicity and fish full-life-cycle tests in support of regulatory programs in both North America and Europe [12]. Indeed, tolerance of adverse conditions and ease of spawning make the fathead minnow ideal for laboratory culture. Brood stock can be maintained in spawning condition year-round, ensuring a constant supply of larval fish for toxicity-testing purposes.

In the natural habitat, spawning occurs from May through August in calm shallows of streams and along shorelines in ponds. Males select and prepare the nest site, defending it from intruders and only allowing persistent females to enter and spawn. Nests are typically located on the underside of submerged stones or branches.

Adult minnows have a moderately compressed body, a short, blunt snout, and a slightly subterminal mouth. Colors are olive to brown on the upper body and silvery white on the lower body, with a dark midlateral stripe. Adults range from 3.5 to 7.5 cm in total length and display strong sexual dimorphism [13]. Nuptial males tend to be larger than females with tubercles on the snout and a prominent pad of spongy rugose tissue on the nape [14,15]. The spongy nape pad, used for scrubbing the nest site, has taste buds capable of chemosensory assessment of egg condition and is believed to produce substances with fungicidal properties [14]. In females, the prominence of the ovipositor is an obvious secondary sex characteristic [8]. Female fathead minnows spawn fractionally every 2 to 16 days and each female can produce up to 10,000 eggs in a three-month breeding season [14] or clutches of 50 to 100 eggs every 3 to 5 days under favorable conditions. Fertilization rates of 80 to 100% are typical for fathead minnows [15].

The fathead minnow reaches sexual maturity within four to five months post-hatch and is a fractional spawner (i.e., release of eggs occurs at intervals, usually over several days or weeks). This allows more smaller and immature eggs to be carried in a limited abdominal cavity space, as the intervals enable the smaller eggs to mature; and once shed, eggs mature at different times, which may help avoid complete loss of a season's spawning to predators. Consequently, gametogenesis in both sexes occurs asynchronously (meaning different gametogenic maturation stages are present simultaneously [16]). In rare cases the occurrence of spontaneous hermaphrodism/intersex in fathead minnows is reported. The term *hermaphrodism* is erroneously used indiscriminately and synonymously with *ovo–testis*, *testis–ova*, and *intersex* in much of the literature, and it describes the simultaneous occurrence of male and female germ cells in the gonad of a fish. However, for improved understanding in this book and in keeping with the most recent literature, a distinction is made between naturally occurring juvenile and adult hermaphrodism (e.g., a fish with primary simultaneous or subsequent functional bisexuality) and the spontaneous or xenobiotic-induced intersex condition, in which clearly sexually differentiated gonochoristic fish (e.g., trout, fathead minnow, medaka) present with ovo–testis/testis–ova features (Figure 2.2) in the respective gonad.

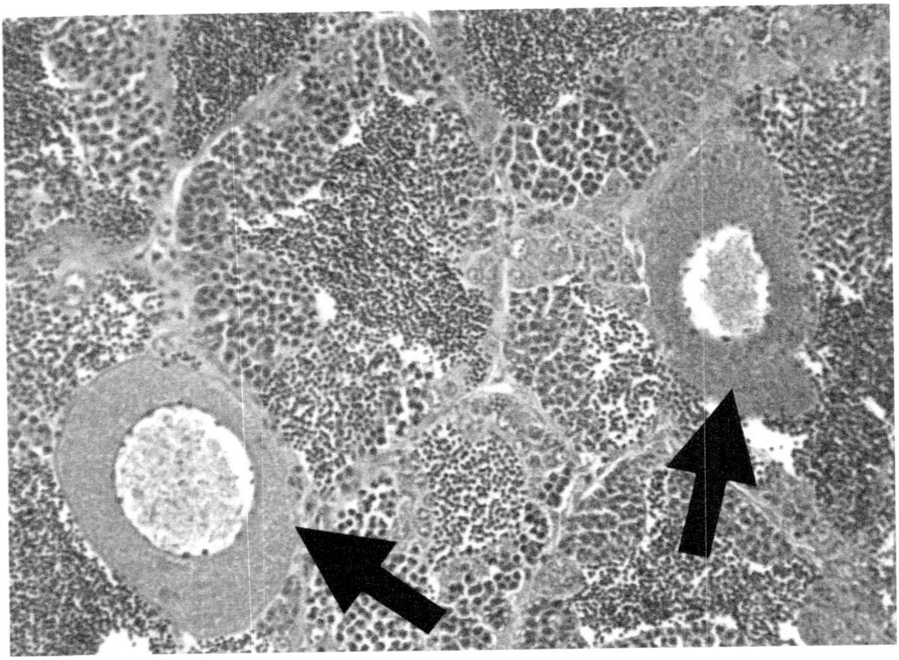

FIGURE 2.2 Testis of a healthy untreated male adult fathead minnow of male external appearance (dark color, nuptial tubercles, and dorsal fat pad; no ovipositor) containing oocytes (arrows). (From [17], with permission.)

2.2 MEDAKA (*Oryzias latipes*)

The Japanese medaka (Figure 2.3), *Oryzias latipes* [family Oryziatidae (Oryziidae), formerly regarded as a member of the family Cyprinodontidae], the brown wild type [19], inhabits fresh and brackish waters from India throughout southeast Asia, Timor, Sulawesi, Luzon, and Japan [20]. In Japan, the breeding season of the medaka extends from mid-April to late September [21]. In the outdoor culture, the period for attainment of sexual maturity depends on the season in which fry are raised. Fry that hatch in May and early June may reach maturity in August of the same year. However, fry that hatch in August may reach maturity in the following spring. Spawning is daily and year-round under optimal artificial conditions, even during the winter season; however, maintenance of an optimal photoperiod and temperature regime is key to successful continuous breeding and "normal" gonadal histopathology [22]. Indeed, a long photoperiod is essential for triggering the hypothalamic secretion of luteinizing hormone–releasing hormone (LH-RH) to stimulate the activity of the pituitary to produce gonadotropic hormone and consequently, promote ovarian development (i.e., recrudescence following sexual regression) [23].

The Japanese medaka has a relatively well-characterized process of gonadal differentiation and development. They are oviparous aquarium fish and have been

FIGURE 2.3 Medaka (*Oryzias latipes*) d-rR strain. Upper, female; lower, male. (From [18], with permission.) (See insert for color representation.)

used as an attractive model system to study sexual differentiation [24–28]. Adult medaka are small (2.5 to 5 cm in length) and sexually dimorphic; however, when determining the sex of a medaka, the use of a low-power microscope is recommended [29]. The dorsal fin of male fish has a cleft, which is absent in females (for an outline of anal and dorsal fins, see Figure 2.4).

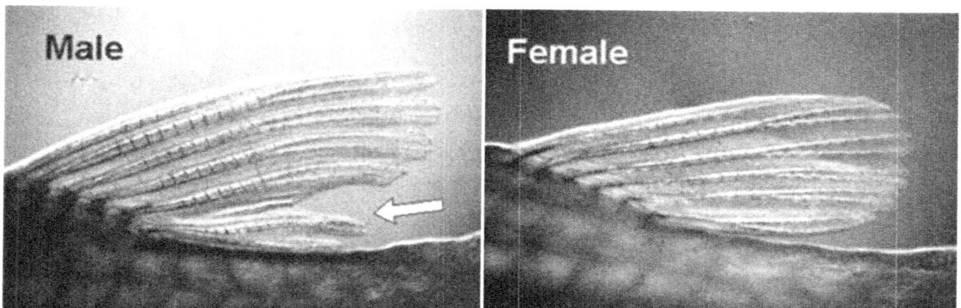

FIGURE 2.4 Dorsal fins of male and female medaka at 100 days post-hatch. A notch of a dorsal fin (white arrow), which is not observed in females, indicates the male phenotype. (From [30], with permission.)

The medaka is considered an ideal model organism for studying reproductive toxicity, due to its short life cycle (maturation within 6 to 8 weeks post-hatch) and the fact that females can spawn 10 to 40 eggs daily under an optimal photoperiod, temperature, and food regime [31]. The generation time is short, two to three months, comparable to that of zebrafish. As with zebrafish, the timing of spawning can be controlled by regulating light conditions during a 24-hour period. Currently, more than 10 inbred strains have been established [32], and about 100 natural mutant strains are maintained [33]. In this context it is important to note that the use of specific strains (outbred or inbred) can be crucial with regard to interpretation of the histopathological effects of endocrine-active compounds [e.g., the incidences of intersex (ovo–testis/testis–ova)] [31,34]. Medaka are also asynchronous spawners, meaning that at least two different size groups of oocytes are present in a female at the same time, the larger group usually being more homogeneous (representing the eggs for the next spawning season) [15]. As in many other fish, the sex of medaka is XX/XY-coded and can be reversed *phenotypically* if androgens or estrogens are applied during a sensitive period of embryonic or juvenile development [35]. Gonadal differentiation and morphology have been well charâcterized [36,37] and thus allow distinction of xenobiotic-induced histopathological effects [38] from a range of histopathological changes normally occurring in untreated fish. Medaka are considered sensitive to endocrine-modulating compounds and are well known for displaying intersex in gonadal tissue when exposed to such compounds [31,39]. However, there is a paucity of information regarding the "natural" incidence of intersex, a topic of more recent debate. Indeed, background incidences of ovo–testis may be strain-specific (i.e., many Japanese and some U.S. groups report 0% incidence of ovo–testis in control fish) [29,39], whereas high incidences (intersex control fish in 15 of 41 toxicological studies) have been observed from medaka strains employed in other American laboratories [40].

2.3 ZEBRAFISH (*Danio rerio*)

The Indian and southeast Asian (Burma) zebrafish is an oviparous fish (family: Cyprinidae) and is considered an attractive study model in a wide range of scientific areas, such as basic reproduction biology, genomics [42,43], and toxicology [44]. Adult zebrafish can reach a length of up to 4 to 5 cm and exhibit only subtle secondary sex characteristics (i.e., the female having a more pronounced belly and the male having slightly stronger yellowish coloration). The male zebrafish has a larger anal fin than that of females, and females typically have a rounded, fuller body contour (Figure 2.5). The secondary sexual characteristics in zebrafish are typically not used as a quantifiable endpoint. Under optimal conditions, spawning can be induced every 3 to 4 days (van der Ven, personal communication), and a female can produce an average of 50 to 400 eggs [15,42]. Zebrafish are free spawners, and they prefer fine-leaved plants as a spawning substrate. The zebrafish has a relatively short generation time, reaching reproductive maturity three to four months after fertilization, making it a suitable test species for partial- and full-life-cycle studies and the evaluation of xenobiotic exposure and assessment of endocrine-modulating effects. In addition, the zebrafish

FIGURE 2.5 Female (bottom) and male (top) zebrafish, *Danio rerio*. (From [41], with permission.) (See insert for color representation.)

is recommended as a test species in a number of existing test protocols (e.g., various OECD and ISO guidelines) [15]. Currently, 20 wild-type and about 57 mutant strains have been entered in the zebrafish database (www.zfin.org). Spontaneous intersex is observed at low incidences in zebrafish, although as with fathead minnows and medaka, the influence of the strain on the incidence of intersex is at present unknown. However, in contrast to the fathead minnow and medaka, there is currently an ongoing discussion as to whether or not zebrafish undergo a transitional period in development referred to as *juvenile hermaphrodism* (see Chapter 3). If present, strain-related juvenile hermaphrodism has the potential to contribute to low incidences of spontaneous intersex, however, this remains to be determined.

2.4 OTHER FISH SPECIES

Other fish species frequently used for a range of laboratory experiments and basic research are:

1. The platyfish (*Xiphophorus maculates*, Figure 2.6A), a cyprinodont belonging to the family Poeciliidae, is approximately 4 to 7 cm in length and native to fresh waters of Central America. This species is viviparous, with a short generation time, and reproduces continually, making it an ideal multigenerational test species [45]. Males with the gonopodium (copulatory organ, see the arrow in Figure 2.6A) are slightly larger than females and brighter in coloration.

FIGURE 2.6 A: male (top) with gonopodium (arrow) and female (bottom) platyfish, Florida introduced, *Xiphophorus variatus* (from [53]); B: male (top) and female (bottom) guppy, *Poecilia reticulata* (from [54]); C: male three-spined stickleback, *Gasterosteus aculeatus*, in breeding colors (from [55]); D: eelpout, *Zoarces viviparous*, of unknown sex (from [56], with permission.) (See insert for color representation.)

However, it is important to note that in addition to endocrine-active compounds, environmental factors such as the ambient pH may have a tremendous influence on the sex ratio of the developing population (e.g., platyfish and other poecilids) [46]. Indeed, in *Xiphophorus helleri,* the development of male monosex populations (100%) or a nearly complete female population (>98%) is obtained at an acid pH (6.2) or at a slightly basic value (pH 7.8), respectively [47], emphasizing the need to understand the underlying factors determining sexual development in small fish before any inferences can be made with regard to endocrine-active compounds.

2. The guppy (*Poecilia reticulata*, Figure 2.6B), a viviparous fish also belonging to the family Poeciliidae, is native from Brazil to Central America. Males with the gonopodium (copulatory organ) are smaller than females (3 cm compared to 6 cm) and brilliantly colored. The guppy is an omnivorous fish easy to handle, with a short reproduction time and year-round breeding capacities, making it a suitable laboratory fish species. The hormonal pathways that control the expression of male guppies' sexual characteristics are not fully understood yet; however, all present knowledge suggests that sperm production, body coloration, and courtship behavior are controlled by androgens [48].

3. The three-spined stickleback (*Gasterostreus aculeatus*, Figure 2.6C) is widely distributed throughout the northern hemisphere and is found in lakes, rivers, ponds, estuaries, and the sea. The most common species present, in general, three dorsal spines, two pelvic spines, and a single anal spine. Adult fish are usually from 4 to 6 cm in length but may reach a maximum length of 11 cm. Colors vary due to location and environmental and physiological conditions. Freshwater fish can be olive green above, fading to bronze or silver below; an overall bronze with gray mottling above; an overall bronze/green; or a bright brassy color. During the breeding season, males become highly colored, to attract as many females as possible, and develop a red breast. Breeding takes place from March to August. Each female lays from 10 to 100 eggs and the male guards these and the resulting fry, which hatch after up to 20 days. The threespine stickleback is probably one of the most studied fish by fishery biologists and behavioral biologists [49,50].
4. The eelpout (*Zoarces viviparous*, Figure 2.6D), a viviparous fish belonging to the family Perciformes, is distributed worldwide from the intertidal zone (at the north and south poles) down to 5000 m. The eelpout inhabits rocky shores under stones, within algae and in tidal pools, and between tidemarks down to 40 m, and may remain out of the water under rocks or seaweed. Adult eelpout can reach a length of 52.0 cm and a maximum age of 10 years. Mating takes place from August through September with internal fertilization of the eggs. Females give birth to 30 to 400 developed young (35 to 55 mm). Eelpouts have a minimum population doubling time of 4.5 to 14 years [51,52]. Although these fish are not optimal for maintenance under laboratory conditions, this species may be of interest when monitoring seawater samples.

REFERENCES

1. Ankley, G.T. and R.D. Johnson, Small fish models for identifying and assessing the effects of endocrine-disrupting chemicals. *ILAR Journal*, 2004, **45**(4):469–483.
2. Nash, J.P., et al., Long-term exposure to environmental concentrations of the pharmaceutical ethynylestradiol causes reproductive failure in fish. *Environmental Health Perspectives*, 2004, **112**(17):1725–1733.
3. Jobling, S., et al., Altered sexual maturation and gamete production in wild roach (*Rutilus rutilus*) living in rivers that receive treated sewage effluents. *Biology of Reproduction*, 2002, **66**:272–281.
4. Jobling, S., et al., Widespread sexual disruption in wild fish. *Environmental Science and Technology*, 1998, **32**:2498–2506.
5. Palace, V.P., et al., Induction of vitellogenin and histologic effects in wild fathead minnows from a lake experimentally treated with the synthetic estrogen, ethynylestradiol. *Water Quality Research Journal of Canada*, 2002, **37**(3):637–650.

6. Wolf, J.C., et al., Qualitative and quantitative histomorphologic assessment of fathead minnow *Pimephales promelas* gonads as an endpoint for evaluating endocrine-active compounds: A pilot methodology study. *Toxicologic Pathology*, 2004, **32**(5):600–612.
7. Smith, R.J.F., Seasonal changes in the histology of the gonads and dorsal skin of the fathead minnow, *Pimephales promelas. Canadian Journal of Zoology*, 1978, **56**(10): 2103–2109.
8. Sohoni, P., et al., Reproductive effects of long-term exposure to bisphenol A in the fathead minnow (*Pimephales promelas*). *Environmental Science and Technology*, 2001, **35**(14):2917–2925.
9. Cardwell, J.R., et al., Effect of dominance status on sex hormone levels in laboratory and wild-spawning male trout. *General and Comparative Endocrinology*, 1996, **101**(3): 333–341.
10. Orlando, E.F., et al., Endocrine-disrupting effects of cattle feedlot effluent on an aquatic sentinel species, the fathead minnow. *Environmental Health Perspectives*, 2004, **112**(3):353.
11. Palmer, David A., Wildlife International, Ltd., 8598 Commerce Drive, Easton, Maryland, http://www.wildlifeinternational.com.
12. Jensen, K.M., et al., Aspects of basic reproductive biology and endocrinology in the fathead minnow (*Pimephales promelas*). *Comparative Biochemistry and Physiology, Part C: Toxicology and Pharmacology*, 2001, **128**(1):127.
13. U.S. EPA, Revised draft detailed review paper on fish screening assays for endocrine disruption. Battelle, Columbus, OH, 2002.
14. Yonkos, L.T., A.S. Kane, and R. Reimschuessel, Fathead minnow histology atlas: World Wide Web outreach and utilization. *Marine Environmental Research*, 2000, **50**(1):312.
15. U.S. EPA, Draft detailed review paper on a fish two-generation toxicity test. Battelle, Columbus, OH, 2002.
16. U.S. EPA, A short-term test method for assessing the reproductive toxicity of endocrine-disrupting chemicals using the fathead minnow *(Pimephales promelas*). U.S. Environmental Protection Agency, Washington, DC, 2002, pp. 141–144.
17. Rühl-Fehlert, C.I., et al., Spontaneous intersex in a fathead minnow (*Pimephales promelas*). *European Society of Toxicologic Pathology (ESTP) Congress*, Hamburg, Germany, 2003.
18. National Institute for Agro-Environmental Sciences, Department of Environmental Chemistry (NIAES), Japan.
19. Hori, H., et al., http://biol1.bio.nagoya-u.ac.jp:8000/, 1995.
20. Volff, J.N., et al., Governing sex determination in fish: Regulatory putsches and ephemeral dictators. *Sexual Development*, 2007, **1**(2):85.
21. Iwamatsu, T., Stages of normal development in the medaka *Oryzias latipes. Zoological Science*, 1994, **11**:825–839.
22. Koger, C.S., S.J. Teh, and D.E. Hinton, Variations of light and temperature regimes and resulting effects on reproductive parameters in medaka (*Oryzias latipes*). *Biology of Reproduction*, 1999, **61**:1287–1293.
23. Ka-Sing Chan, K., Effect of synthetic luteinizing hormone–releasing hormone (LH-RH) on ovarian development in Japanese medaka, *Oryzias latipes. Canadian Journal of Zoology*, 1977, **55**(1):155–160.
24. Nagahama, Y., The functional morphology of teleost gonads, in *Fish Physiology*, W.S. Hoar, D.J. Randall, and E.M. Donaldson, Editors. Academic Press, San Diego, CA, 1983, pp. 223–275.

25. Nagahama, Y., H. Kagawa, and G. Young, Cellular sources of sex steroids in teleost gonads. *Canadian Journal of Fisheries and Aquatic Sciences*, 1982, **39**:56–64.

26. Nagahama, Y., et al., Sexual plasticity in fish: A possible target of endocrine disruptor action. *Environmental Sciences*, 2004, **11**(1):73–82.

27. Nagahama, Y., Gonadotropin action on gametogenesis and steroidogenesis in teleost gonads. *Zoological Science*, 1987, **4**(2):209–222.

28. Gray, M.A. and C.D. Metcalfe, Induction of testis–ova in Japanese medaka (*Oryzias latipes*) exposed to *p*-nonylphenol. *Environmental Toxicology and Chemistry*, 1997, **16**(5): 1082–1086.

29. Yamamoto, T., Induced intersexes, in *Medaka (Killifish): Biology and Strains*. Keigaku Publishing Co., Tokyo, 1975, pp. 214–222.

30. Chemicals Evaluation and Research Institute. http://w-edcdb.nies.go.jp/SHf/index.html, 2003.

31. Seki, M., et al., Effect of ethinylestradiol on the reproduction and induction of vitellogenin and testis–ova in medaka (*Oryzias latipes*). *Environmental Toxicology and Chemistry*, 2002, **21**(8):1692–1698.

32. Hyodo-Taguchi, Y. and N. Egami, Establishment of inbred strains of the medaka *Oryzias latipes* and the usefulness of the strains for biomedical research. *Zoological Science*, 1985, **2**:305–316.

33. Tomita, H., The lists of the mutants and strains of the medaka, common gambusia, silver crucian carp, goldfish, and golden venus fish maintained in the laboratory of freshwater fish stocks. Nagoya University. *Fish Biology Journal of Medaka*, 1992, **4**:45–48.

34. Gray, M.A., A.J. Niimi, and C.D. Metcalfe, Factors affecting the development of testis–ova in medaka, *Oryzias latipes,* exposed to octylphenol. *Environmental Toxicology and Chemistry*, 1999, **18**(8):1835–1842.

35. Papoulias, D.M., D.B. Noltie, and D.E. Tillitt, An in vivo model fish system to test chemical effects on sexual differentiation and development: Exposure to ethinyl estradiol. *Aquatic Toxicology*, 1999, **48**(1):37–50.

36. Hamaguchi, S., Sex differentiation of germ cells and their supporting cells in *Oryzias latipes. Fish Biology Journal of Medaka*, 1992, **4**:11–17.

37. Grier, H.J., Sperm development in the teleost *Oryzias latipes. Cell and Tissue Research*, 1976, **168**:419–431.

38. Scholz, S. and H.O. Gutzeit, 17α-Ethinylestradiol affects reproduction, sexual differentiation and aromatase gene expression of the medaka (*Oryzias latipes*). *Aquatic Toxicology*, 2000, **50**(4):363–373.

39. Koger, C.S., S.J. Teh, and D.E. Hinton, Determining the sensitive developmental stages of intersex induction in medaka (*Oryzias latipes*) exposed to 17β-estradiol or testosterone. *Marine Environmental Research*, 2000, **50**(1–5):201–206.

40. Grim, K.C., et al., Intersex in Japanese medaka (*Oryzias latipes*) used as negative controls in toxicologic bioassays: A review of 54 cases from 41 studies. *Environmental Toxicology and Chemistry*, 2007, **26**(8):1636–1643.

41. Schartl, M., *Physiological Chemistry*, University of Würzburg, Germany.

42. Laale, H.W., The biology and use of zebrafish, *Brachydanio rerio*, in fisheries research: A literature review. *Journal of Fish Biology*, 1977, **10**(2):121–173.

43. Teh, C., S. Parinov, and V. Korzh, New ways to admire zebrafish: Progress in functional genomics research methodology. *Biotechniques*, 2005, **38**(6):897–906.

44. Spitsbergen, J.M. and M.L. Kent, The state of the art of the zebrafish model for toxicology and toxicologic pathology research: Advantages and limitations. *Toxicologic Pathology*, 2003, **31**(Suppl.):62–87.

45. Kinnberg, K., B. Korsgaard, and P. Bjerregaard, Concentration-dependent effects of nonylphenol on testis structure in adult platyfish, *Xiphophorus maculatus*. *Marine Environmental Research*, 2000, **50**(1–5):169–173.

46. Baroiller, J.-F., Y. Guiguen, and A. Fostier, Endocrine and environmental aspects of sex differentiation in fish. *Cellular and Molecular Life Sciences*, 1999, **55**:910–931.

47. Rubin, D.A., Effect of pH on sex ratio in cichlids and a poeciliid (Teleostei). *Copeia*, 1985, **1**:233–235.

48. Baatrup, E. and M. Junge, Antiandrogenic pesticides disrupt sexual characteristics in the adult male guppy (*Poecilia reticulata*). *Environmental Health Perspectives*, 2001, **109** (10):1063–1070.

49. The Aquarium Project. http://web.ukonline.co.uk/aquarium/pages/threespinestickleback.html, 2004.

50. Nelson, J.S., http://www2.biology.ualberta.ca/uamz.hp/culea.html, 2004.

51. Andriashev, A.P., Zoarcidae, in *Fishes of the North-eastern Atlantic and the Mediterranean*, P.J.P. Whitehead et al., Editors. UNESCO, Paris, 1986, pp. 1130–1150.

52. FishBase. http://ichtyonb1.mnhn.fr/, 2007.

53. Shafland, Paul, Florida Fish and Wildlife Commission, www.myfwc.com.

54. Cisnetti, Federico, picture licensed under the GNU Free Documentation License.

55. Spaans, Piet, picture licensed under the GNU Free Documentation License.

56. Gercken, Jens, Institut für Angewandte Ökologie Forschungsgesellschaft mbH, http://www.ifaoe.de/.

CHAPTER 3

Sexual Determination, Differentiation, and Gonadal Development

Few reviews deal with early sex development in fish: Yamamoto [1,2], van der Kraak et al. [3], Nakamura et al. [4,5], Piferrer et al. [6,7], and Devlin and Nagahama [8], to name the most important. While the understanding of sexual development has been crucial since the advent of aquaculture and fundamental biology, this understanding has found application in and become important for other fields of research only recently (e.g., ecotoxicology, environmental risk assessment). For example, Gimeno et al. [9–13] studied the impact of endocrine-active chemicals on critical life stages in carp. Similar studies were carried out in roach (*Rutilus rutilus*) [14–17], fathead minnow [18–21], medaka [17,22–54], and zebrafish [55–70]. Rather than reiterating already well-reviewed knowledge [8], in this chapter we deal with an overview of progress since the late 1990s and identify knowledge gaps or current points of discussion.

Sexual determination and differentiation relate to the events that occur during development and allow expression of genetic sex into the corresponding phenotypic sex (see Figure 3.1). Gonadal sex determination involves the primary controls that influence the course of gonadal sex differentiation [8]. Thus, it would be mistaken to assume that the processes above involve only sex chromosomal systems, as sex determination in fish has presented with an unforeseen plasticity and variety of regulatory mechanisms, including environmental factors (e.g., population density, temperature, xenobiotics) [7,17,71,72]. The color of sex-linked patterns in Japanese medaka has made it possible to study the mechanisms of sex determination and differentiation. Yamamoto [2,73] demonstrated that in most fish the steroidal estrogens and androgens are the natural inducers for sexual differentiation.

Sexual determination comprises the genetic events (oocyte fertilization by a spermatozoid with interchange of chromosome material) that lead to male or female gonadal development. The genetic sex of an individual is determined at the time of fertilization and dictates the direction of the development of the primitive gonad [74].

Sexual differentiation relates to the subsequent events that allow expression of the genetic sex into the appropriate phenotypic sex, whereby the primordial

Histological Analysis of Endocrine-Disruptive Effects in Small Laboratory Fish
By Daniel R. Dietrich and Heiko O. Krieger

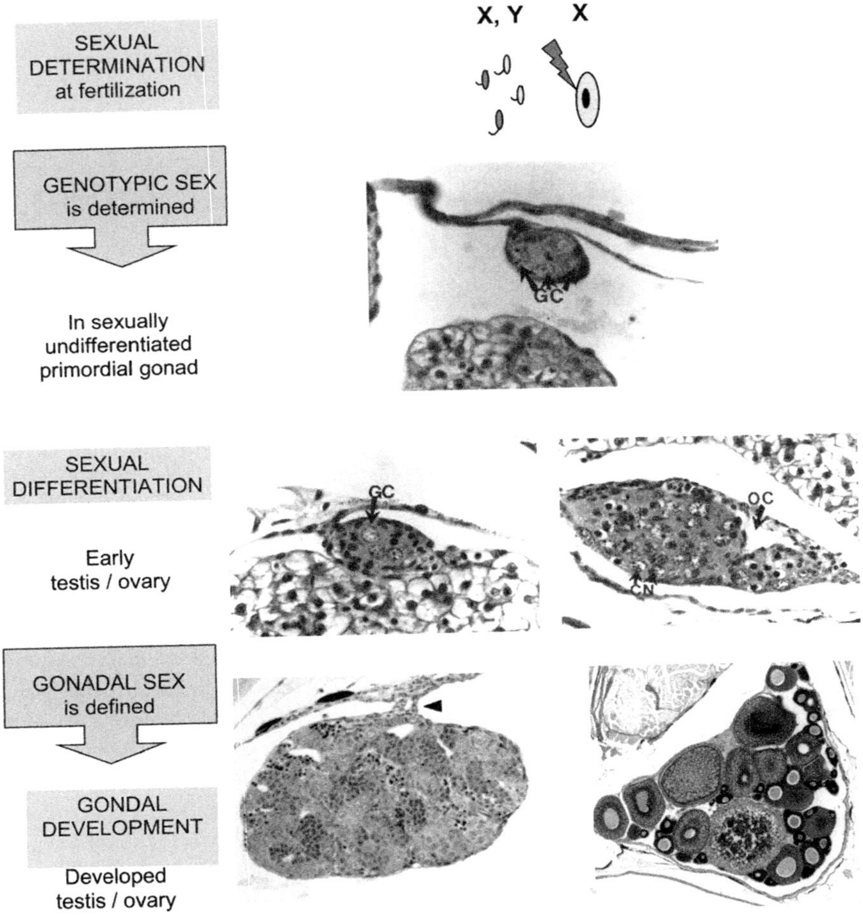

FIGURE 3.1 Sexual determination, differentiation, and gonadal development. GC, germ cells; OC, ovarian cavity; CN, chromatin–nucleolus phase of oocytes. Arrowhead: gonadal duct. (From [75] and [18], with permission.)

(undifferentiated) gonad differentiates into a testis or ovary. This early gonadal development encompasses all the events that take place in the primordial gonad, including the migration of primordial germ cells, the establishment of gonadal ridges, and the differentiation of the gonad proper into testes or ovaries [5,6]. At later stages, secondary sexual characteristics develop due to hormones secreted by the respective gonads, thereby resulting in the sex-specific expression of characteristic features (see Chapter 2).

The timing of the sexual differentiation processes in fish is species dependent, as shown in Figures 3.1 and 3.2, for a few well-studied fish species. In contrast, the duration of sexual differentiation is not only fish species–inherent but also dependent on environmental parameters (e.g., temperature, feed availability, predation stress) (see Figure 3.3) [76–78]. Sexual differentiation in the rainbow trout (*Oncorhynchus mykiss*),

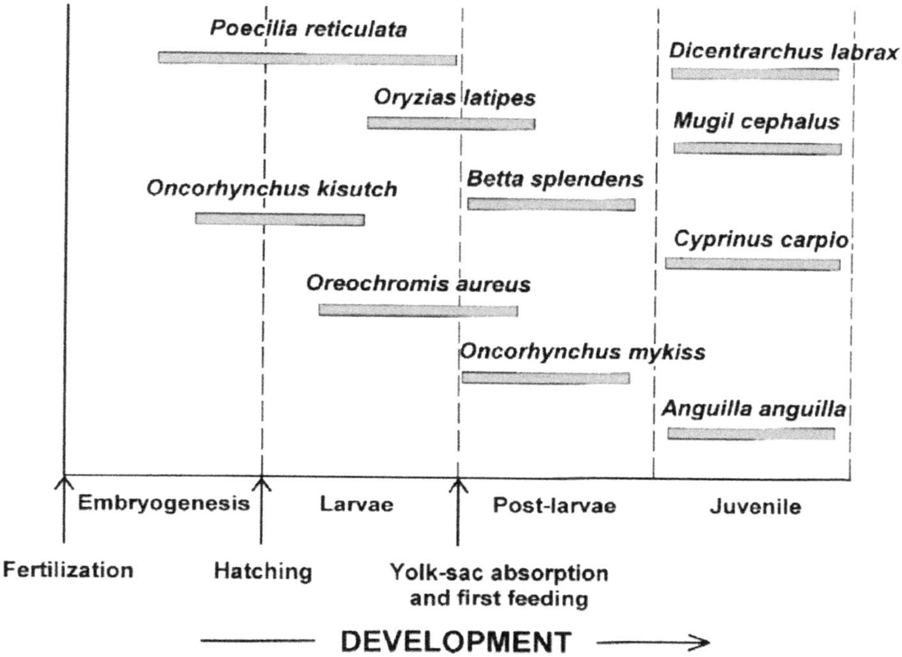

FIGURE 3.2 Timing and duration of sexual differentiation in selected teleost species. The labile period for effective sex steroid treatment in relation to development is highlighted. (Adapted from [6], with permission.)

carp (*Cyprinus carpio*), and roach begins at over 50 days post-hatching (dph), at the end of the first year in the golden mullet (*Mugil auratus*), and in some species [e.g., medaka and coho salmon (*Oncorhynchus kisutch*)] sexual differentiation occurs during the embryonic or larval period [8].

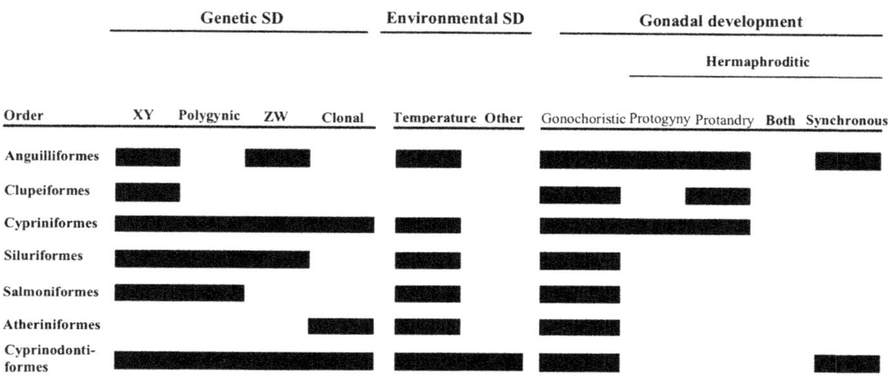

FIGURE 3.3 Factors influencing sexual differentiation of selected fish orders. (Modified from [8], with permission.)

3.1 PRIMORDIAL GERM CELLS IN THE PRIMORDIAL (PRIMARY) GONAD

As in all vertebrates, the fish primordial gonad (i.e., prior to sexual differentiation) is composed of both somatic and germ cells. Primordial germ cells (PGCs) have the ability to become oogonia or spermatogonia in the ovary and testis, respectively. After hatching, PGCs surrounded by one or two somatic cells are usually found at the dorsal wall of the coelomic cavity (e.g., in zebrafish they are located close to the distal end of the swimbladder) [58,62]. Subsequently, both somatic and germ cells proliferate to form the early (primary), still undifferentiated gonads.

At this stage, the gonad is generally suspended from the dorsal peritoneal wall by a thin sheet of somatic cells [79]. The germ cells continue a gradual increase in number, resulting in a gradual enlargement of the gonad. Following initial PGC proliferation, the primordial germ cells mature to spermatogonia or oogonia (the earliest differential stage of gametogenesis), as detailed below. Whether the somatic cells of the gonad in fish are of dual origin, as in the case in mammals (with the cortex originating from the peritoneal wall and the medulla originating from the mesonephric blastema), or whether it is of unitary origin is a subject of some controversy [5]. It is, however, of importance that the developmental timing of gonad formation in fish species can vary greatly (see Figure 3.2).

3.1.1 Differentiation and Number of PGCs

Various Japanese authors have studied the differentiation of primordial germ cells in medaka during normal embryonic development. Satoh and Egami [80] discovered that this occurs at the time of hatching, when no sexual differences are observable in the somatic cells. Histologically (routine light microscopy) it is not possible to determine sex in the primordial (undifferentiated) gonads on the basis of the morphology of their germ cells, as spermatogonia and oogonia cannot be distinguished. Therefore, sexing at this stage generally rests on the somatic gonadal characteristics (see Section 3.2). However, it appears that PGCs of newly hatched genetic female medaka fry undergo mitosis and meiosis more frequently than do the PGCs of genetic males, resulting in a higher *number of PGCs* in female than in male medaka gonads [81]. Therefore, for medaka and tilapia (*Oreochromis mossambicus*), the number of germ cells may be indicative of the genetic sex and thus type of gonad, since germ cells in putative ovaries appear to outnumber those in putative testes [5].

Sexing the early, indifferent gonads on the basis of the morphology of their germ cells is not possible, but the distinction between spermatogonia and oogonia in fish generally rests on the somatic gonadal characteristics. In differentiated gonochorists (see Section 3.2) and in fathead minnows, medaka, and zebrafish, transition from the morphologically indifferent gonad into the sexually differentiated gonad generally occurs earlier in female than in male fish. It is known, however, that although the fundamental processes and mechanisms underlying the genesis and migration of PGCs are probably universal in fish, the initial distribution of PGCs in early-stage embryos and the pathways of their migration prior to entering the gonad-forming dorsal mesentery area are largely different between zebrafish [82–87] and medaka

[22,88,89]. For example, in medaka, the abundance of PGCs also appears to be regulated by maternal factors [90], emphasizing that multiple factors are involved in the regulation of PGC formation, abundance, and maintenance. Indirect evidence is provided from an in-bred rainbow trout family with a somatic gonadal tissue but absent germ cell development, suggesting that PGC formation and/or maintenance is under genetic control [91], thus substantiating the findings reported for medaka. In outbred fish species such as carp and salmonids, sex-specific differences in the number of PGCs have not been observed [92], highlighting the difficulties with generalizations on sex differentiation and development in fish species.

Besides the number of germ cells, the *location of meiotic germ cells* may also be indicative of the type of gonad: Meiotic germ cells in the center of the gonad are indicative of ovarian differentiation, while in presumptive testes, germ and somatic cells are scattered within the primitive gonad [93]. Moreover, some sex specificity is observed with regard to the *timing of germ cells in entering meiosis* (see Figure 3.4). Indeed, although it is presently unclear whether this can be generalized from medaka and fathead minnows to all teleosts, mitosis and meiosis of germ cells apparently take place earlier in ovaries than in testes [5]. For example, in the fathead minnow, ovarian germ cells and testicular germ cells enter into meiosis around 10 to 25 dph and 90 to 120 dph, respectively [93].

A reduced number of PGCs, that is, the average number of PGCs per histological gonad cross section, was observed in the gonads of male carp that had been exposed to 17β-estradiol (E_2) and a weak estrogenic-active chemical [4-*tert*-pentylphenol

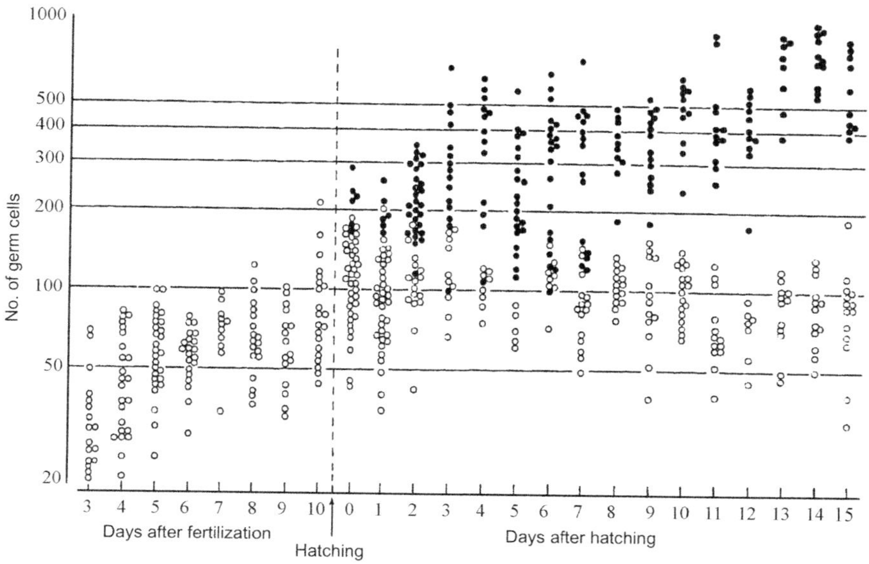

FIGURE 3.4 Changes in the number of germ cells in medaka, *Oryzias latipes*, during normal embryonic development. Open circles, embryos and fry containing no germ cells in the meiotic prophase; filled circles, fry containing germ cells in the meiotic prophase. (From [80], with permission.)

(4-TPP)] [11,13,94]. The severity of the effects appeared to be related to the estrogenic potency of the test substance. A maximum exposure of 86 or 110 days starting either at the embryo, yolk sac, or feeding larvae stage, 24 dph or 50 dph, respectively, thus including the active period of germ cell proliferation, resulted in the complete inhibition of PGC proliferation (see Figure 3.5) and the formation of filiform testes. Inhibition of PGC proliferation by EDCs during early gonadal development thus may suppress spermatogenesis, eventually leading to reduced production of spermatozoa or complete sterilization, thereby reducing the reproductive capacity of the species affected [95].

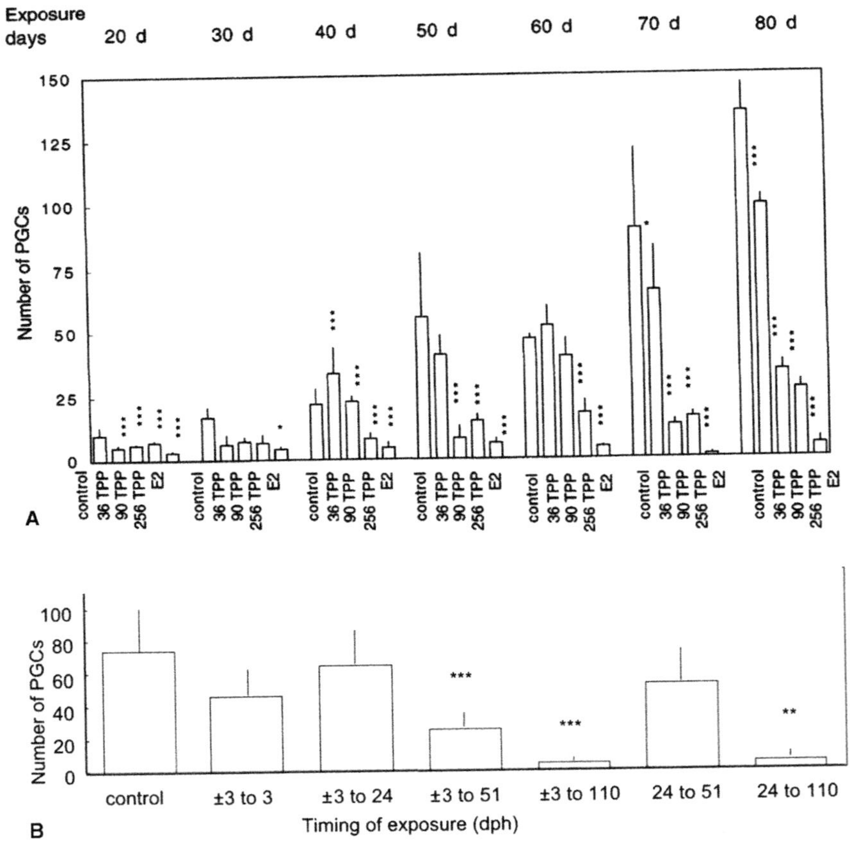

FIGURE 3.5 Effect of the duration and of exposure (A) and timing (B) on number of primordial germ cells PGCs: (A) number of PGCs counted by histological testis cross section of carp exposed 50 dph to E_2 or TPP for various time periods; (B) number of PGCs counted by histological testis cross section of carp exposed to 140 μg/L TPP for various time periods. Fish were examined at 110 dph. Start of the exposure at ±3 dph with either −3, 0, or 3 dph for embryos, yolk sac larvae, or feeding larvae, respectively, or at 24 dph with fingerling carp. Shown are mean values ± standard deviation; significantly different PGCs from controls indicated as **, $p < 0.01$; ***, $p < 0.001$ (two-sided t-test). (From [11,13], with permission.)

3.1.2 Molecular Markers of PGCs

In hatched fish, PGCs can be recognized histologically by their large size, low nucleus/cytoplasm ratio, clear nuclear borders, and granular nuclear chromatin [70]. PGCs of both sexes have common characteristics, as demonstrated elegantly for carp by Parmentier and Timmermans [92]: Sperm-specific antigenic determinants were found to be present on the PGCs of all individuals from hatching onward, but were absent from somatic cells. In those gonads that developed into testis, these antigenic determinants remained present during differentiation up to the mature spermatozoa, whereas in females they are expressed only until the oogonia and early prophase stages, and disappear in growing follicular oocytes. The study of PGC emergence and development during the embryonic stages became possible after the *vasa* gene was identified as a molecular marker of PGCs in fish [70,85–87,90,96,97]. For instance, in zebrafish and medaka, the maternal *vasa* RNA localizes in the cleavage plane at the two- to four-cell-stage embryo, but spreads to the cytoplasm of the PGCs during the blastula stage [85,89]. At the gastrula stage, maternally derived *vasa* transcripts in zebrafish and medaka appear to diminish, while during late gastrulation some of the cells located around the posterior region of the embryonic shield appear to retain or even increase *vasa* expression and the PGCs start to migrate and coalesce with somatic gonadal cells [98]. It is interesting to note that two isoforms of *vasa* homologs were identified in tilapia and have been demonstrated to be the product of differential cleavage products. These isoforms were expressed differentially during germ cell differentiation, dependent on germ cell differentiation but not on genotypic sex [97]. Thus, the *vasa* protein isoforms appear to play a major role in the developmental distinction between germ cells and somatic cells.

Overall, current knowledge regarding PGC formation and development, especially with regard to the influence of maternally derived gene products, suggests that any adverse interaction (e.g., via chemicals) could disturb proper gonadal development as foreseen by the genetic sex. Indeed, 10^{-7} M concentrations of endosulfan and nonylphenol have been demonstrated to alter significantly the somite-specific domain aggregation but not the number of PGCs in developing zebrafish embryos [72], possibly altering the biochemical communication between the PGCs and the mesodermal signaling center. Whether or not such mechanisms of xenobiotic interaction could also be responsible for the occurrence of intersex (see Figures 3.14 to 3.17) in genetic males [71,99,100] is subject to future investigations. It is of importance to note, however, that the influence of sex determination on sex differentiation appears to vary among fish species. For example, application of aromatase inhibitors and antiestrogens had little effect on ovary development of medaka [41], whereas similar treatment during early development of coho salmon, rainbow trout, and tilapia resulted in abrogation of estradiol synthesis and thus masculinization of all fish [101–104]. Recent publications on the expression of cytochrome P450 aromatase genes (*CYP19a* and *CYP19b*) in the brain and early gonad of zebrafish [105,106], as well as the importance of sex-steroids in the sex differentiation of medaka, zebrafish, and fathead minnow (see below), suggest that sex-determination mechanisms do not initiate the process of sex cell differentiation but, depending on the fish species, are intricately involved in steps leading to initiation of steroid synthesis and gonadal differentiation.

3.2 REPRODUCTIVE STRATEGIES

There exist considerable differences among fish species in their reproductive strategies and thus mode of morphological gonad differentiation [2,8,107,108]; see Figure 3.6.

Gonochorism Gonochorism in which the undifferentiated gonads develop as either testes or ovaries, is the most common difference [109,110]. Two developmental extremes of gonochorism exist:

- *Differentiated gonochorist*. PGCs differentiate directly either into oocytes and thus to the development of ovaries, or into spermatogonia and therefore into testes. The phenotypic sex of a differentiated gonochoristic fish is more stable, and spontaneous intersex or sex reversal is not expected to occur [2].
- *Undifferentiated gonochorist* (transient or nonfunctional hermaphrodites). PGCs of all animals (both genetic sexes) appear to differentiate at first into an ovary-like gonad (nonfunctional female phase, gonads containing early-stage oocytes), and then at some point testicular differentiation takes over in approximately 50% of the animals (male genetic sex), and the oocytes present regress; that is, the nonfunctional hermaphroditic gonads resolve into functional ovaries or testes.

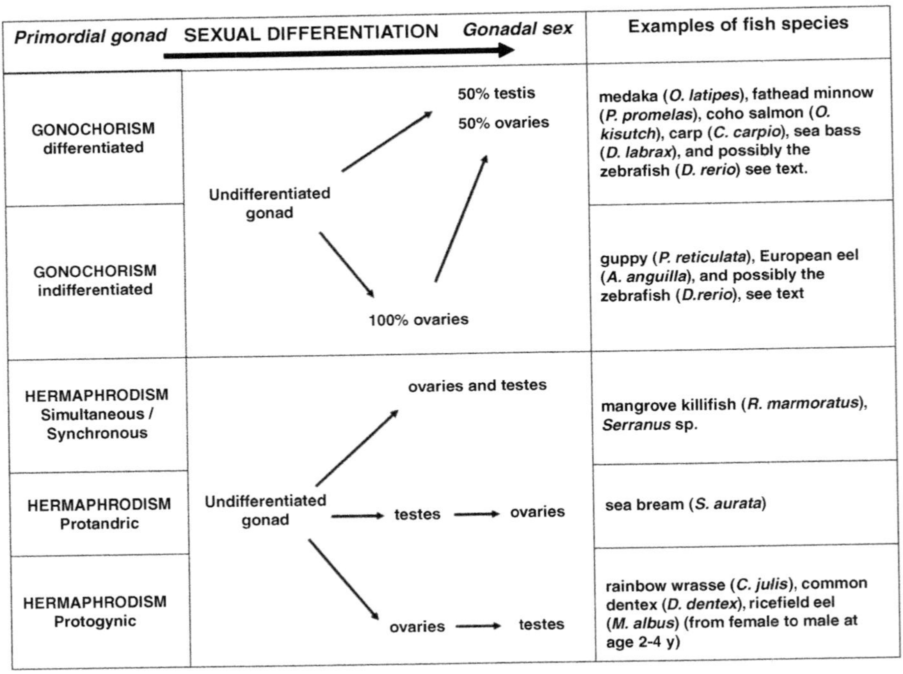

FIGURE 3.6 Reproductive strategies and associated pathways of gonadal differentiation.

Hermaphrodism Besides the two forms of gonochorism noted above, various types of hermaphrodism, a normal sexual condition where an individual produces mature gametes of both sexes, exist in various ways and occurs in some fish species (see Figure 3.6). This includes simultaneous (synchronous) hermaphrodism (both gonad types occurring at the same time within an organism) as well as sequential hermaphrodism, that is, protandric species that first develop into functional males and then turn into functional females, or protogynic species that first develop into functional females prior to changing into functional males [7,8]. Sequential hermaphrodism occurs in several hundreds of the more than 20,000 fish species worldwide, representing several families from various habitats [111].

Intersex (see Figures 3.14 to 3.17) Intersex, which is clearly distinct from hermaphrodism, refers to the presence of female and male gonadal tissue in the same gonad of gonochoristic fish individuals [112]. It is thought to occur spontaneously and sporadically, either as the result of a particular genetic constitution or from a particular endocrine condition [113] and is considered to be an accidental or pathological phenomenon [112].

3.3 DIFFERENTIATION OF THE PRIMORDIAL GONAD INTO OVARY OR TESTIS

Somatic changes occur during sexual differentiation, in addition to germ cells entering into meiosis that make it possible to distinguish a developing ovary or testis. Yet the developmental period when morphological differentiation of the gonad takes place varies greatly among teleost species (Figures 3.2 and 3.3). However, the fact that not all fish (teleosts) gonads seem to have a dual corticomedullary primordial gonad such as cartilaginous fish and higher vertebrates do, but rather present with a primordial gonad resembling cortical tissue, makes it difficult to distinguish when ovarian and testicular differentiation has started [114]. Whether the somatic cells of the gonad in fish are of dual origin, as is the case in mammals (with the cortex originating from the peritoneal wall and the medulla originating from the mesonephric blastema), or whether it is of unitary origin is a subject of controversy [5].

Obviously, PGCs have to exchange information with the adjacent somatic cell layer of the early gonad, which is derived from the cortex epithelial layer. This exchange of information is largely under endocrine control and driven by the synthesis of steroids with their respective receptors not only in the developing gonad proper and adjacent somatic tissue, but also in other organs (e.g., the brain) [105,106,115].

Ovarian Differentiation The ovarian differentiation is characterized by the entry of some germ cells into meiosis and by the concomitant formation of the ovarian cavity [79] (see also Section 4.1.2). In tilapia this can be as early as 20 dph. Van Aerle et al. [93] noticed the presence of meiotic cells located in the center of the gonad in fathead minnow as of 10 to 25 dph, with somatic cells located at the periphery of the gonad. Concomitantly with the start of oogenesis, formation of the ovarian cavity may take

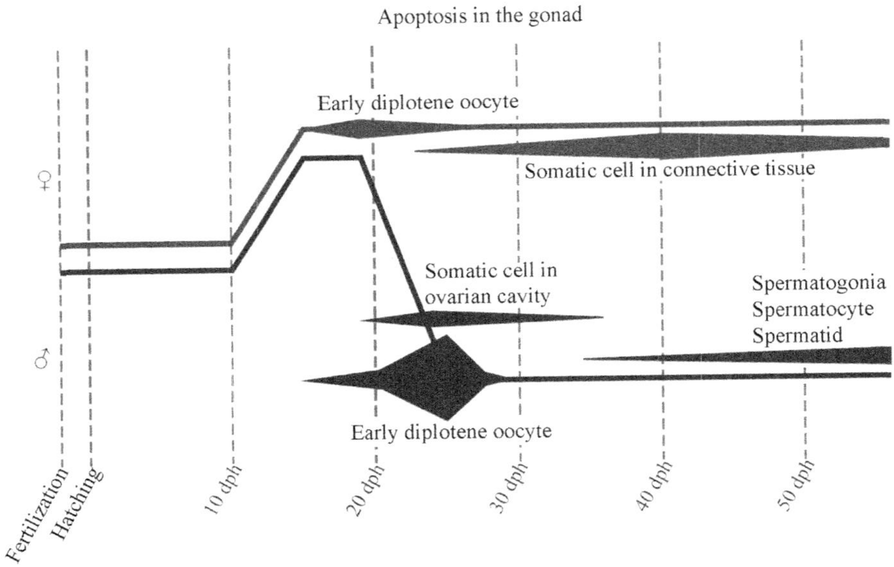

FIGURE 3.7 Apoptosis (width of diamonds represents number of apoptotic cells) during the gonadal differentiation processes of developing zebrafish. (From [61], with permission.)

place via the proliferation of peritoneal somatic cells and concurrent degeneration, probably by an apoptotic process (Figure 3.7) of the medullary somatic cells [5]. Formation of the ovarian cavity can also occur after the start of oogenesis, for instance in the mosquitofish (*Gambusia affinis*), or it may occur in a rudimentary form as in salmonids, which ovulate directly into the coelomic cavity. In differentiated gonochorists (i.e., also in fathead minnows, medaka, and zebrafish), transition from the morphologically indifferent gonad into the sexually differentiated gonad generally occurs earlier in female than in male fish.

Testicular Differentiation Conversely, male sex differentiation occurs later and is characterized by the appearance of spermatocysts containing spermatogonia B and later stages of spermatogenesis and differentiation of the vascular system of the testis and the sperm ducts. In the fathead minnow, van Aerle et al. [93] mentioned that as of 25 dph, somatic cells can be observed scattered throughout the gonad. As of 60 dph the sperm ducts become apparent and the testis lobules start to form. The germ cells (spermatogonia) do not enter in meiosis until 90 to 120 dph.

It is important to note at this point that differentiated gonochorists are expected to provide for a low incidence of spontaneous intersex in adult animals, as no period of "transitional hermaphrodism" is encountered during male gonad differentiation (see Figure 3.6). Indeed, 0% incidence of intersex in medaka has been reported by a Japanese laboratory for a period of 30 years [73] and by many American laboratories [33,116], which would support the hypothesis above, whereas only one report from a single American laboratory observed high incidences of intersex in medaka (J. Wolf, personal communication). Similar contradictory findings have been reported for carp:

whereas Parmentier and Timmermans [92] reported that gonad development is of the differentiated gonochoristic type, Davies and Takahashima [117] observed that carp gonads develop through an initial female phase (i.e., they would be undifferentiated gonochorists). A question thus arises as to whether these differences could stem from the use of different strains or different rearing conditions.

Gonad Differentiation in Medaka, Fathead Minnows, and Zebrafish With regard to gonad differentiation of small laboratory fish species, both medaka and fathead minnow are differentiated gonochorists. An overview of the gonad development of these two species is given in the Atlas of Medaka Gonadal Histology of the CERI Institute in Japan [118] and on the Medaka fish homepage [119], in the Digital Atlas of Fathead Minnow Normal Histology in Maryland [120], and in the presentation by van Aerle et al. [93], respectively. The issue of the type of gonad differentiation is more controversial in the case of zebrafish (see also Figure 3.6).

Zebrafish Takahashi [62] reported that zebrafish develops as an undifferentiated gonochorist, with the gonads of all individuals developing initially as nonfunctional ovaries containing primary growth stage oocytes. This observation was corroborated by the findings of Maack and Segner [58], although the time frame when this juvenile hermaphroditic phase occurred differed between the two studies (10 to 25 dph in the the study of Takahashi [62], 40 to 70 days post-fertilization (approximately 36 to 66 dph) in the study of Maack and Segner [5,58,61–64]). Transition from the early ovary into the testis (see also Figure 3.7) appears to occur through a gradual decrease in the number and size of oocytes, oocyte degeneration, and proliferation of somatic tissue [58,62,66]. With progressive disappearance of oocytes from these altered gonads, clusters of gonial germ cells become prominent, and at a later stage, spermatocytes.

In contrast to the reports of Takahashi, Maack and Segner, and Oern et al. [58,62,66], van der Ven and colleagues found no evidence of transient juvenile hermaphrodism in zebrafish and classified this species as differentiated gonochorist [56,121]. A question arises as to how to explain these contradictory reports. One possible explanation could be a strain difference, although recently, Maack et al. [69] did not observe differences in gonad development of a heterozygous and homozygous strain. The presence of strain variations, however, is suggested by the varying incidences of intersex (testis–ova) or other forms of partial or full hermaphrodism (Figure 3.8) in testis of zebrafish. For example, an intersex ratio of 1 : 350 was found in the RIVM (National Institute of Public Health and the Environment, The Netherlands) stock of zebrafish [121], versus an incidence of up to 1 : 4 reported by Takahashi [62]. The presence of strain differences gains further support by the findings of Hsiao and Tsai [122]. Indeed, the latter authors reported the existence of two phenotypic variations of male development: one with and one without transient occurrence of diplotene oocytes. Depending on the ratio of these male phenotypes in a given strain of zebrafish, a lower or higher number of genetically male fish would go through juvenile hermaphrodism. At present, one cannot determine which underlying mechanism provides the predominance of one or the other male phenotype, or whether the(se) mechanism(s) depend(s) on environmental factors or on genetic polymorphism.

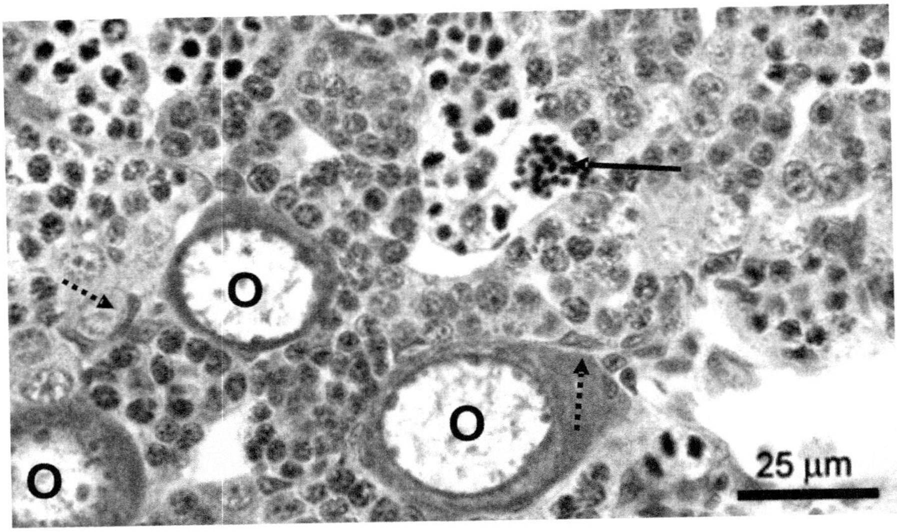

FIGURE 3.8 Testis–ova (intersex) in the developing testis of a male zebrafish with the clustered organization (spermatocysts), showing both oocytes (O) and various stages of spermatogenesis, including mature sperm (arrow). Sertoli cells (spotted arrows) can also be discerned. (Courtesy of L. van der Ven and P. Wester, RIVM, Bilthoven, The Netherlands.)

Other possible explanations of the contradictory observations on zebrafish gonadal development are experimental differences, such as inadequate developmental windows studied, insufficient numbers of fish analyzed, or results confounded by pathological processes. It is important to note that all of the evidence brought forth to support or refute the existence of juvenile hermaphrodism in zebrafish was based on morphological assessment. However, some additional evidence exists that could support the existence of juvenile hermaphrodism in zebrafish. Uchida et al. [61] investigated the frequency of apoptotic oocytes in the gonads of developing zebrafish of a wild-type strain and of a genetic all-female strain. Both strains showed gonadal differentiation between 15 and 30 dph; thus, the time frame was comparable to the strain used in the study of Takahashi [62]. In the all-female strain, the number of apoptotic oocytes increased until day 19 post-hatching, but declined drastically thereafter (Figure 3.7). An increasing number of apoptotic oocytes until day 19 post-hatching were also present in the wild-type strain. Then, however, two distinct patterns of apoptosis developed: Whereas some individuals behaved like fish of the all-female strain and showed a drastic decline in the number of apoptotic oocytes in their gonads, other individuals of the wild-type strain showed a continued, strong increase in the number of apoptotic oocytes up to a climax at 25 dph, and by 29 dph (5 weeks post-fertilization), all oocytes had disappeared from the gonads.

As it takes place in the genetic female strain as well as in all individuals of the wild-type strain, the authors interpret the increasing number of apoptotic oocytes until 19 dph to reflect normal oocyte loss during developmental differentiation of the early ovary, once a given status of differentiation of the ovary is reached. In the females in this study [61], this was obviously the case around 20 dph, as after that, apoptotic oocyte loss strongly declines in all individuals (genetic females) as well as in one fraction of individuals of the wild-type strain. Other individuals of the wild-type strain presented with sustained apoptotic oocyte loss after 19 dph (Figure 3.7). These fish were considered to represent genetic males, with their gonads undergoing transition from the early ovary into the testis and therefore eliminating the oocytes completely. These findings [61] could also reflect the presence of the male phenotype with transient occurrence of diplotene oocytes, as described by Hsiao and Tsai [122].

Knowledge of the gonadal expression of key steroidogenic enzymes, particularly of aromatase, could provide some clarity as to the mode of gonadal differentiation in zebrafish, as analysis of aromatase expression can be valuable to distinguish genetic males from genetic females [5]. Fenske and Segner [68] analyzed expression of the gonad-specific aromatase isoform cytochrome P450 19a (cyp19a) in developing and adult zebrafish. In adult zebrafish, cyp19a transcript abundance shows a dimorphic pattern, with male fish showing significantly lower cyp19a expression than that in female fish. In developing zebrafish, a distinction according to sex is not possible; thus, only the intra-cohort variation of cyp19a expression may be taken as an indicator of sex-specific aromatase expression. At day 36 post-fertilization (in the study of Fenske and Segner [68], this age corresponded to the juvenile "protogynic" stage), low intra-cohort variation of cyp19a mRNA levels was found. However, at day 71 post-fertilization, when differentiation of ovaries and testis was completed, variation of cyp19a transcript abundance was much more pronounced, which could be taken as an indication that individuals with female-like aromatase expression and individuals with male-like aromatase expression were both present. When zebrafish were exposed to the aromatase inhibitor fadrozole, 100% masculinization was induced (i.e., all 71-day-old fish showed gonads with testicular morphology and expressed low cyp19a levels). Corroborating evidence was reported by Uchida et al. [123], showing that treatment of genetic all-female zebrafish with fadrozole resulted in 100% gonadal masculinization, with the induced sex reversal accompanied by oocyte apoptosis.

Results from estrogen-exposure experiments may provide further support to the hypothesis of zebrafish as an undifferentiated gonochorist. When zebrafish were exposed to 3 ng/L 17α-ethinylestradiol (EE_2) from fertilization up to the reproductive stage, 100% gonadal feminization occurred [68]. Histological examination of the gonads revealed two forms of ovarian morphology: In some fish, mature ovaries containing all oocyte stages were observed, whereas in other individuals, the ovaries were immature, containing only or mainly avitellogenic primary growth stage oocytes. When, after chronic EE_2 exposure, fish were transferred into clean water, in addition to mature females, fish with mature testis were also found. The question is: If there was 100% feminization at termination of the EE_2 treatment, where are the male fish coming from, and was there sufficient evidence demonstrating that the females were fully functional (i.e., produced viable and fertilizable oocytes)? These observations—100% feminization, including both mature and immature ovaries at the end of chronic EE_2 exposure, and appearance of testis after recovery in clean water—could be explained based on the assumption that zebrafish develop as undifferentiated gonochorists, males passing through an ovarian stage during development and developing gonads with testicular morphology only after this ovarian stage (see Figure 3.6). Exogenous EE_2 would disturb the endogenous balance of steroid sex hormones as established by gender-specific aromatase levels, and would thereby prevent the gonads of genetic males from undergoing a transition from the early ovarian stage into the testis stage. The "immature" ovaries observed in adult EE_2-exposed zebrafish would then represent male fish that were prevented from normal testicular differentiation. If the exogenous EE_2 is then removed, male gonad differentiation resumes, explaining the presence of male fish in a previously 100% feminized cohort. The results described here for zebrafish exposed to estrogens during their full life cycle, from fertilization until reproduction, agrees quite well with the results of Hill and Janz [63] from a partial life-cycle exposure: Treatment of juvenile zebrafish with EE_2 concentrations of 1, 10, and 100 ng/L from 2 to 60 dph resulted in a clear sex shift toward 100% phenotypic female fish (having ovaries) [63]. However, continued rearing of these fish to adulthood (160 dph) demonstrated that these fish presented no significant departure from the 1 male:1 female ratio, suggesting that the original "genetic" determination of the gonad had been reestablished, thus possibly also providing some evidence for gonochoristic sex differentiation in zebrafish. This is further corroborated by the findings of Nash et al. [124], who reported that long-term exposure to EE_2 suppresses male pattern sexual differentiation and arrests development rather than producing functional females.

The observations above are not restricted to zebrafish, for which it is not yet clear whether they should be considered undifferentiated or differentiated gonochorists, but also hold true for established differentiated gonochorists (e.g., medaka and fathead minnow), whereby long-term exposure to EE_2 suppresses male pattern sexual differentiation, induces the occurrence of intersex, and arrests the development of testes rather than producing functional females.

3.4 GONADAL DUCT FORMATION

The formation of gonadal ducts is dependent, among other factors, on the appropriate balance of sex steroids during sexual differentiation [74] (see Sections 3.1 and 3.2). In sexually indifferent mammals and birds, both male and female embryos have two sets of paired ducts: the Müllerian (female) and Wolffian (male) ducts [125]. Every sexually undifferentiated individual would develop a female gonad without the influence of the *SRY* gene, located on the human Y chromosome, in combination with an up-regulation of *SOX9*, located on human chromosome 17, whose expression controls the development and differentiation of the testis proper [126,127]. Thus, *SRY* has been suggested as the testis-determining gene in mammals [128] and a number of other species [129], but expression appears not to be limited to the gonadal ridge or early development [130,131]. SOX9 is a gene assumed to be regulated by *SRY* (Figure 3.9), functioning in the male sex-determination pathway [132]. In consequence of the up-regulation of *SOX9* and testicular development, Leydig cells produce testicular androgens which induce the Wolffian ducts to become the epidydimus, *vas deferens*, seminal vesicles, and ejaculatory ducts, while Sertoli cells produce Müllerian-inhibiting substance (or anti-Müllerian hormone, *amh*), which induces regression of the Müllerian ducts

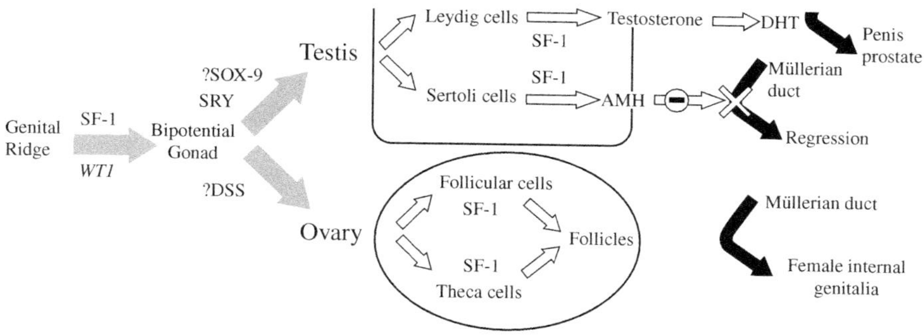

FIGURE 3.9 Ontogeny of expression and functions affecting the fetal mammalian gonads. (Adapted and modified from [128].)

(see Figure 3.9). In females, as a result of the down-regulation of *SOX9* and the absence of the *SRY* gene and *amh,* the Wolffian ducts degenerate and the Müllerian duct gives rise to the oviducts, uterus, and upper vagina.* However, although most of the above holds true for mammals [125,133], this is not true for fish, as the *SRY* gene is absent in fish [111] (see Sections 3.1 and 3.2).

In fish, gametes are discharged into the ambient water through gonadal ducts as in most vertebrates. Thus, gonadal ducts serve to collect eggs and sperm and transport the gametes to the cloaca (Figure 3.10). Consequently, they link the ovarian cavity and central lumen to the urogenital pore and carry the gametes (oocytes in the ovarian cavity and sperm released in the central lumen) to their appropriate internal or external destinations [134,135].

Female Gonadal Duct Ovarian ducts are present in most teleosts and are a direct posterior extension of the ovarian wall and lumen rather than a modified Müllerian funnel (oviduct) as seen in other vertebrate taxa. In teleosts the embryology of the ovary and ducts reveals that the ovary has a central cavity continuous with the oviduct. Except for murenoids (family Muraenidae), the ovaries of teleosts present with an ovarian cavity [113]. The ovarian cavity can be formed in various ways, giving rise to various cavity types. The *para-ovarian cavity* is formed from the most dorsal portion of the coelomic epithelium and the lateral walls of the gonadal rudiments (Figure 3.11), such as in carp, fathead minnows, and zebrafish. The internal epithelium of the endovarian cavity (formed inside the gonad) has an exclusively gonadal origin (Figure 3.11) [113]. In salmonids, the ovarian cavity is formed in only a rudimentary fashion and ovarian ducts are incomplete (Figure 3.10). In this case, ovulation involves the deterioration of intraovarian connective tissue, rupture of the ovarian wall, and release of the eggs into the ovarian central cavity, from which the eggs are expelled via muscular contractions through the urogenital papilla [134,137]. The ovarian ducts penetrate abdominal pores and terminate independently at the

* Secondary sexual characteristics differ from sex accessory structures, as they do not participate in gamete production and transport, but are essential for courtship and mating (e.g., gonopodium, nuptial tubercles) [131].

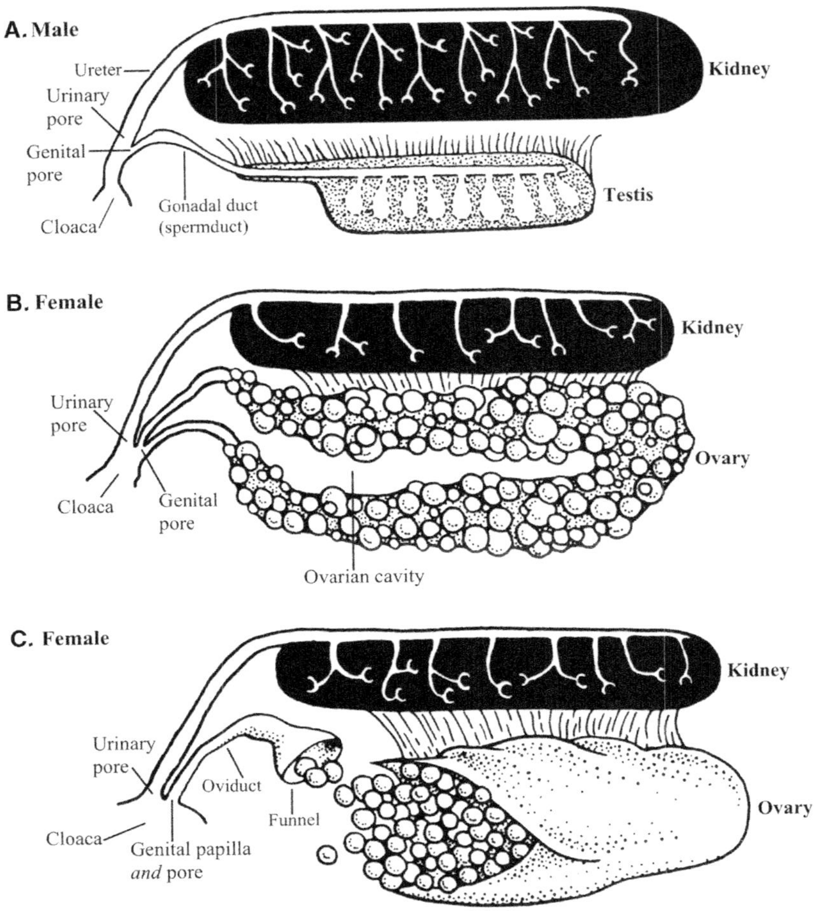

FIGURE 3.10 Schematic representation (lateral views) of gonads and gonadal ducts in fish: (A) male urogenital system; (B) female urogenital system; (C) rudimentary ovarian cavity as observed, for example, in salmonids. (Modified from [136], with permission.)

genital pore, situated between the anal and urinary pores. In the yellow perch (one fused ovary), the single ovarian duct appears to be a narrowing of the posterior ovary, which terminates in the abdominal pore.

On transverse/cross section, the (para-ovarian) ovarian cavity and ducts (two points of attachments) are unequivocally discernible, allowing for the rapid identification of ovarian differentiation and detection effects of estrogens or other sex steroids on gonadals differentiation (examples in Table 3.1). It is important to note that only transverse/cross sections will allow the detection of subtle but permanent changes in gonadal duct formation (see Figure 3.12, Chapters 7 and 8, and Appendix), whereas longitudinal sections (employed for histological detection of EDC- induced changes in sex ratios [63,138]) will not.

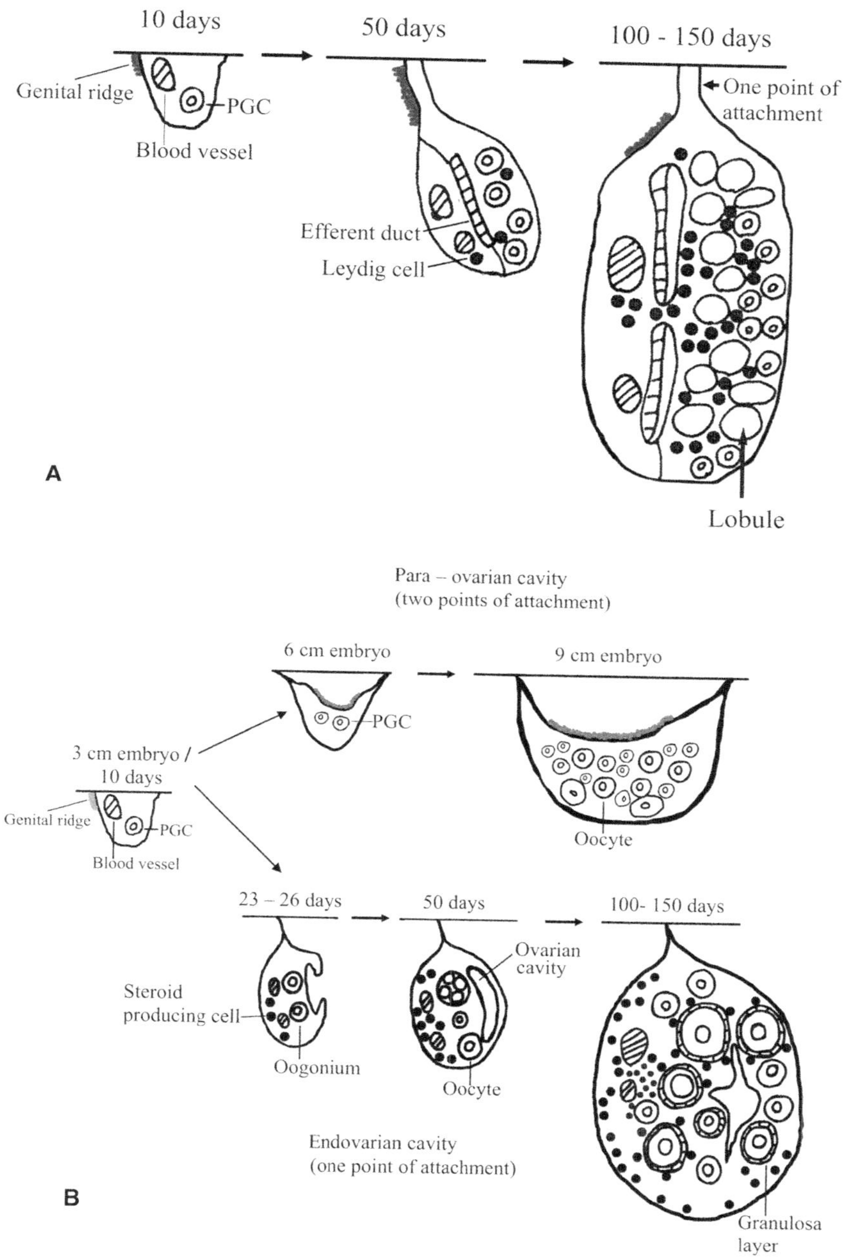

FIGURE 3.11 Gonadal differentiation patterns (cross-sectional views): (A) male testicular differentiation as observed in European perch (*Perca fluviatilis*, Percidae); (B) female ovarian differentiation as observed in Nile tilapia (*Oreochromis niloticus*, Cichlidae, upper panel) and European perch (lower panel). (Redrawn from [5] and [113].)

TABLE 3.1 Effects of Sex Steroids on the Gonadal Duct/Ovarian Cavity Formation[a]

Fish Species	Exposure Duration and Life Stage	Compound and Concentration	Effects Observed at the Level of the Gonadal Ducts/Ovarian Cavity	Refs.
Tilapia (*Tilapia mossambica*)	Daily oral administration of E_2, examination at 30, 40, 60, 70 dph)	E_2 (50 μg/g diet) from 6 to 25 dph	Complete feminization of gonad of genetic males, similar to ovaries of females	[79]
		E_2 (50 μg/g diet) from 6 to 15 dph	No signs of modifications, no intersex gonad	
		E_2 (50 μg/g diet) from 11 to 20 dph	Incomplete feminization of genetic males, intersex gonads (testis in posterior region, ovary in anterior region including ovarian cavity)	
	Daily oral administration for 19 or 44 days from 7 days after hatching	17α-methyltestosterone (MT), 50 μg/g for 19 d	No ovarian cavity in females, but testes with well-formed efferent ducts	[79]
		MT 1000 μg/g for 19 days	No effect in ovaries, testes with ovarian cavity	
		MT 1000 μg/g for 44 days	Females with expanded efferent ducts, no effect on germ cells, males with ovarian cavity and oocytes	
Nile tilapia (*Oreochromis niloticus*)	Juvenile female tilapia	MT 50 μg/g diet for	100% masculinization, suppression of cyp450 cholesterol-side-chain cleavage, 3β-hydroxysteroid dehydrogenase, and aromatase in the gonads	[155]
Roach (*Rutilus rutilus*)	During sexual differentiation, exposure starts at 50 dph for 50 or 100 days	Treated sewage effluent (100% contains 37 ng/L estrone, 5.9 ng/L E_2, < 1 ng/L EE_2)	Observations at 100 dph: Increased incidence of "male" fish with feminized gonadal ducts (two points of attachments to the mesentery: 100% incidence at 100% effluent, 70% at 47.4% effluent) Observations at 350 dph: 100% incidence at 100% effluent	[14]

European sea bass (*Dicentrarchus labrax*)	Oral administration for 100 days	MT 10 mg/kg feed: 126–226 dpf 226–236 dpf 326–426 dpf	Observations at 3 years of age: Permanent masculinization, increased proportion of males with functional ducts producing sperm at 3 years, sperm capable of fertilizing ova	[156]
Carp (*Cyprinus carpio*; genetic males)	During sexual differentiation, exposure starts at 50 dph	10 μg/L E_2 from 50 dph	Observations at 70 dph: All testes with oviduct as of 20 days of exposure	[11,94]
Medaka (*Oryzias latipes*) d-rR strain	d-rR strain, nano-injection exposure	0.5–2.5 ng EE_2 per egg 0.005–0.05 ng EE_2 per egg	Observations in adult fish: Sex-reversed males had typical female ducts with separate opening for the genital and urinary tracts Non-sex-reversed males had normally fused urinary and gonadal ducts, forming a single urogenital opening	[160]
Zebrafish (*Danio rerio*)	3 weeks' exposure, starting at various life stages: embryos, juveniles, adults	E_2, at 5, 25, and 100 ng/L 0 to 21 dpf or 21 to 42 dpf	Observations at 42 dpf: Increased proportion of presumptive males with feminized gonadal ducts (two points of attachment forming an ovarian cavity)	[151]
Fathead minnow (*Pimephales promelas*)	Various water exposure durations during sexual differentiation	EE_2 10 ng/L 0–5 dph 5–10 dph 10–15 dph 15–10 dph 0–20 dph	Observations at 100 dph: <10% males with ovarian-like cavity 40% males with ovarian-like cavity 65% males with ovarian-like cavity 45% males with ovarian-like cavity 20% males with ovarian-like cavity	[18]

(*continued*)

TABLE 3.1 *(Continued)*

Fish Species	Exposure Duration and Life Stage	Compound and Concentration	Effects Observed at the Level of the Gonadal Ducts/Ovarian Cavity	Refs.
	During sexual differentiation, exposure starts at embryolarval stage	4-*tert*-pentylphenol: 56, 180, and 560 μg/L	In male fish exposed to: 180 μg/L: mixed gonads (predominately testes with a scattering of primary oocytes); formation of an ovarian-like cavity 560 μg/L: complete absence of testicular development	[161]
	During sexual differentiation, exposure starts in early blastula embryos until 107 dph	10 ng/L EE_2	No fish with male gonadotype Sexually undifferentiated fish with two mesentarial attachments and ovarian-like cavity	[161]
	During sexual differentiation, exposure starts in early blastula embryos until 114 dph	Methyldihydrotestosterone: 0, 0.10, 0.32, and 1.0 μg/L	Mixed sex gonads (10–25%) and smaller testes or dysplasia of gonadal tissue Suppression of ovarian development	[162]

[a] This is not a complete list; more examples are given in Chapter 6.

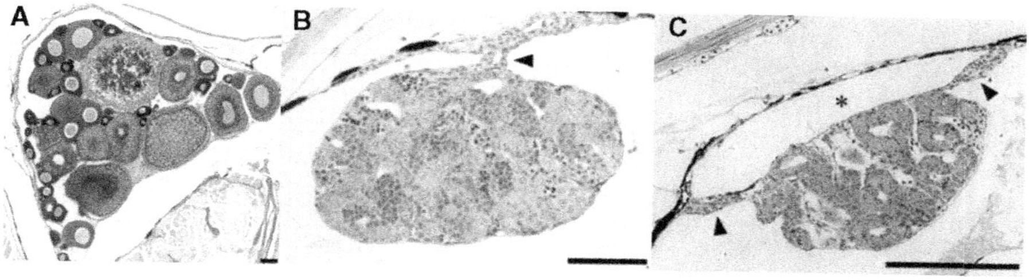

FIGURE 3.12 Histology of the gonad in control and EE_2-exposed fathead minnow at 100 dph (transverse/cross sections): (a) normal female, depicting the ovarian cavity; (b) normal male, depicting the sperm duct (arrowhead); (c), male with fully formed ovarian-like cavity (*) attached to the peritoneal wall by duct cells (arrowhead). Scale bar = 100 μm. (From [18], with permission.)

Male Gonadal Duct Sperm ducts are well developed in fish (chondrostei and teleostei) but are lacking in cyclostomes and are secondarily lost in salmonids [139]. The sperm duct (*vas deferens*) originates from the mesonephric system; however, in teleosts there is no connection between the mesonephros and the gonad at maturity; that is, unlike most other vertebrates, the sperm ducts of teleosts are anatomically separate from the nephric duct or Wolffian duct [114,134]. The efferent ducts of each spermatogenic lobule (*vasa efferentia*; see Section 5.1.1) carry mature sperm toward the *vas deferens* (Figure 3.12). The sperm ducts may be internalized within the testes or be evident along the dorsal posterior region of each testis. The two sperm ducts (*vasa deferentia*) fuse posterior to the main sperm duct, lobular outgrowths from the sperm duct (probably having additional secretory functions) may be present, or, as in goldfish (*Carassius auratus*), the duct is modified to provide an accessory sperm storage "organ," analogous to seminal vesicles in other vertebrate species. The main sperm duct generally terminates at the genital pore, which exits the body independently between the more anterior anal pore and the more posterior urinary pore, whereby the genital and urinary pores are often associated with an externally prominent urogenital papilla [134].

Effect of Sex Steroids on Gonadal Ducts In fish there is ample evidence that sex steroids induce gonoduct differentiation [5,8,17,125,140–142]. Indeed, using genetically controlled all-female and all-male Nile tilapia, Nakamura et al. [142] demonstrated that endogenous estrogen acts as an ovarian inducer and that the lack of steroid hormone, including androgen, is important for testicular differentiation. Similar earlier and most recent findings corroborate the interpretation of Nakamura et al. in that in all the species investigated, estrogenic exposure of genetic or presumed males resulted in the development of female gonadal ducts (Figure 3.12), albeit not in all cases including an ovarian cavity, and the development of the oviduct in females exposed to estrogens tended to occur earlier (see Table 3.1 and Chapters 4 and 6). This is not surprising since estrogens have an essential function in stimulating growth of the female reproductive tract (oviduct), with E_2 stimulating ovarian mitoses and growth (see Chapter 4). The oviduct is the target organ for estrogens and progesterone in most vertebrates, including oviparous species. The liver is considered the primary target organ for estrogens in oviparous vertebrates (fish, amphibians, reptiles, birds), as the liver is primarily responsible for synthesis of the future yolk proteins (e.g., vitellogenin) upon estrogenic stimulation [18,66,143–151]. However, contrary to the interpretation of Nakamura et al. [142], several authors demonstrated that exposure to androgens [17,152] or aromatase inhibitors [68,102,104,153,154] can provide an all-male phenotype or a high predilection for male gonads in various protogynic fish species (Table 3.1). Moreover, a partial or even all-male phenotype can be achieved in tilapia via temperature-induced reduced aromatase expression or xenobiotic-mediated aromatase inhibition, even in genetic all-female tilapia [76–78,155]. The latter thus suggest that in some fish species, the ratio of estrogen to androgen concentrations, rather than a single hormone (estrogen or androgen), may be critical in guiding gonadal differentiation toward a female or a male pathway. It is important to note that for many fish species (e.g., carp [11,12], bass [140,156], zebrafish [157]), EDC-induced phenotypic gonadal differentiation, once fully established, appears permanent. On the other hand, in cases where gonads have not differentiated fully, cessation of EDC exposure may lead to reestablishment of the genetically determined phenotypic sex and thus gonad differentiation [157–159]. The latter appears, however, to depend on the duration and concentration of EDC exposure, and foremost on whether EDC exposure covered the critical period of gonadal differentiation (see Section 3.6).

3.5 ENDOCRINOLOGY: INFLUENCE ON GONADOGENESIS

A wealth of evidence is available suggesting that sex steroids are inducers of sexual differentiation in teleosts [6–8,17,141]. Past experience with estrogen treatment of salmonid fish in aquaculture has aimed at either feminizing fish directly using high levels of estrogens (direct feminization), or indirectly, by producing phenotypic and functional males with genetic females via androgen treatment, thus producing neomales that can be paired with normal females to produce an all-female stock ([163]; for an in-depth review, see Piferrer [6]).

Based on these findings, one is led to assume that treatment of eggs or juveniles of all fish with either estrogens or androgens will lead to the feminization or masculinization of 100% of the treated fish, respectively. Indeed, estrogen (e.g., EE_2, E_2) [63] or androgen [100 ng/L nonaromatizable methyldihydrotestosterone (MDHT)] [164] treatment during sexual differentiation of zebrafish, medaka [23,33,45,116,165,166], and fathead minnows [18,19,167] will lead to an increased incidence of phenotypical and functional female fish (see also Section 6.2.1 and Figure 3.12). However, 100% feminization or masculinization was not always reached; rather, the degree (incidence) of feminization or masculinization of genetic male or female, respectively, appeared largely dependent on the steroid dose, type of steroid, route and duration of application, time of application (developmental stage of the fish), and on the definition of males and females. Indeed, in some studies males with undifferentiated testes have been categorized as females, resulting in the reporting of a skewed sex ratio [63,64]. This has been a slightly misleading interpretation, as, for example, in (un)differentiated gonochorists such as the zebrafish, estrogen treatment will arrest sexual differentiation and overall development in males, giving an erroneous impression of early ovary tissue (present in both sexes; see above), while cessation of exposure results in fully functional male fish [63,124,157,159].

Indeed, the proportion of undifferentiated juveniles appeared to depend largely on the actual size of the zebrafish; that is, in small (head–tail length) 42-dph zebrafish, a higher number of undifferentiated juveniles were present than in large 42-dph zebrafish [168]. Moreover, despite the use of high concentrations of the aromatase inhibitor tamoxifen (up to 27 μM), no overt masculinization of the F_1 zebrafish generation was observed (a decrease in hatching, survival, body weight, and length as well as feminization), whereas treatment with 1.0 nM E_2 resulted in a 72.1% female and 27.9% undifferentiated F_1, demonstrating that in this cohort of zebrafish, sex differentiation can be influenced only partially by EDCs, [168]. Similarly, the degree of masculinization achieved with aromatizable methyltestosterone in fathead minnows [19] and zebrafish [169] or with testosterone or methylandrostenediol in medaka [33,41], rarely reached 100%. Furthermore, higher doses of androgens appeared to induce feminization rather than a masculinization of presumably genetic males, probably largely dependent on the degree of aromatase expression in the developing gonads at the time of exposure during larval development [68,101,106,170–172]. This implies, however, that either germ cells or somatic cells in the developing gonad express aromatase and other essential enzymes for steroid biosynthesis [173]. In the developed ovary, oocytes do not express aromatase; rather, this enzyme expression appears restricted to granulosa and theca cells [5] (see Chapter 4). This would imply that steroid-producing cells would have to develop earlier (morphological differentiation) than, for example, the differentiating step from primordial germ cells to primary oocytes (cytological differentiation). This indeed appears to be the case, as aromatase activity in the brain was reported to occur before sexual differentiation of the gonad in zebrafish [105,106] and medaka [174], thus possibly explaining that 100% masculinization was achieved in zebrafish, medaka, and tilapia treated with the aromatase inhibitor fadrozole [68,102,123,174]. Furthermore, Piferrer et al. [104] were able to demonstrate that a brief exposure of genetic female salmon to an aromatase inhibitor during sexual differentiation resulted in nearly complete development of the fish to normal functional males. Similarly, 100% masculinization was achieved in zebrafish treated with 100 ng/L tributyltin (TBT), an aromatase inhibitor, from 0 to 70 dph [154]. However, this TBT dose, as well as the lower dose (10 ng/L TBT), also induced 100% abnormal sperm, suggesting that other mechanisms besides sexual differentiation were severely affected by TBT. Furthermore, the finding that lower doses of TBT did not induce 100% sex reversal may suggest that sex reversal, whether induced directly by high doses of estrogens or androgens or indirectly via the inhibition of aromatases, is a high (supraphysiological)-dose effect. The high dose may represent a concentration that is several orders of magnitude higher than the normal physiological concentrations present and thus may also have toxic effects beyond interaction with the respective receptors and enzymes. Contrary to the latter findings, studies using fadrozole in nonsexually differentiated juvenile fathead minnows [19] and fertilized eggs of medaka [41] had no effect on the sex ratio, suggesting that the start, duration, and dose of aromatase inhibitor exposure may be critical for achieving morphological sex reversal. Contrary to the findings with estrogens, androgens, and aromatase inhibitors, histological and ultrastructural studies in several fish species, including the medaka and guppy of either sex, suggest that steroid-hormone biosynthesis and steroid-producing cells in the gonad appear after the completion of gonadal sex differentiation [41,80]. Last but not least, van den Hurk and Slof [175] and van den Hurk and Lambert [176] have reported that progesterone treatment of medaka and rainbow trout does not influence sexual differentiation, suggesting that the necessary enzymes for conversion of progesterone, testosterone, and/or estradiol are not present during sexual differentiation. In contrast, gene expression of 11β-hydroxylase, an enzyme necessary for the synthesis of the androgen 11-ketotestosterone, was found prior to sex differentiation in genotypic male rainbow trout [103,177], reinforcing the importance of androgens in sexual differentiation in rainbow trout and possibly all salmonids.

It must be emphasized however, that at present there is an extreme paucity of toxicodynamic data demonstrating the concentration of xenobiotics and natural hormones at the site of interaction (receptors and enzymes) and during the various phases of sexual differentiation. Although the data above appear somewhat controversial, it could suggest that at least in fathead minnows [19], medaka [41], and zebrafish [61], possibly with the exception of salmon [104], sex steroids are not synthesized in the developing embryo adjacent to or within the developing gonad and appear not initially involved in the primary sexual differentiation of primordial germ cells into ovaries or testes. However, current data on brain aromatases also strongly suggest involvement of sex steroids in the initial steps of sexual differentiation.

Indeed, the brain isoform cyp19b is expressed in zebrafish mainly in the hypothalamus, pituitary, and sensory control tissues [178]. The production of estrogen in these regions apparently influences neuroendocrine functions, sexual behavior, and sexual differentiation. In addition, sexually dimorphic expression of brain aromatase is observed in developing zebrafish, whereas it is absent for ovarian aromatase expression [106], supporting the findings of Tong et al. [178], although it remains to be proven that brain-synthesized E_2 is produced in sufficient quantities to be able to reach the gonadal ridge and thus has influence on the developing gonad proper. Recent studies of zebrafish development suggest that maternal P450aromB (cyp19b) is enriched in oocytes, then rapidly degraded post-fertilization, while embryonic P450aromB (cyp19b) expression begins as early as 5 hours post-fertilization, reaching its maximal expression (10-fold) at 72 hours post-fertilization [105]. Conversely, embryonic P450aromA (cyp19a) expression begins at 48 hours post-fertilization. Studies have also shown that an up-regulation of brain aromatase (cyp19b) resulting from estradiol treatment occurs as early as 24 hours post-fertilization, in zebrafish [179], whereas ovarian aromatase is not expressed in developing (0- to 14-day) medaka fry [174]. However, in zebrafish, embryonic P450aromB (cyp19b) but not embryonic P450aromA (cyp19a) was inducible with estrogens [105].

Similar to cyp19a, the cyp19b promoter has a potential binding site for steroidogenic factor 1 (SF-1), involved in the formation of the gonads [180,181] (see Figure 3.9).

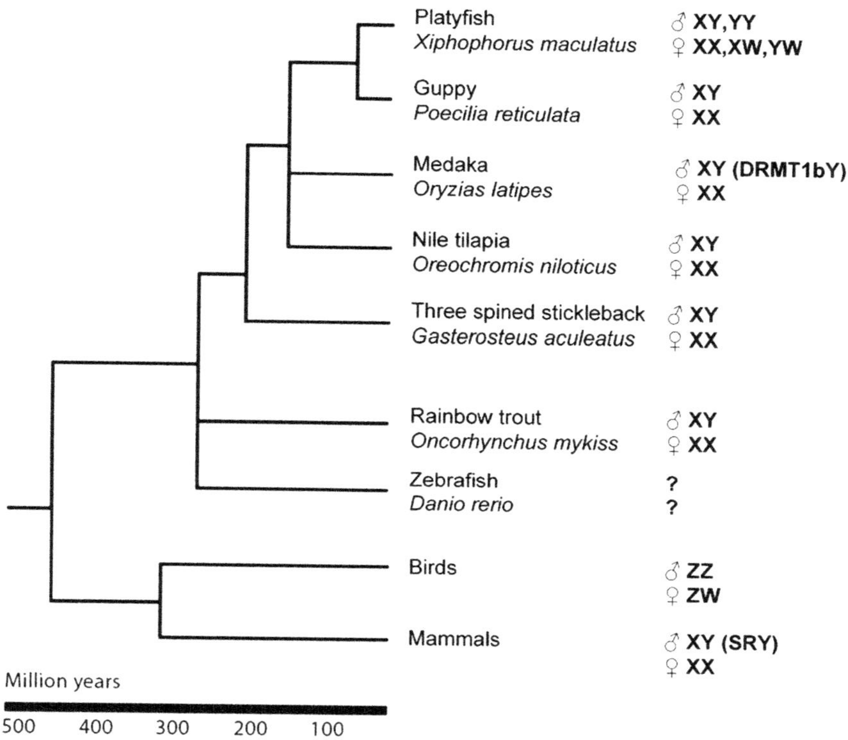

FIGURE 3.13 Major fish models and their sex determination systems. Known master sex determining genes are indicated. (Redrawn and modified from [111].)

SOX9 binding sites have also been identified in the cyp19a and cyp19b promoters of medaka [174], and the expression pattern of *SOX9* is similar to that of aromatase [182,183]. This could suggest that aromatase is downstream of *SRY*-like *SOX* genes in the sex-determining pathway [132,183–205] and therefore, along with estrogen production, is important in sex determination, sex differentiation, and sex reversal in fish [153]. In the absence of the *SRY* gene in fish [111], the regulation of *SOX9* is complicated by genome duplication [198] and subfunctionalization, largely depending on the fish species. For example, one copy of *SOX9a* is expressed and up-regulated in testis of undifferentiated male zebrafish and precedes the expression of *amh*, while another copy of *SOX9* (i.e., *SOX9b*, is expressed in zebrafish ovaries [198,203]. However, despite the current understanding of *SOX9* in zebrafish, no master sex regulator (Figure 3.13) is known for this as for most other fish species [206]. Notable exceptions to the latter, albeit without suggesting exclusivity or completeness, include the guppy, three-spined stickleback (*Gasterosteus aculeatus*), and medaka [111,206]. Indeed, in medaka the X and Y chromosomes are homomorphic (i.e., morphologically indistinguishable); however, Matsuda et al. [54,207] and Nanda et al. [208,209] were able to visualize medaka sex chromosomes via fluorescent *(in situ)* hybridization (FISH). More important, the medaka is the only nonmammalian vertebrate for which a master sex-determining gene (*dmrt1bY*; see Figure 3.13) at the molecular level is known [210–213]. *Dmrt1bY or dmy* is a Y-specific copy of the autosomal *dmrt1* gene. Consequently, males have, in addition to the two autosomal allelic copies of *dmrt1*, another copy linked to the Y-chromosome (*dmrt1bY*), whereas female medaka possess only the two autosomal allelic copies [207,209]. Consistent with a master sex-determining gene, mutations in the *dmrt1bY* gene and thus reduced expression of *dmrt1bY* in XY medaka embryos results in phenotypical female XY medaka [207,214]. The latter results would suggest that early testis development requires *DMY* expression above a certain threshold level, which if not attained will result in functional phenotypic females. However, to complicate issues, Nanda and co-workers [208] also found genotypic XX medaka males with normal phenotype and uncompromized fertility, but lacking *dmrt1bY*, which were capable of breeding with XX females. The latter coupling did, however, bring about a higher incidence of female offspring. All of the above suggests that *dmrt1bY* may not be solely responsible for male development, but that still unidentified autosomal modifiers may play a similarly important role in sex determination.

The data above suggest three issues:

1. A very early critical window in teleost (here medaka and fathead minnows, possibly also zebrafish) development exists, where sex steroids have no or only limited influence on sexual differentiation of the gonads, possibly due to limited synthesis and lack of sufficient transport from the brain to the gonads, and lack of expression of the necessary steroid synthetic pathways and hormone receptors *in situ*. This is supported by the recent discovery of the *DMY* gene in medaka, the testis-determining gene in medaka considered equivalent to the mammalian *SRY* gene, whose expression in mammals triggers testicular differentiation prior to the production of steroids [207] and the fact that once the critical window is surpassed, even extreme concentrations of steroids [41,218] will not evolve in complete sex reversal (see below).
2. A subsequent critical window in development exists, where the presence of the respective hormone receptors renders the presumptive ovary or testes susceptible to sex reversal by high concentrations of sex steroids or structurally/functionally analogous compounds.
3. The germ cells in the undifferentiated gonad have a bi-potential nature, in that they can differentiate into gametes different from the predetermined sex upon continued high-dose stimulus with a sex steroid or sex steroid analog.

Indeed, it must be kept in mind that all experiments to date using exposure of eggs, embryos, or juvenile fish to steroids or steroid analogs or active compounds were carried out generally with rather high doses of these compounds (0.2 ng/L to 1.2 mg/L concentrations in ambient water; see Chapter 6). Relative to the normal plasma steroid concentrations expected in developmental stages (probably below pg/mL compared to the 300 pg/mL E_2 in adult male fathead minnows [170,219,220] and used as an approximation in view of lacking corresponding values in juvenile fish), the exogenously applied steroids would exceed the endogenously present concentrations by several orders of magnitude, especially when considering that steroids (e.g., 16 or 64 ng/L EE_2)

demonstrated bioconcentration factors (BCFs; fish/ambient water concentration) above 600 [167]. Indeed, using this BCF and an assumed ambient EE_2 concentration of 5 ng/L and 80% uptake, exposed male fathead minnows would end up with a plasma concentration for EE_2 of 2400 pg/mL at steady state. Consequently, *environmentally relevant* steroid concentrations represent excessive doses from the perspective of the adult small fish and more so for the embryo and the developing juvenile fish. It thus must be expected that these high exogenously applied steroid concentrations would overrun any normal physiologic hormonal signaling within the hypothalamic–pituitary–gonadal axis [221], and therefore describe a high–dose effect of these steroids and androgens. Consequently, these experiments may not be representative of the subtle hormonal changes involved in sexual differentiation. To better understand the effects of steroids on sexual differentiation, *in situ* determinations of steroid levels would be necessary. Indeed, when measured in blood or whole-body homogenate, the natural steroid concentrations will not reflect the local levels that are delivered in a paracrine way (i.e., the hormone-producing cell is directly adjacent or close to the target germ cell), and much higher levels can be expected in this microenvironment than in blood. Furthermore, binding proteins can act as docking proteins, thus optimizing delivery of the steroid hormone to the receptor. This mechanism may work efficiently for the natural compound but not for analogs, which may bind with lower affinities to both receptors and steroid-binding hormones, and to reach a similar effect may therefore need much higher concentrations than those of natural compounds.

3.6 CRITICAL PERIOD OF SEXUAL DIFFERENTIATION IN DEVELOPING FISH

Using varying developmental stages (fertilized eggs to months post-hatch) and treatment of these stages with short- or long-term high concentrations of steroids (primarily estrogens; more rarely, nonaromatizable androgens and aromatase inhibitors), numerous investigators have tried to define the *critical period of sexual differentiation* (reviewed by Piferrer [6]). This period was commonly defined as the time between the earliest detection of gonadal differentiation and the time when both ovaries and testes are fully differentiated, such that application of high doses of steroids later than the end of the critical period of sexual differentiation would not result in complete phenotypic sex reversal.

Correspondingly, for *zebrafish* the critical period of sexual differentiation was suggested to lie between 20 and 60 days post-hatch (dph), as development of ovaries was detected at approximately 23 to 25 dph [58,61], and fully developed testes were observed at 60 dph [66]. Indeed, exposure of juvenile zebrafish with EE_2 concentrations ranging from 1, 10, and 100 ng/L from 2 to 60 dph resulted in a clear sex shift toward 100% phenotypic female fish (having ovaries), as expected [63]. However, continued rearing of these fish to adulthood (160 dph) demonstrated that these fish presented no significant departure from the 1 male:1 female ratio. The latter data were corroborated by van der Ven and co-workers [168], whereby exposure of embryos up to 42 dph with 1 nM E2 resulted in a feminization degree of 54.2 to 72.1%, 0.0 to 9.5% males, and 27.9 to 36.3% undifferentiated fish, suggesting that cessation of exposure and continued rearing would have reestablished a nearly 1 male: 1 female ratio, as was also reported by Schafers and co-workers [159] in a similar experiment. This suggests that exposure of zebrafish to strong estrogens during sexual differentiation and early gametogenesis does not irreversibly alter phenotypic sex, and furthermore, that sex steroids are involved only secondarily in the sexual differentiation of zebrafish gonads, as suggested earlier (see Section 3.5). In contrast, life-long exposure to high concentrations of sex steroids (>10 ng/L EE_2) nearly irreversibly inhibits reproduction, whereby reproduction inhibition is induced at various levels of interaction: decreased survival, growth, gamete production, and increased pathological changes secondary to increased vitellogenin synthesis in presumptive males—however, not via an increased incidence of intersex (<0.1%) [159,168]. Conversely, Kinnberg and co-workers [222] reported no intersex gonads in fish of the control group, whereas intersex gonads were observed in 1, 1, or 12% of the fish in the groups exposed to 16, 64, or 202 µg/L prochloraz, respectively, for 60 days as of 24 hours post-fertilization. In addition, a dose-dependent increase was observed in the proportion of males and undifferentiated fish. Whether the latter is really contradictory to the findings of van der Ven et al. [168], or whether this may be a reflection of the genetic background of the zebrafish cohort employed (see Sections 3.2 and 3.7) and/or a compound specific trait, cannot be determined at this time.

In *medaka* the normal period of earliest ovary and testicular differentiation lies between 4–7 and 35 dph, respectively [50,223,224]. It is important to note, however, that to achieve complete sex reversal with steroids in medaka, the route of steroid application (oral dosing vs. ambient water vs. egg injection), the type of steroid, and the dose and duration of steroid application are crucial. Indeed, nearly complete sex reversal can be achieved with injection of high doses (0.5 to 2.5 ng/egg) of EE_2 [116], high oral doses (ppm) of E_2 or estrone in the feed with newly hatched fry [45,73], short (4 to 10 days) immersion exposure of fertilized eggs to 100 µg/L E_2 [27,165], or prolonged exposure (one or two months) of newly hatched sexually undifferentiated juveniles to high concentrations (0.01 to 1.66 µg/L) of E_2 [23] or EE_2 (100 ng/L) [166] in the ambient water (see also Section 3.4 and Table 3.1). It is interesting to note that lowering the concentration of EE_2 to 10 ng/L did not induce sex reversal [166], while the application of 10 ng/L E_2 for a much shorter period still had a 100% reversal effect [23]. Furthermore, Yamamoto [45,73] and Iwamatsu [27,165] reported that lower doses of estrogens result in a high incidence of intersex and in a lower number, or an absence, of sexually reversed males, respectively. Conversely, 1 to 14 days of post-fertilization exposure of medaka to 5.5 µg/L *o,p*-DDT resulted in a fivefold-increased expression of brain aromatase (cyp19b) as well as in a complete male-to-female sex reversal [174]. At least for the medaka, these data suggest that extreme steroid doses are necessary to induce sex reversal and that this can be achieved within the period of the newly fertilized egg and the latest point of sexual differentiation of the male and female gonads. Low doses (ng/L) of steroids appear to promote intersex and thus appear to have a different mechanistic pathway (although possibly employing similar or the same receptors as those involved in sex reversal) than that of high doses. However, medaka exposed to intermediate concentrations of EE_2 (10 ng/L) starting at hatching until sexual maturity induced not only intersex but also fibrosis of the gonad and apoptosis of spermatogenic cells in male fish as well as hepatotoxicity and renal toxicity in both sexes [225].

In *fathead minnows*, up to 10 dph the primitive gonad is characterized by undifferentiated primordial germ cells (PGCs) surrounded by a thin layer of somatic tissue. The first signs of ovarian differentiation are observed at 10 dph, with continued differentiation and development up to a full functional ovary at approximately 90 to 100 dph with vitellogenic follicles [226]. In males, presumptive testes are observed at 10 dph, characterized by a diffuse assembly of somatic and germ cells and a "sperm duct," yet meiosis of spermatogonia and differentiation are not observed until 90 dph [226]. Similar to the findings in medaka, fathead minnows treated with an intermediate concentration (10 ng/L) of EE_2 [18] in the ambient water at various timeranges throughout development did not result in sexreversal. The primary findings were a disruption in sperm duct development (presence of an ovarian-like cavity), with the highest incidence (64.3%) in fish exposed during from 10 to 15 dph, and the presence of increased vitellogenin levels in male fish. In contrast, Länge et al. [167] found skewed sex ratios as well as intersex (84% female: 5% male, 11% male with testis–ova) and full sex reversal in fathead minnows exposed from newly fertilized embryos to 56 and 172 dph with 4 ng/L EE_2, respectively, while skewed sex ratios (65% female: 35% male) were observed at 1 ng/L EE_2 and 172 dph. Furthermore, exposure of nonsexually differentiated juvenile fathead minnows for 21 days to high concentrations (25 to 100 μg/L) of the highly specific aromatase inhibitor fadrozole [19] had no effect on the sex ratio. The data above corroborate previous findings in medaka: Chronic exposure to intermediate concentrations of estrogens can result in sex reversal while short exposures induce transient changes (reversible) only. Consequently, it can be assumed that fathead minnows, similar to medaka, have a male determinant gene whose signaling during sexual differentiation of the gonads can be overridden by high concentrations of external estrogens in order to produce the sex reversal observed. Furthermore, short-term exposures to lower doses (ng/L) of exogenous steroids appear to promote intersex and thus appear to have a different mechanistic pathway, although possibly employing similar or the same receptors, than that of the high doses. Consequently, the "critical period of sexual differentiation" in fathead minnows is defined primarily by the effects that can be superimposed on normal sexual development by high concentrations of estrogens relative to the absent or undetectable endogenous steroids present in embryos and juvenile fish.

3.7 BI-POTENTIALITY OF GERM CELLS IN ADULT FISH

The observation that short-term exposures to varying (< 10 ng/L to 30 μg/L) concentrations of estrogens can induce intersex (= ovo–testis = testis–ova) in juvenile medaka or fathead minnows (see above) is also reported for adult medaka treated over a short period of time with high concentrations (30 μg/L) of E_2 [29,42,99]. Unfortunately, little information is available with regard to the hormonal induction of intersex in adult zebrafish and fathead minnows, with the exception that 21 days of treatment of adult zebrafish with high doses of EE_2 (0.25 to 85 ng/L) did not result in the generation of oocytes or ovarian-like tissue in the testes of exposed male zebrafish [227], nor did treatment of zebrafish embryos with 1 nM E_2 up to 42 dph [168]. Similar results were

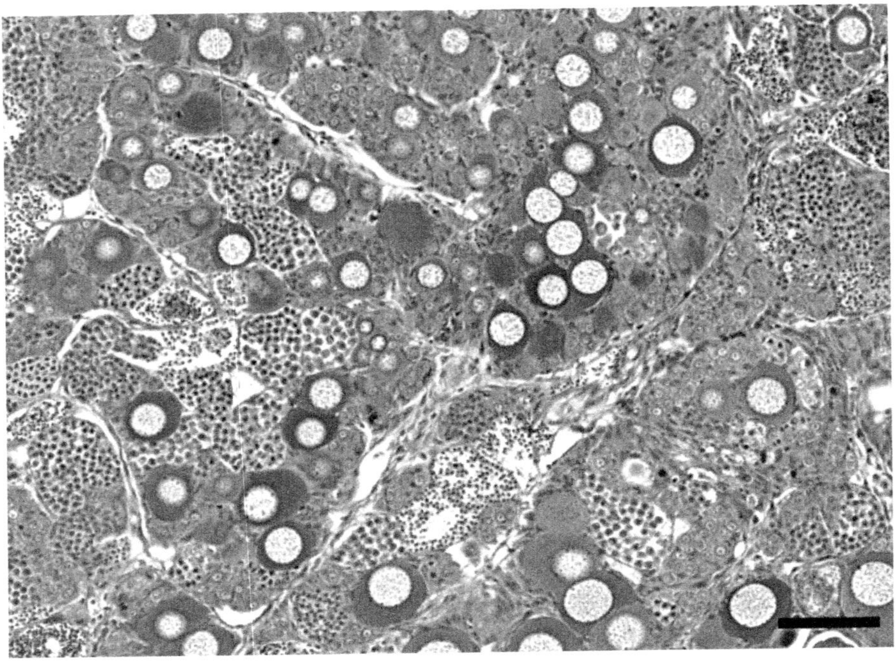

FIGURE 3.14 Testis–ova in a male medaka. Scale bar = 50 μm. (Courtesy of J. W. Fournie, U.S. EPA, Gulf Breeze, Florida.) (See insert for color representation.)

reported by van den Belt et al. [228] using concentrations of 10 to 25 ng/L EE_2 for 24 days. Conversely to the generation of intersex, increased incidences of hermaphrodism (fully functional ovaries and testes within the same animal) were found in the F_1 generation treated continuously with EE_2 in a multigenerational study [124]). Intersex (see also Chapters 2 and 6) is, for example, the development of oocytes within fully developed functional testes (Figures 3.14 to 3.17) following treatment with high concentrations of estrogens, aromatizable androgens, or hormonally active xenobiotics [229], as was demonstrated for medaka by Balch and co-workers [230]. The development of intersex suggests that within adult testicular tissue, relatively undifferentiated gonial stages are permanently present that can be induced to differentiate along a female pathway given a sufficient dose of female steroids. Indeed, intersex phenomena have frequently been observed in roach and gudgeon (*Gobio gobio*) in the UK and Denmark [15,16,231,232]: however, always in locations in close proximity to sewage treatment plants and thus in situations where high ambient concentrations of steroids (E_2, estriol, estrone, and EE_2) are expected and have been confirmed analytically [233]. These observations could lead to the erroneous conclusion that sewage treatment plant effluents and specifically natural and synthetic estrogens as well as mimics thereof are causally related to the appearance of intersex in roach and gudgeon. Indeed, Jobling et al. [15] and van Aerle et al. [232] reported intersex in fish at reference sites (i.e., apparently pristine rivers and still waters, respectively),

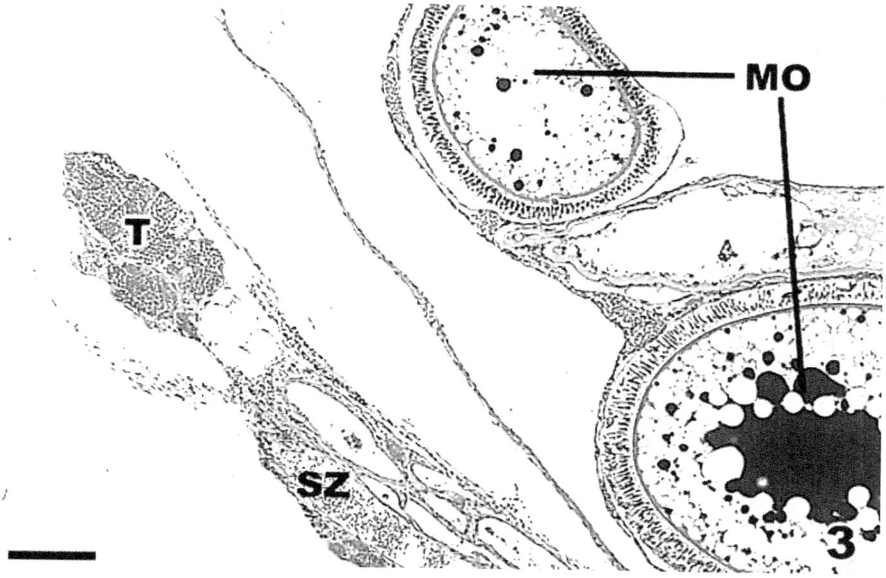

FIGURE 3.15 Gonad of a medaka (age 5 months) exposed as fry (at 7 dph) to 100 μg/L testosterone, showing both mature oocytes (MOs) and mature spermatozoa (SZ). Scale bar = 88 μm. (From [33], with permission.)

suggesting that a low background incidence of intersex could indeed be a normal phenomenon, as has recently been reported for zebrafish in captivity [168]. In view of this, the question must be asked whether there are populations with a naturally high background incidence of intersex as well as populations whose genetic predisposition provides a higher susceptibility to intersex formation. Indeed, the recently reported intersex phenomena in roach, with 4.5 to 5% at reference sites and 6.7 to 6.5% at sewage effluent–impacted sites, where the highest incidence of 26.5% was found at the site with the highest EDC concentration in the effluent, does suggest that EDCs may have an enhancing effect on the development of intersex, especially, probably, when the background (genetic susceptibility) is already high [16]. The latter interpretation is also supported by the fact that intersex phenomena in roach were not inducible via estrogen exposure in the laboratory [234]. Moreover, this clearly suggests that genetic predisposition for the formation of intersex is an important area that must be understood very well before inferences can be made to the intersex-inducing effects of estrogens in roach.

Shibata and Hamaguchi [99] described in detail the genesis of intersex within the testicular tissue in medaka. These authors were able to demonstrate that at least in the medaka, spermatogonia B are involved primarily in sexual differentiation into oocytes when stimulated with extreme doses of estrogens. These experiments would thus also suggest that spermatogonia A cannot be influenced and thus represent the genetically determined (DMY) subset of gonial cells that are formed during the

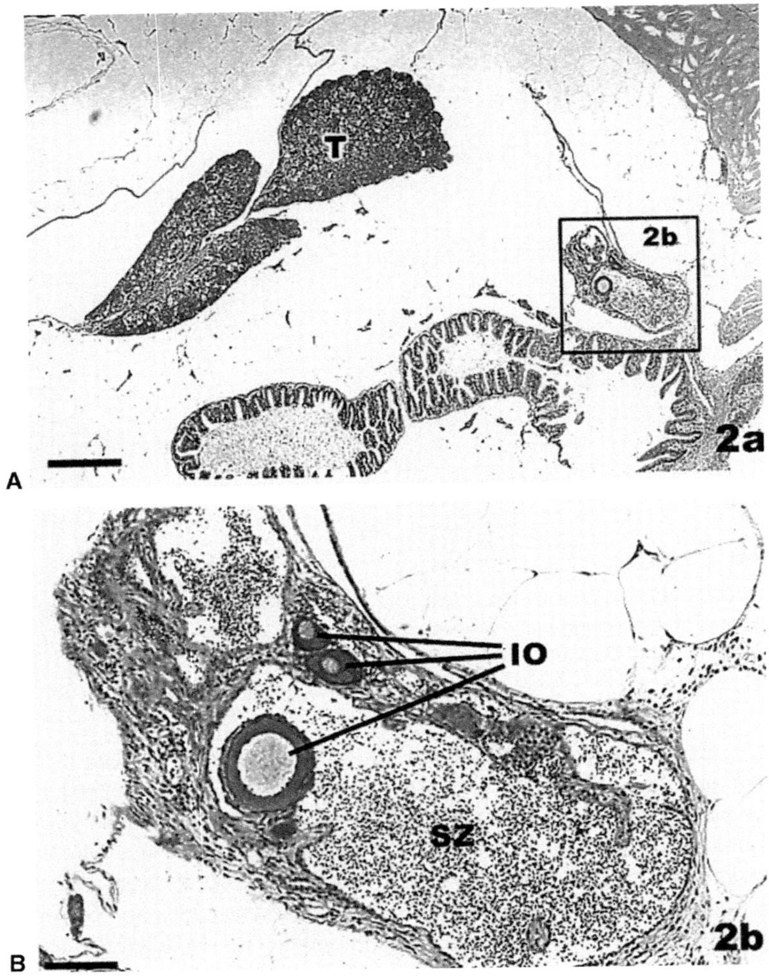

FIGURE 3.16 (A) Gonad of a medaka (age 5 months) exposed as fry to 15 μg/L estradiol for 6 days with normal testis morphology (T). However, several immature oocytes are present within the testicular duct system (insert B). Scale bar = 218 μm. (B) Immature oocytes (IOs) are shown within the duct system of a medaka fry exposed to 15 μg/L estradiol, along with mature spermatozoa (SZ). Scale bar = 44 μm. (From [33], with permission.)

sexual development of medaka (see Sections 3.4 to 3.6). Support for this suggestion is found in experiments by Metcalfe et al. [218], who were able to show that despite exposure of medaka to 100 ng/L E_2 from hatch to 100 days post-hatch, the 100-dph fish with intersex gonads had typical testicular tissue in the anterior and primarily ovarian tissue in the posterior end of the gonad (Figure 3.17). As the germ cells in the gonad of the male medaka develop in an anterior-to-posterior fashion, this indicates that the stimulus of the high dose of E_2 was merely able to

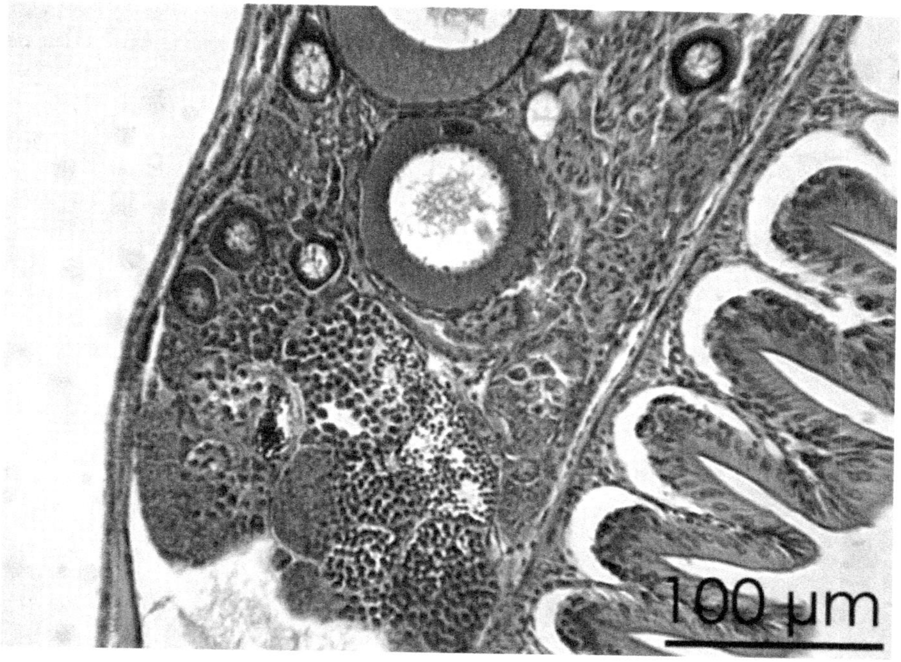

FIGURE 3.17 Testis–ova in the gonads of a male medaka (103 dph) exposed to 100 μg/L nonylphenol for 100 days. (Courtesy of D. Janz, University of Saskatchewan, Saskatoon, Canada, [235].) (See insert for color representation.)

influence gonial differentiation over a given period of time, whereas genetic determination took over at a later time, irrespective of the E_2 concentration present. Conversely, this suggests that intersex phenomena in adult fish probably represent a transient effect, largely dependent on the duration and concentration of exogenous steroid exposure.

REFERENCES

1. Yamamoto, K., Studies on the formation of fish eggs: Annual cycle in the development of the ovarian eggs in the flounder, *Liopsetta obscura*. *Journal of the Faculty of Science, Hokkaido University, Series VI, Zoology*, 1956, **12**:362–373.
2. Yamamoto, T.-O., Sex differentiation, in *Fish Physiology*, W. S. Hoar, and D. J. Randall, Editors. Academic Press, New York, 1969, pp. 117–175.
3. Van der Kraak, G., et al., Effects of LH-RH and des-Gly10[D-Ala6]LH-RH-ethylamide on plasma gonadotropin levels and oocyte maturation in adult female coho salmon (*Oncorhynchus kisutch*). *General and Comparative Endocrinology*, 1983, **49**(3): 470–476.

4. Nakamura, M., Effects of estradiol-17β on gonadal sex differentiation in two species of salmonids, the masu salmon *Oncorhynchus masou*, and the chum salmon *O. keta*. *Aquaculture*, 1984, **43**(1–3):83–90.

5. Nakamura, M., et al., Gonadal sex differentiation in teleost fish. *Journal of Experimental Zoology*, 1998, **281**(5):362–372.

6. Piferrer, F., Endocrine sex control strategies for the feminization of teleost fish. *Aquaculture*, 2001, **197**(1–4):229–281.

7. Piferrer, F., et al., Genetic, endocrine, and environmental components of sex determination and differentiation in the European sea bass (*Dicentrarchus labrax L.*). *General and Comparative Endocrinology*, 2005, **142**(1–2):102.

8. Devlin, R.H. and Y. Nagahama, Sex determination and sex differentiation in fish: An overview of genetic, physiological, and environmental influences. *Aquaculture*, 2002, **208**(3–4):191–366.

9. Gimeno, S., *The Estrogenicity of Alkylphenols in the Aquatic Environment: An Approach for Risk Assessment using Various Life Stages of all Male Carp Populations.* PhD thesis, Utrecht University, Utrecht, The Netherlands, 1997, Chap. 1.

10. Gimeno, S., Gonadal sexual differentiation as histological endpoint. *OECD Technical Workshop on Gonadal Histology of Small Laboratory Fish*, Bilthoven, The Netherlands, 2002.

11. Gimeno, S., et al., Feminisation of young males of the common carp, *Cyprinus carpio*, exposed to 4-*tert*-pentylphenol during sexual differentiation. *Aquatic Toxicology*, 1998, **43**(2–3):77–92.

12. Gimeno, S., et al., Demasculinisation of sexually mature male common carp, *Cyprinus carpio*, exposed to 4-*tert*-pentylphenol during spermatogenesis. *Aquatic Toxicology*, 1998, **43**(2–3):93–109.

13. Gimeno, S., et al., Disruption of sexual differentiation in genetic male common carp (*Cyprinus carpio*) exposed to an alkylphenol during different life stages. *Environmental Science and Technology*, 1997, **31**(10):2884–2890.

14. Rodgers-Gray, T.P., et al., Exposure of juvenile roach (*Rutilus rutilus*) to treated sewage effluent induces dose-dependent and persistent disruption in gonadal duct development. *Environmental Science and Technology*, 2001, **35**(3):462–470.

15. Jobling, S., et al., Altered sexual maturation and gamete production in wild roach (*Rutilus rutilus*) living in rivers that receive treated sewage effluents. *Biology of Reproduction*, 2002, **66**:272–281.

16. Bjerregaard, L.B., B. Korsgaard, and P. Bjerregaard, Intersex in wild roach (*Rutilus rutilus*) from Danish sewage effluent-receiving streams. *Ecotoxicology and Environmental Safety*, 2006, **64**(3):321–328.

17. Nagahama, Y., et al., Sexual plasticity in fish: A possible target of endocrine disruptor action. *Environmental Sciences*, 2004, **11**(1):73–82.

18. van Aerle, R., et al., Windows of sensitivity for the estrogenic effects of ethinylestradiol in early life-stages of fathead minnow, *Pimephales promelas*. *Ecotoxicology*, 2002, **11**(6):423–434.

19. Zerulla, M., et al., Morphological sex reversal upon short-term exposure to endocrine modulators in juvenile fathead minnow (*Pimephales promelas*). *Toxicology Letters*, 2002, **131**:51–63.

20. Rühl-Fehlert, C.I., et al., Spontaneous intersex in a fathead minnow (*Pimephales promelas*). *European Society of Toxicologic Pathology (ESTP) Congress*, Hamburg, Germany, 2003.

21. Kidd, K.A., et al., Collapse of a fish population after exposure to a synthetic estrogen. *Proceeding of the National Academy of Science of the USA*, 2007, **104**(21): 8897–8901.

22. Hamaguchi, S., Sex differentiation of germ cells and their supporting cells in *Oryzias latipes*. *Fish Biology Journal of Medaka*, 1992, **4**:11–17.

23. Nimrod, A.C. and W.H. Benson, Reproduction and development of Japanese medaka following an early life stage exposure to estrogens. *Aquatic Toxicology*, **44**(1–2):1998, 141–156.

24. Davis, C.R., M.S. Okihiro, and D.E. Hinton, Effects of husbandry practices, gender, and normal physiological variation on growth and reproduction of Japanese medaka, *Oryzias latipes*. *Aquatic Toxicology*, 2002, **60**(3–4):185–201.

25. Foran, C.M., et al., Influence of exposure history on vitellogenin induction in medaka: A case of 'estrogen memory'? *Marine Environmental Research*, 2000, **50**(1–5):196.

26. Iwamatsu, T., Stages of normal development in the medaka *Oryzias latipes*. *Zoological Science*, 1994, **11**:825–839.

27. Iwamatsu, T., et al., Sex reversal of medaka (*Oryzias latipes*) by a short exposure of embryos to sex steroids. *Zoological Science*, 2000, **17**:15.

28. Kang, I.J., et al., Effects of bisphenol A on the reproduction of Japanese medaka (*Oryzias latipes*). *Environmental Toxicology and Chemistry*, 2002, **21**(11):2394–2400.

29. Kang, I.J., et al., Effect of 17β-estradiol on the reproduction of Japanese medaka (*Oryzias latipes*). *Chemosphere*, 2002, **47**(1):71–80.

30. Kawahara, T. and I. Yamashita, Sexual differentiation of gonadal cells in the medaka *Oryzias latipes*. *Zoological Science*, **14**(Suppl.):1997, 70.

31. Kiparissis, Y., et al., Effects of the antiandrogens, vinclozolin and cyproterone acetate on gonadal development in the Japanese medaka (*Oryzias latipes*). *Aquatic Toxicology*, 2003, **63**(4):391–403.

32. Koger, C.S., S.J. Teh, and D.E. Hinton, Variations of light and temperature regimes and resulting effects on reproductive parameters in medaka (*Oryzias latipes*). *Biology of Reproduction*, 1999, **61**:1287–1293.

33. Koger, C.S., S.J. Teh, and D.E. Hinton, Determining the sensitive developmental stages of intersex induction in medaka (*Oryzias latipes*) exposed to 17β-estradiol or testosterone. *Marine Environmental Research*, 2000, **50**(1–5):201–206.

34. Metcalfe, T.L., et al., Gonadal development and endocrine responses in Japanese medaka (*Oryzias latipes*) exposed to *o,p*-DDT in water or through maternal transfer. *Environmental Toxicology and Chemistry*, 2000, **19**(7):1893–1900.

35. Patyna, P.J., et al., A proposed multigeneration protocol for Japanese medaka (*Oryzias latipes*) to evaluate effects of endocrine disruptors. *Science of the Total Environment*, 1999, **233**(1–3):211–220.

36. Wester, P.W. and J.H. Canton, Histopathological study of *Oryzias latipes* (medaka) after long-term β-hexachlorocyclohexane exposure. *Aquatic Toxicology*, 1986, **9**(1): 21–45.

37. Wester, P.W., Histopathological study of *Poecilia reticulata* (guppy) and *Oryzias latipes* (medaka) following methyl bromide and sodium bromide exposure, in *Pathology* University of Utrecht, Utrecht, The Netherlands, 1988, p. 208.
38. Wester, P.W., J.H. Canton, and J.A.M.A. Dormans, Pathological effects in freshwater fish *Poecilia reticulata* (guppy) and *Oryzias latipes* (medaka) following methyl bromide and sodium bromide exposure. *Aquatic Toxicology*, 1988, **12**(4):323–344.
39. Yokota, H., et al., Effect of bisphenol A on the early life stage in Japanese medaka (*Oryzias latipes*). *Environmental Toxicology and Chemistry*, 2000, **19**(7):1925–1930.
40. Yokota, H., et al., Life-cycle toxicity of 4-nonylphenol to medaka (*Oryzias latipes*). *Environmental Toxicology and Chemistry*, 2001, **20**(11):2552–2560.
41. Kawahara, T. and I. Yamashita, Estrogen-independent ovary formation in the medaka fish, *Oryzias latipes*. *Zoological Science*, 2000, **17**(1):65–68.
42. Gray, M.A., A.J. Niimi, and C.D. Metcalfe, Factors affecting the development of testis–ova in medaka, *Oryzias latipes*, exposed to octylphenol. *Environmental Toxicology and Chemistry*, 1999, **18**(8):1835–1842.
43. Iwamatsu, T., et al., Oogenesis in the medaka *Oryzias latipes*: Stages of oocyte development. *Zoological Science*, 1988, **5**:353–373.
44. Yoshikawa, H. and M. Oguri, Gonadal sex differentiation in the medaka, *Oryzias latipes*, with special regard to the gradient of the differentiation of testes. *Bulletin of the Japanese Society of Science and Fisheries*, 1979, **45**:1115–1121.
45. Yamamoto, T., A further study of induction of functional sex reversal in genotypic males of the medaka (*Oryzias latipes*) and progenies of sex reversals. *Genetics*, 1959, **44**:739–757.
46. Yamamoto, K., Progenies of sex-reversal females mated with sex-reversal males in the medaka, *Oryzias latipes*. *Journal of Experimental Zoology*, 1961, **146**:163–179.
47. Morinaga, C., et al., Mutations affecting gonadal development in medaka, *Oryzias latipes*. *Mechanisms of Development*, 2004, **121**(7–8):829.
48. Winkler, C., et al., Developmentally regulated and non-sex-specific expression of autosomal *dmrt* genes in embryos of the medaka fish (*Oryzias latipes*). *Mechanisms of Development*, 2004, **121**(7–8):997–1005.
49. Wester, P.W. and J.H. Canton, Histopathology of *P. reticulata* (guppy) and *O. latipes* (medaka) in toxicity testing of some environmental contaminants. *Aquatic Toxicology*, 1988, **11**(3–4):426.
50. Hartley, W.R., et al., Gonadal development in Japanese medaka (*Oryzias latipes*) exposed to 17β-estradiol. *Marine Environmental Research*, 1998, **46**(1–5):145–148.
51. Kobayashi, D., et al., Steroidogenesis in the ovarian follicles of the medaka (*Oryzias latipes*) during vitellogenesis and oocyte maturation. *Zoological Science*, 1996, **13**(6):921–927.
52. Nakamoto, M., et al., Testicular type Sox9 is not involved in sex determination but might be in the development of testicular structures in the medaka *Oryzias latipes*. *Biochemical and Biophysical Research Communications*, 2005, **333**(3):729–736.
53. Kondo, M., et al., Molecular cloning and characterization of *DMRT* genes from the medaka *Oryzias latipes* and the platyfish *Xiphophorus maculatus*. *Gene*, 2002, **295** (2):213–222.
54. Matsuda, M., Sex determination in fish: Lessons from the sex-determining gene of the teleost medaka, *Oryzias latipes*. *Development, Growth and Differentiation*, 2003, **45** (5–6):397–403.

55. van der Ven, L.T.M. and P.W. Wester, Normal histology and effects of hormonal active agents of reproductive organs of the zebrafish (*Danio rerio*). *OECD Technical Workshop on Gonadal Histology of Small Laboratory Fish as a Tool for Endocrine Disruption Screening and Testing*, Bilthoven, The Netherlands, 2002.

56. van der Ven, L.T.M., P.W. Wester, and J.G. Vos, Histopathology as a tool for the evaluation of endocrine disruption in zebrafish (*Danio rerio*). *Environmental Toxicology and Chemistry*, 2003, **22**(4):908–913.

57. Maack, G. and M. Fenske, Gonad development of zebrafish exposed to low levels of ethynylestradiol during different life stages. *Journal of Fish Pathology*, 1999, **55** (Suppl.):245–246.

58. Maack, G. and H. Segner, Morphological development of the gonads in the zebrafish (*Danio rerio*). *Journal of Fish Biology*, 2003, **62**(4): 895–906.

59. van der Ven, L.T.M., et al., Comparative histopathology of zebrafish gonads after disruption of the sex hormone system. *SETAC Europe 12th Annual Meeting*, Vienna, Austria, 2002.

60. Oern, S., et al., The impact on reproduction of an orally administered mixture of selected PCBs in zebrafish (*Danio rerio*). *Archives of Environmental Contamination and Toxicology*, 1998, **35**(1):52–57.

61. Uchida, D., et al., Oocyte apoptosis during the transition from ovary-like tissue to testes during sex differentiation of juvenile zebrafish. *Journal of Experimental Biology*, 2002, **205**:711–718.

62. Takahashi, H., Juvenile hermaphroditism in the zebrafish *Brachydanio rerio*. *Bulletin of the Faculty of Fisheries, Hokkaido University*, 1977, **28**:57–65.

63. Hill, L. Jr. and D.M. Janz, Developmental estrogenic exposure in zebrafish (*Danio rerio*): I. Effects on sex ratio and breeding success. *Aquatic Toxicology*, 2003, **63**(4): 417–429.

64. Weber, L.P., R.L. Hill Jr., and D.M., Janz, Developmental estrogenic exposure in zebrafish (*Danio rerio*): II. Histological evaluation of gametogenesis and organ toxicity. *Aquatic Toxicology*, 2003, **63**(4):431–446.

65. Pelegri, F. and S. Schulte-Merker, A gynogenesis-based screen for maternal-effect genes in the zebrafish, *Danio rerio*. *Methods in Cell Biology*, 1999, **60**:1–20.

66. Oern, S., et al., Gonad development and vitellogenin production in zebrafish (*Danio rerio*) exposed to ethinylestradiol and methyltestosterone. *Aquatic Toxicology*, 2003, **65** (4):397–411.

67. Ball, J.S., et al., Assessing the effects of oestrogen exposure in zebrafish during early life on the expression of genes controlling gonadal sex differentiation and development. *CREDO Cluster Workshop on Ecological Relevance of Chemically-Induced Endocrine Disruption in Wildlife*, University of Exeter, UK, 2004.

68. Fenske, M. and H. Segner, Aromatase modulation alters gonadal differentiation in developing zebrafish (*Danio rerio*). *Aquatic Toxicology*, 2004, **67**(2):105–126.

69. Maack, G., H. Segner, and C.R. Tyler, Ontogeny of sexual differentiation in different strains of zebrafish (*Danio rerio*). *Fish Physiology and Biochemistry*, 2003, **28**(1–4):125–128.

70. Nagai, T., E. Yamaha, and K. Arai, Histological differentiation of primordial germ cells in zebrafish. *Zoological Science*, 2001, **18**(2):215–223.

71. Afonso, L.O.B., et al., Y-chromosomal DNA markers for discrimination of chemical substance and effluent effects on sexual differentiation in salmon. *Environmental Health Perspectives*, 2002, **110**(9):881–887.

72. Willey, J.B. and P.H. Krone, Effects of endosulfan and nonylphenol on the primordial germ cell population in pre-larval zebrafish embryos. *Aquatic Toxicology*, 2001, **54** (1–2): 113.
73. Yamamoto, T., Induced intersexes, in *Medaka (Killifish): Biology and Strains*. Keigaku Publishing Co., Tokyo, 1975, pp. 214–222.
74. Wilson, J.D., F.W. George, and J.E. Griffin, The hormonal control of sexual development. *Science*, 1981, **211**(4488):1278–1284.
75. Uguz, C., Histological evaluation of gonadal differentiation in fathead minnows (*Pimephales promelas*). *Tissue Cell*, 2008, **40**(4):299–306.
76. Baroiller, J.F. and H. D'Cotta, Environment and sex determination in farmed fish. *Comparative Biochemistry and Physiology, C: Toxicology and Pharmacology*, 2001, **130**(4):399–409.
77. D'Cotta, H., et al., Search for genes involved in the temperature-induced gonadal sex differentiation in the tilapia *Oreochromis niloticus*. *Journal of Experimental Zoology*, 2001, **290**(6):574–585.
78. D'Cotta, H., et al., Aromatase plays a key role during normal and temperature-induced sex differentiation of tilapia *Oreochromis niloticus*. *Molecular Reproduction and Development*, 2001, **59**(3):265–276.
79. Nakamura, M. and H. Takahashi, Gonadal sex differentation in *Tilapia mossambica* with special regard to the time of oestrogen treatment effective in inducing feminization of genetic fishes. *Bulletin of the Faculty of Fisheries, Hokkaido University*, 1973, **24**:1–23.
80. Satoh, N. and N. Egami, Sex differentation of germ cells in the teleost *Oryzias latipes* during normal embryonic development. *Journal of Embryology and Experimental Morphology*, 1972, **28**(2):385–395.
81. Quirk, J.G. and J.B. Hamilton, Number of germ cells in known male and known female genotypes of vertebrates embryos (*Oryzias latipes*). *Science*, 1973, **180** (4089):963–964.
82. Weidinger, G., et al., Dead end, a novel vertebrate germ plasm component, is required for zebrafish primordial germ cell migration and survival. *Current Biology*, 2003, **13**(16): 1429–1434.
83. Weidinger, G., et al., Identification of tissues and patterning events required for distinct steps in early migration of zebrafish primordial germ cells. *Development*, 1999, **126**(23): 5295–5307.
84. Weidinger, G., et al., Regulation of zebrafish primordial germ cell migration by attraction towards an intermediate target. *Development*, 2002, **129**(1):25–36.
85. Braat, A.K., et al., *Vasa* protein expression and localization in the zebrafish. *Mechanisms of Development*, 2000, **95**(1–2):271–274.
86. Olsen, L.C., R. Aasland, and A. Fjose, A *vasa*-like gene in zebrafish identifies putative primordial germ cells. *Mechanisms of Development*, 1997, **66**:95–105.
87. Yoon, C., K. Kawakami, and N. Hopkins, Zebrafish *vasa* homologue RNA is localized in cleavage planes of 2- and 4-cell-stage embryos and is expressed in primordial germ cells. *Development*, 1997, **124**:3157–3166.
88. Shinomiya, A., et al., The *vasa*-like gene, *olvas*, identifies the migration path of primordial germ cells during embryonic body formation stage in medaka, *Oryzias latipes*. *Development, Growth and Differentiation*, 2000, **42**(4):317–326.

89. Tanaka, M., et al., Establishment of medaka (*Oryzias latipes*) transgenic lines with the expression of green fluorescent protein fluorescence exclusively in germ cells: a useful model to monitor germ cells in a live vertebrate. *Proceedings of the National Academy of Sciences (USA)*, 2001, **98**(5):2544.

90. Sasado, T., et al., Mutations affecting early distribution of primordial germ cells in medaka (*Oryzias latipes*) embryo. *Mechanisms of Development*, 2004, **121**(7–8):817.

91. Herman, R.L. and H.L. Kincaid, Gonadal dysgenesis in rainbow trout Salmo gairdneri. *Diseases of Aquatic Organisms*, 1986, **1**:227–228.

92. Parmentier, H.L. and L.P.M. Timmermans, The differentiation of germ cells and gonads during development of carp, *Cyprinus carpio*. A study with anti-carp sperm monoclonal antibodies. *Journal of Embryology and Experimental Morphology*, 1985, **90**:13–32.

93. van Aerle, R., T.J. Runnalls, and C.R., Tyler, Ontogeny of gonadal sex development relative to growth in fathead minnow. *Journal of Fish Biology*, 2004, **64**(2):355.

94. Gimeno, S., et al., Feminization of male carp. *Nature*, 1996, **384**,(6606):221–222.

95. Biava, C.G., E.A. Smuckler, and D. Whorton, The testicular morphology of individuals exposed to dibromochloropropane. *Experimental and Molecular Pathology*, 1978, **29**(3):448–458.

96. Krovel, A.V. and L.C. Olsen, Sexual dimorphic expression pattern of a splice variant of zebrafish *vasa* during gonadal development. *Developmental Biology*, 2004, **271**(1):190.

97. Kobayashi, T., H. Kajiura-Kobayashi, and Y. Nagahama, Two isoforms of *vasa* homologs in a teleost fish: Their differential expression during germ cell differentiation. *Mechanisms of Development*, 2002, **111**(1–2):167.

98. Yoshizaki, G., et al., Primordial germ cells: The blueprint for a piscine life. *Fish Physiology and Biochemistry*, 2002, **26**(1):3–12.

99. Shibata, N. and S. Hamaguchi, Evidence for the sexual bipotentiality of spermatogonia in the fish, *Oryzias latipes*. *Journal of Experimental Biology*, 1988, **245**(1):71–77.

100. Krisfalusi, M. and J.J. Nagler, Induction of gonadal intersex in genotypic male rainbow trout (*Oncorhynchus mykiss*) embryos following immersion in estradiol-17β. *Molecular Reproduction and Development*, 2000, **56**(4):495–501.

101. Afonso, L.O.B., et al., The effect of the aromatase inhibitor fadrozole and two polynuclear aromatic hydrocarbons on sex steroid secretion by ovarian follicles of coho salmon. *General and Comparative Endocrinology*, 1997, **106**(2):169–174.

102. Afonso, L.O.B., G.J. Wassermann, and R. Terezinha de Oliveira, Sex reversal in Nile tilapia (*Oreochromis niloticus*) using a nonsteroidal aromatase inhibitor. *Journal of Experimental Zoology*, 2001, **290**(1):177–181.

103. Guiguen, Y., et al., Steroids and gonadal sex differentiation in the rainbow trout, *Oncorhynchus mykiss*. *Proceedings of the Sixth International Symposium on the Reproductive Physiology of Fish.* Institute of Marine Research, Bergen, Norway, 2000.

104. Piferrer, F., et al., Brief treatment with an aromatase inhibitor during sex differentiation causes chromosomally female salmon to develop as normal functional males. *Journal of Experimental Zoology*, 1994, **270**(3):255–262.

105. Sawyer, S.J., K.A. Gerstner, and G.V. Callard, Real-time PCR analysis of cytochrome P450 aromatase expression in zebrafish: Gene specific tissue distribution, sex differences,

developmental programming, and estrogen regulation. *General and Comparative Endocrinology*, 2006, **147**(2):108–117.

106. Trant, J.M., et al., Developmental expression of cytochrome P450 aromatase genes (*CYP19a* and *CYP19b*) in zebrafish fry (*Danio rerio*). *Journal of Experimental Zoology*, 2001, **290**(5):475–483.
107. Colombo, G. and G. Grandi, Gonad sex differentiation of *Anguilla anguilla* by sex steroids. *International Review of Hydrobiology*, 1990, **75**(6): 763–773.
108. Colombo, G. and G. Grandi, Sex differentiation in the European eel: Histological analysis of the effects of sex steroids on the gonad. *Journal of Fish Biology*, 1995, **47**(3): 394–413.
109. Hoar, W.S., D.J. Randall, and E.M. Donaldson, *Fish Physiology,* Vol. 4, *Reproduction,* Part B, *Behaviour and Fertility*. Academic Press, San Diego, CA, 1983.
110. Hoar, W.S., D.J. Randall, and E.M. Donaldson, *Fish Physiology*, Vol. 4, *Reproduction*, Part A, *Endocrine Tissues and Hormones*. Academic Press, San Diego, CA, 1983.
111. Volff, J.N., et al., Governing sex determination in fish: Regulatory putsches and ephemeral dictators. *Sexual Development*, 2007, **1**(2):85.
112. Atz, J.W., Intersexuality in fishes, in *Intersexuality in Vertebrates Including Man*, C. N. Armstron and A. J. Marshall, Editors. Academic Press, New York, 1964, pp. 145–232.
113. Lepori, N.G., Gonadal sex differentiation in teleosts, in *Sex Differentiation, Hermaphroditism and Intersexuality in Vertebrates Including Man*, N. G. Lepori, Editor. Piccin Medical Books, Padua, Italy, 1980, pp. 13–37.
114. Nagahama, Y., The functional morphology of teleost gonads, in *Fish Physiology*, W.S. Hoar, D.J. Randall, and E. M. Donaldson, Editors. Academic Press, San Diego, CA, 1983, pp. 223–275.
115. Matsuoka, M.P., et al., Real-time PCR analysis of ovary- and brain-type aromatase gene expression during Atlantic halibut (*Hippoglossus hippoglossus*) development. *Comparative Biochemistry and Physiology, Part B: Biochemistry and Molecular Biology*, 2006, **144**(1):128.
116. Papoulias, D.M., D.B. Noltie, and D.E. Tillitt, An in vivo model fish system to test chemical effects on sexual differentiation and development: Exposure to ethinyl estradiol. *Aquatic Toxicology*, 1999, **48**(1): 37–50.
117. Davies, P.R. and F. Takahashima, Sex differentiation in common carp, *Cyprinus carpio*. *Journal of the Tokyo University of Fisheries*, 1980, **66**:191–199.
118. Chemicals Evaluation and Research Institute, Atlas of medaka gonadal histology. http://w-edcdb.nies.go.jp/SHf/index.html, 2003.
119. Hori, H., et al., Medakafish homepage. http://biol1.bio.nagoya-u.ac.jp:8000/, 1995; accessed Jan. 27, 2007.
120. Yonkos, L.T., R. Reimschuessel, and A.S. Kane. Digital atlas of fathead minnow normal histology, http://aquaticpath.umd.edu/fhm/. 2003.
121. van der Ven, L.T.M. and P.W. Wester, Toxicological pathology atlas of small laboratory fish with special reference to normal histology and effects of endocrine disruption in the zebrafish (*Danio rerio*). http://www.rivm.nl/fishtoxpat, 2003.
122. Hsiao, C.-D. and H.-J. Tsai, Transgenic zebrafish with fluorescent germ cell: A useful tool to visualize germ cell proliferation and juvenile hermaphrodism in vivo. *Developmental Biology*, 2003, **262**(2):313–323.

123. Uchida, D., et al., An aromatase inhibitor or high water temperature induce oocyte apoptosis and depletion of P450 aromatase activity in the gonad of genetic female zebrafish during sex reversal. *Comparative Biochemistry and Physiology, Part B: Biochemistry and Molecular Biology*, 2004, **137**(1):11–20.

124. Nash, J.P., et al., Long-term exposure to environmental concentrations of the pharmaceutical ethynylestradiol causes reproductive failure in fish. *Environmental Health Perspectives*, 2004, **112**(17):1725–1733.

125. Adkins-Regan, E., Hormones and sexual differentiation, in *Hormones and Reproduction in Fishes, Amphibians, and Reptiles*, D.O. Norris and R.E. Jones, Editors. Plenum Press, New York, 1987, pp. 1–27.

126. Waters, P.D., M.C. Wallis, and J.A.M. Graves, Mammalian sex: Origin and evolution of the Y chromosome and *SRY*. *Seminars in Cell and Developmental Biology*, 2007, **18**(3):389.

127. Ferguson-Smith, M., The evolution of sex chromosomes and sex determination in vertebrates and the key role of *DMRT1*. *Sexual Development*, 2007, **1**(1):2.

128. Schafer, A.J., Sex determination and its pathology in man. *Advances in Genetics*, 1995, **33**:275–329.

129. Pamilo, P. and R.J. O'Neill, Evolution of the SRY genes. *Molecular Biology and Evolution*, 1997, **14**(1):49–55.

130. Ito, M., et al., Investigation of the fate of SRY-expressing cells using an in vivo Cre/loxP system. *Development, Growth and Differentiation*, 2006, **48**(1):41–47.

131. Meyer, B., Sex in the worm: Counting and compensating X chromosome dose. *Trends in Genetics*, 2000, **16**(6):247–253.

132. Koopman, P., M. Bullejos, and J. Bowles, Regulation of male sexual development by *Sry* and *Sox9*. *Journal of Experimental Zoology*, 2001, **290**(5):463–474.

133. Schartl, M., Developmental biology in fish at the University of Würzburg: The Winkler Lab. http://www.biozentrum.uni-wuerzburg.de/pc1/winkler/home_en.htm, 2004.

134. Redding, J.M. and R. Patino, Reproductive systems: Gross functional anatomy, in *The Handbook of Experimental Animals: The Laboratory Fish*, G. K. Ostrander, Editor. Academic Press, San Diego, CA, 2000, pp. 261–272.

135. Redding, M.J. and R. Patino, Reproductive physiology, in *The Physiology of Fishes*. Marine Science Series., CRC Press, Boca Raton, FL, 1993, pp. 503–534.

136. Hoar, W.S. and D.J. Randall, Reproduction, in *Fish Physiology*, Vol. III, *Reproduction and Growth, Bioluminiscence, Pigments, and Poisons*. Academic Press, New York, 1969, pp. 1–72.

137. Harder, W., *Anatomy of Fishes*, Vol. 2. E. Schweizerbart'sche Verlagsbuchhandlung, Stuttgart, Germany, 1975.

138. Petersen, G.I., et al., Suitability of zebrafish as test organism for detection of endocrine distriputing chemicals. Nordic Council of Ministers. *TemaNord*, 2001, **597**:5–63.

139. Norris, D.O., Regulation of male gonaducts and sex accessory structures, in *Hormones and Reproduction in Fishes, Amphibians and Reptiles*, D. O. Norri and R. E. Jones, Editors. Plenum, Press, New York, 1987, pp. 327–354.

140. Blazquez, M., et al., Structural and functional aspects of early exposure to estradiol 17β and 17α-ethynylestradiol on the gonads of the gonochoristic teleost *Dicentrarchus labrax*. *Fish Physiology and Biochemistry*, 1998, **18**(1):37–47.

141. Nagahama, Y., Endocrine regulation of gametogenesis in fish. *International Journal of Developmental Biology*, 1994, **38**:217–229.

142. Nakamura, M., R.K. Bhandari, and M. Higa, The role estrogens play in sex differentiation and sex changes of fish. *Fish Physiology and Biochemistry*, 2003, **28**(4): 113–117.

143. Seki, M., et al., Effect of ethinylestradiol on the reproduction and induction of vitellogenin and testis–ova in medaka (*Oryzias latipes*). *Environmental Toxicology and Chemistry*, 2002, **21**(8):1692–1698.

144. Goodbred, S.L., et al., Reconnaissance of 17β-estradiol, 11-ketotestosterone, vitellogenin, and gonad histopathology in common carp of United States streams: Potential for contaminant-induced endocrine disruption. *Geological Survey Open-File Report 96-627*. USGS, Washington, DC, pp. 1997, 1–47.

145. Carlson, D.B. and D.E. Williams, Sex-specific vitellogenin production in immature rainbow trout. *Environmental Toxicology and Chemistry*, 1999, **18**(10):2361–2363.

146. Palmer, B.D., et al., Vitellogenin as a biomarker for xenobiotic estrogens in an amphibian model system. *Environmental Toxicology and Chemistry*, 1998, **17**(1):30–36.

147. Korte, J.J., et al., Fathead minnow vitellogenin: Complementary DNA sequence and messenger RNA and protein expression after 17β-estradiol treatment. *Environmental Toxicology and Chemistry*, 2000, **18**(4):972–981.

148. Cheek, A.O., et al., Experimental evaluation of vitellogenin as a predicitve biomarker for reproductive disruption. *Environmental Health Perspectives*, 2001, **109**(7):681–690.

149. Pawlowski, S., et al., Effects of 17[alpha]-ethinylestradiol in a fathead minnow (*Pimephales promelas*) gonadal recrudescence assay. *Ecotoxicology and Environmental Safety*, 2004, **57**(3):330.

150. Arukwe, A. and A. Goksoyr, Eggshell and egg yolk proteins in fish: Hepatic proteins for the next generation: oogenetic, population, and evolutionary implications of endocrine disruption. *Comparative Hepatology*, 2003, **2**(1):4.

151. Brion, F., et al., Impacts of 17beta-estradiol, including environmentally relevant concentrations, on reproduction after exposure during embryo-larval-, juvenile- and adult-life stages in zebrafish (*Danio rerio*). *Aquatic Toxicology*, 2004, **68**(3):193–217.

152. Piferrer, F., I.J. Baker, and E.M. Donaldson, Effects of natural, synthetic, aromatizable, and nonaromatizable androgens in inducing male sex differentiation in genotypic female chinook salmon (*Oncorhynchus tshawytscha*). *General and Comparative Endocrinology*, 1993, **91**(1):59–65.

153. Kitano, T., et al., Aromatase inhibitor and 17alpha-methyltestosterone cause sex-reversal from genetic females to phenotypic males and suppression of P450 aromatase gene expression in Japanese flounder (*Paralichthys olivaceus*). *Molecular Reproduction and Development*, 2000, **56**(1):1–5.

154. McAllister, B.G. and D.E. Kime, Early life exposure to environmental levels of the aromatase inhibitor tributyltin causes masculinisation and irreversible sperm damage in zebrafish (*Danio rerio*). *Aquatic Toxicology*, 2003, **65**(3):309–316.

155. Bhandari, R.K., et al., Suppression of steroidogenic enzyme expression during androgen-induced sex reversal in Nile tilapia (*Oreochromis niloticus*). *General and Comparative Endocrinology*, 2006, **145**(1):20–24.

156. Blazquez, M., et al., Development of sex control techniques for European sea bass (*Dicentrarchus labrax L.*) aquaculture: Effects of dietary 17α-methyltestosterone prior to sex differentiation. *Aquaculture*, 1995, **135**(4):329–342.

157. Fenske, M., et al., An environmentally relevant concentration of estrogen induces arrest of male gonad development in zebrafish, *Danio rerio*. *Environmental Toxicology and Chemistry*, 2005, **24**(5):1088–1098.

158. van den Belt, K., et al., Effects of ethynylestradiol on the reproductive physiology in zebrafish (*Danio rerio*): Time dependency and reversibility. *Environmental Toxicology and Chemistry*, 2002, **21**(4):767–775.

159. Schafers, C., et al., Concentration- and time-dependent effects of the synthetic estrogen, 17alpha-ethinylestradiol, on reproductive capabilities of the zebrafish, *Danio rerio*. *Journal of Toxicology and Environmental Health A*, 2007, **70**(9):768–779.

160. Papoulias, D.M., D.B. Noltie, and D.E. Tillitt, An in vivo model fish system to test chemical effects on sexual differentiation and development: Exposure to ethinyl estradiol. *Aquatic Toxicology*, 2000, **48**(1):37–50.

161. Panter, G.H., et al., Development of chronic tests for endocrine active chemicals: Part 1. An extended fish early-life stage test for oestrogenic active chemicals in the fathead minnow (*Pimephales promelas*). *Aquatic Toxicology*, 2006, **77**(3):279–290.

162. Bogers, R., et al., Development of chronic tests for endocrine active chemicals: Part 2: An extended fish early-life stage test with an androgenic chemical in the fathead minnow (*Pimephales promelas*). *Aquatic Toxicology*, 2006, **80**(2):119–130.

163. Hunter, G.A. and E.M. Donaldson, Hormonal sex control and its application to fish culture, in *Fish Physiology*,, W.S. Hoar, D.J. Randall, and E.M Donaldson, Editors. Academic Press, San Diego, CA, 1983, pp. 223–303.

164. Wester, P.W., et al., Identification of endocrine disruptive effects in the aquatic environment: A partial life cycle assay in zebrafish., Rijksinstituut voor Volksgezondheid en Milieu, Bilthoven, The Netherlands, 2003, p. 112.

165. Iwamatsu, T., Convenient method for sex reversal in a freshwater teleost, the medaka. *Journal of Experimental Zoology*, 1999, **283**(2):210–214.

166. Scholz, S. and H.O. Gutzeit, 17α-Ethinylestradiol affects reproduction, sexual differentiation and aromatase gene expression of the medaka (*Oryzias latipes*). *Aquatic Toxicology*, 2000, **50**(4):363–373.

167. Länge, R., et al., Effects of the synthetic estrogen 17α-ethynylestradiol on the life-cycle of the fathead minnow (*Pimephales promelas*). *Environmental Toxicology and Chemistry*, 2001, **20**(6):1216–1227.

168. van der Ven, L.T., et al., Effects of the estrogen agonist 17beta-estradiol and antagonist tamoxifen in a partial life-cycle assay with zebrafish (*Danio rerio*). *Environmental Toxicology Chemistry*, 2007, **26**(1):92–99.

169. Petersen, G.I., et al., Baseline test studies and development of ELISA. *Nordic Council of Ministers. TemaNord*, 2000, **555**:35–46.

170. Ankley, G.T., et al., Evaluation of the aromatase inhibitor fadrozole in a short-term reproduction assay with the fathead minnow (*Pimephales promelas*). *Toxicological Sciences*, 2002, **67**:121–130.

171. Afonso, L.O.B., G.K. Iwama, and E.M. Donaldson, Effects of the aromatase inhibitor fadrozole on plasma sex steroid secretion and ovulation rate in female coho salmon,

Oncorhynchus kisutch, close to final maturation. *General and Comparative Endocrinology*, 1999, **113**(2):221–229.

172. Afonso, L.O.B., et al., Effects of the aromatase inhibitor fadrozole on plasma sex steroid and oocyte maturation in female coho salmon (*Oncorhynchus kisutch*) during vitellogenesis. *Fish Physiology and Biochemistry*, 1999, **20**:231–241.
173. Nagahama, Y., Gonadotropin action on gametogenesis and steroidogenesis in teleost gonads. *Zoological Science*, 1987, **4**(2):209–222.
174. Kuhl, A.J., S. Manning, and M. Brouwer, Brain aromatase in Japanese medaka (*Oryzias latipes*): Molecular characterization and role in xenoestrogen-induced sex reversal. *Journal of Steroid Biochemistry and Molecular Biology*, 2005, **96**(1):67.
175. van den Hurk, R. and G.H. Slof, A morphological experimental study of gonadal sex differentiation in the rainbow trout, *Salmo gairdneri*. *Cell Tissue Research*, 1981, **218**(3):487–497.
176. van den Hurk, R. and J.G.D. Lambert. Temperature and steroid effects on gonadal sex differentiation in rainbow trout. *Proceedings of the Second International Symposium on the Reproductive Physiology Fish*, Wageningen, The Netherlands, 1982.
177. Liu, S., et al., Expression of cytochromeP45011b (11β-hydroxylase) gene during gonadal differentiation and spermatogenesis in rainbow trout, *Oncorhynchus mykiss*. *Journal of Steroid Biochemistry and Molecular Biology*, 2000, **75**(4–5): 291–298.
178. Tong, S.-K., et al., Phylogeny, expression and enzyme activity of zebrafish *cyp19* (P450 aromatase) genes. *Journal of Steroid Biochemistry and Molecular Biology*, 2001, **79**(1–5):299–303.
179. Kishida, M. and G.V. Callard, Distinct cytochrome P450 aromatase isoforms in zebrafish (*Danio rerio*) brain and ovary are differentially programmed and estrogen regulated during early development. *Endocrinology*, 2001, **142**(2):740–750.
180. Shen, W.-H., et al., Nuclear receptor steroidogenic factor 1 regulates the mullerian inhibiting substance gene: A link to the sex determination cascade. *Cell*, 1994, **77**(5): 651–661.
181. Tanaka, M., et al., Structure and promoter analysis of the cytochrome P-450 aromatase gene of the teleost fish, medaka (*Oryzias latipes*). *Journal of Biochemistry*, 1995, **117**(4): 719–725.
182. Ito, M., et al., A rainbow trout *SRY*-type gene expressed in pituitary glands. *FEBS Letters*, 1995, **377**(1):37–40.
183. Vriz, S., et al., Zygotic expression of the zebrafish *Sox-19*, an HMG box-containing gene, suggests an involvement in central nervous system development. *Brain Research: Molecular Brain Research*, 1996, **40**(2):221–228.
184. Grier, H.J., Ultrastructural identification of the Sertoli cell in the testis of the northern pike, *Esox lucius*. *American Journal of Anatomy*, 1977, **149**:283–288.
185. Gao, S., et al., Molecular cloning, expression of Sox5 and its down-regulation of *Dmrt1* transcription in zebrafish. *Journal of Experimental Zoology Part B: Molecular and Developmental Evolution*, 2005, **304**(5):476–483.
186. Wang, R., et al., Molecular cloning and expression of *Sox17* in gonads during sex reversal in the rice field eel, a teleost fish with a characteristic of natural sex transformation. *Biochemical and Biophysical Research Communications*, 2003, **303**(2):452.
187. Bagheri-Fam, S., et al., Comparative genomics of the SOX9 region in human and *Fugu rubripes*: Conservation of short regulatory sequence elements within large intergenic regions. *Genomics*, 2001, **78**(1–2):73–82.

188. Chiang, E.F., et al., Two *sox9* genes on duplicated zebrafish chromosomes: Expression of similar transcription activators in distinct sites. *Developmental Biology*, 2001, **231**(1):149–163.

189. Hett, A.K., et al., Characterization of Sox9 in European Atlantic sturgeon (*Acipenser sturio*). *Journal of Heredity*, 2005, **96**(2):150–154.

190. Hsiao, N.-W., et al., Mutagenesis study on the zebra fish *SOX9* high-mobility group: Comparison of sequence and non-sequence specific HMG domains. *Biochemistry*, 2003, **42**(38):11183–11193.

191. Koopman, P., *Sry* and *Sox9*: mammalian testis-determining genes. *Cellular and Molecular Life Sciences*, 1999, **55**(6–7):839–856.

192. Li, M., et al., Zebrafish *sox9b* is an early neural crest marker. *Development: Genes and Evolution*, 2002, **212**:203–206.

193. Takamatsu, N., et al., Rainbow trout *SOX9*: cDNA cloning, gene structure and expression. *Gene*, 1997, **202**(1–2):167–170.

194. Yan, Y.L., et al., A zebrafish *sox9* gene required for cartilage morphogenesis. *Development*, 2002, **129**(21):5065–5079.

195. Yan, Y.L., et al., A pair of *Sox*: Distinct and overlapping functions of zebrafish *sox9* co-orthologs in craniofacial and pectoral fin development. *Development*, 2005, **132** (5):1069–1083.

196. Zhang, G., M.M. Miyamoto, and M.J. Cohn, Lamprey type II collagen and Sox9 reveal an ancient origin of the vertebrate collagenous skeleton. *Proceedings of the National Academy of Sciences USA*, 2006, **103**(9):3180–3185.

197. Hett, A.K. and A. Ludwig, *SRY*-related (*Sox*) genes in the genome of European Atlantic sturgeon (*Acipenser sturio*). *Genome*, 2005, **48**(2):181–186.

198. Cresko, W.A., et al., Genome duplication, subfunction partitioning, and lineage divergence: *Sox9* in stickleback and zebrafish. *Developmental Dynamics: An Official Publication of the American Association of Anatomists*, 2003, **228**(3): 480–489.

199. Lu, H., et al., Two alleles of the Sox9a2 in the rice field eel. *Journal of Experimental Zoology, Part B: Molecular Development and Evolution*, 2003, **299**(1):36–40.

200. Zhou, R., et al., Similar gene structure of two *Sox9a* genes and their expression patterns during gonadal differentiation in a teleost fish, rice field eel (*Monopterus albus*). *Molecular Reproduction and Development*, 2003, **66**(3):211–217.

201. Huang, B., et al., Autosomal XX sex reversal caused by duplication of *SOX9*. *American Journal of Medical Genetics*, 1999, **87**(4):349–353.

202. Yokoi, H., et al., *Sox9* in a teleost fish, medaka (*Oryzias latipes*): evidence for diversified function of *Sox9* in gonad differentiation. *Molecular Reproduction and Development*, 2002, **63**(1):5–16.

203. Rodriguez-Mari, A., et al., Characterization and expression pattern of zebrafish anti-Mullerian hormone (Amh) relative to *sox9a*, *sox9b*, and *cyp19a1a*, during gonad development. *Gene Expression Patterns*, 2005, **5**(5):655–667.

204. Kanda, H., et al., Rainbow trout *Sox24*, a novel member of the *Sox* family, is a transcriptional regulator during oogenesis. *General and Comparative Endocrinology*, 1998, **211**(2):251–257.

205. Yamashita, A., et al., cDNA cloning of a novel rainbow trout *SRY*-type HMG box protein, rtSox23, and its functional analysis. *General and Comparative Endocrinology*, 1998, **209** (1–2):192–200.

206. Traut, W. and H. Winking, Meiotic chromosomes and stages of sex chromosome evolution in fish: Zebrafish, platyfish and guppy. *Chromosome Research*, 2001, **9**(8):659–672.

207. Matsuda, M., et al., DMY is a Y-specific DM-domain gene required for male development in the medaka fish. *Nature*, 2002, **417**:559–562.

208. Nanda, I., et al., Common spontaneous sex-reversed XX males of the medaka *Oryzias latipes*. *Genetics*, 2003, **163**(1):245–251.

209. Nanda, I., et al., A duplicated copy of *DMRT1* in the sex-determining region of the Y chromosome of the medaka, *Oryzias latipes*. *Proceedings of the National Academy of Sciences USA*, 2002, **99**(18):11778–11783.

210. Matsuda, M., Sex determination in the teleost medaka, *Oryzias latipes*. *Annual Review of Genetics*, 2005, **39**:293–307.

211. Schartl, M., A comparative view on sex determination in medaka. *Mechanisms of Development*, 2004, **121**(7–8):639–645.

212. Volff, J.N., M. Kondo, and M. Schartl, Medaka *dmY/dmrt1Y* is not the universal primary sex-determining gene in fish. *Trends in Genetics*, 2003, **19**(4):196–199.

213. Volff, J.-N., et al., Evolutionary dynamics of the DM domain gene family in metazoans. *Journal of Molecular Evolution*, 2003, **57**(Suppl 1) S241–S249.

214. Otake, H., et al., Wild-derived XY sex-reversal mutants in the medaka, *Oryzias latipes*. *Genetics*, 2006, **173**(4):2083–2090.

215. Kumar, S. and S.B. Hedges, A molecular timescale for vertebrate evolution. *Nature*, 1998, **392**(6679):917–920.

216. Steinke, D., W. Salzburger, and A. Meyer, Novel relationships among ten fish model species revealed based on a phylogenomic analysis using ESTs. *Journal of Molecular Evolution*, 2006, **62**(6):772–784.

217. Steinke, D., et al., Many genes in fish have species-specific asymmetric rates of molecular evolution. *BMC Genomics*, 2006, **7**:20.

218. Metcalfe, C.D., et al., Estrogenic potency of chemicals detected in sewage treatment plant effluents as determined by in vivo assays with Japanese medaka (*Oryzias latipes*). *Environmental Toxicology and Chemistry*, 2001, **20**(2):297–308.

219. U.S. EPA, Draft final report on comparative evaluation of fathead minnow assays for detecting endocrine-disrupting chemicals. Batelle, Columbus, OH, 2003.

220. Jensen, K.M., et al., Aspects of basic reproductive biology and endocrinology in the fathead minnow (*Pimephales promelas*). *Comparative Biochemistry and Physiology, Part C: Toxicology and Pharmacology*, 2001, **128**(1):127.

221. Arcand-Hoy, L.D. and W.H. Benson, Fish reproduction: An ecologically relevant indicator of endocrine disruption. *Environmental Toxicology and Chemistry*, 1998, **17**(1):49–57.

222. Kinnberg, et al., Effects of the fungicide prochloraz on the sexual development of zebrafish (*Danio rerio*). *Comparative Biochemistry and Physiology Part C: Toxicologic Pharmacology*, 2007, **145**(2):165–170.

223. Arcand-Hoy, L.D. and W.H. Benson, Toxic responses of the reproductive system, in *Target Organ Toxicity in Marine and Freshwater Teleosts*, D. Schlenk and W.H. Benson, Editors. Taylor & Francis, New York, 2001, pp. 175–202.

224. Gray, M.A. and C.D. Metcalfe, Induction of testis–ova in Japanese medaka (*Oryzias latipes*) exposed to *p*-nonylphenol. *Environmental Toxicology and Chemistry*, 1997, **16**(5):1082–1086.

225. Weber, L.P., et al., Increased kidney, liver, and testicular cell death after chronic exposure to 17alpha-ethinylestradiol in medaka (*Oryzias latipes*). *Environmental Toxicology Chemistry*, 2004, **23**(3):792–797.

226. van Aerle, R., T.J. Runnalls, and C.R. Tyler, Gonadal sex differentiation in fathead minnow, *Pimephales promelas*. *Annual Conference of SETAC Europe*, Vienna, Austria, 2002.

227. Islinger, M., et al., Effects of 17α-ethinylestradiol on the expression of three estrogen-responsive genes and cellular ultrastructure of liver and testes in male zebrafish. *Aquatic Toxicology*, 2003, **62**,(2):85–103.

228. van den Belt, K., R. Verheyen, and H. Witters, Reproductive effects of ethynylestradiol and *4t-octylphenol* on the zebrafish (*Danio rerio*). *Archives of Environmental Contamination and Toxicology*, 2001, **41**(4):458–467.

229. Balch, G. and C.D. Metcalfe, Developmental effects in Japanese medaka (*Oryzias latipes*) exposed to nonylphenol ethoxylates and their degradation products. *Chemosphere*, 2006, **62**(8):1214–1223.

230. Balch, G.C., C.A. Mackenzie, and C.D. Metcalfe, Alterations to gonadal development and reproductive success in Japanese medaka (*Oryzias latipes*) exposed to 17alpha-ethinylestradiol. *Environmental Toxicology and Chemistry*, 2004, **23**(3):782–791.

231. Jobling, S., et al., Widespread sexual disruption in wild fish. *Environmental Science and Technology*, 1998, **32**:2498–2506.

232. van Aerle, R., et al., Sexual disruption in a second species of wild cyprinid fish (the gudgeon, *Gobio gobio*) in United Kingdom freshwaters. *Environmental Toxicology and Chemistry*, 2001, **20**(12):2841–2847.

233. O'Brien, E. and D.R. Dietrich, Hindsight rather than foresight: Reality versus the EU draft guideline on pharmaceuticals in the environment. *Trends in Biotechnology*, 2004, **22**(7):326–330.

234. Liney, K.E., et al., Assessing the sensitivity of different life stages for sexual disruption in roach (*Rutilus rutilus*) exposed to effluents from wastewater treatment works. *Environmental Health Perspectives*, 2005, **113**(10):1299–1307.

235. Janz, D.M., Effects of chronic nonylphenol exposure on medaka (*Oryzias latipes*) gonadal histology. Saskatoon, Saskatchewan, Canada, 2004.

CHAPTER 4

Female Gonad Anatomy and Morphology

4.1 GONADOGENESIS: OVARY

4.1.1 Location and Gross Organization

The female reproductive system of teleosts, unlike that of mammals, is highly variable, reflecting the wide range of reproductive patterns, including viviparity. In most teleosts, the ovary is a hollow bilaterally paired (e.g., zebrafish and fathead minnow) structure (Figure 4.1) attached to the dorsal peritoneum (e.g., zebrafish, fathead minnow, medaka). However, in some species (e.g., medaka) only one ovary develops [1–3], or paired structures become fused into a single solid organ during early development (e.g., yellow perch, *Perca flavescens*). Oogenesis occurs within the germinal epithelia associated with the lamellar folds of the ovary (see Chapter 3). Lamellar walls are composed of germinal and follicular epithelia supported by connective tissue stroma.

Ovaries completely filled with mature eggs in some instances account for over 50% of a fish's body weight. In nonspawning conditions, the ovary may regress to a small fraction of its mature size [4].

The ovarian tissue is tubular or saclike. Its coloration ranges from clear to pink, yellow, and orange, depending on the pigmentation of the eggs. Typical egg diameters lie between 0.4 and 3 mm, but in some salmonid species, the eggs may be larger than 5 mm [5]. The eggs are covered by the chorion, which may be smooth or modified for adhesion or attachment to the substrate or to other eggs (e.g., as with medaka eggs) [1]. Ovulated eggs stay in the ovarian lumen, the ovocoel, until spawning. More information and intricate details on fish oocytes, their development, and their hormonal regulation may be found in a recent book specifically dedicated to the understanding of fish oocytes [6]. Therefore, only the main characteristics important for understanding EDC-mediated histopathological changes in the ovaries of the zebrafish, fathead minnow, and medaka, are covered here.

Ovarian ducts are present in most teleosts (see Section 3.3) and are a direct posterior extension of the ovarian wall and lumen rather than a modified Müllerian funnel (oviduct), as seen in other vertebrate taxa. The ovarian ducts penetrate abdominal

Histological Analysis of Endocrine-Disruptive Effects in Small Laboratory Fish
By Daniel R. Dietrich and Heiko O. Krieger

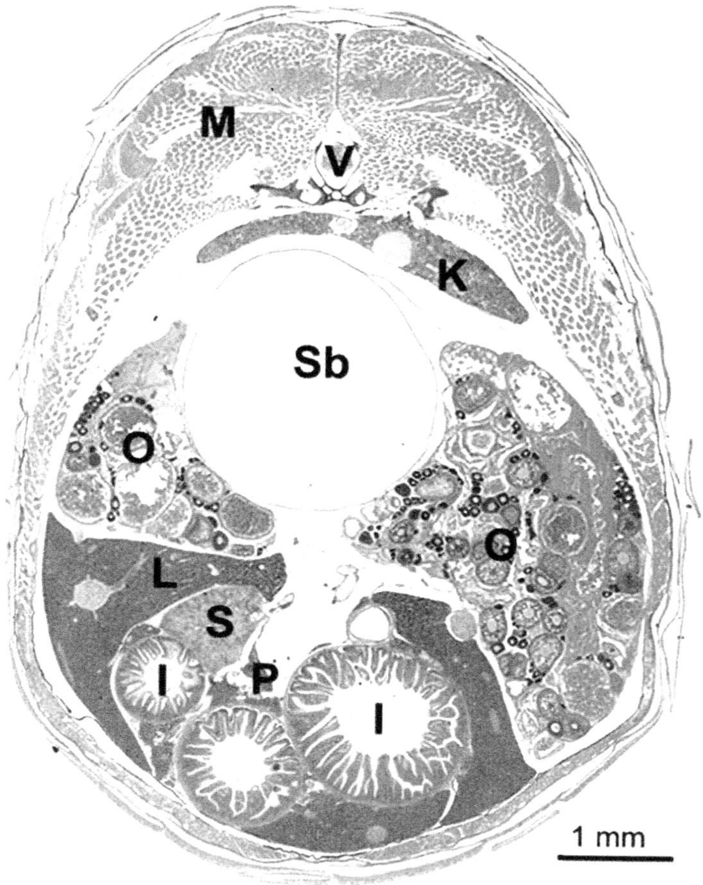

FIGURE 4.1 Relative size and location of organs in a female adult zebrafish. M, muscle; V, vertebra; K, kidney; Sb, swimbladder; O, ovary; L, liver; S, spleen; P, pancreas; I, intestine. (Courtesy of L. van der Ven and P. Wester, RIVM, Bilthoven, The Netherlands). (See insert for color representation.)

pores and terminate independently at the genital pore, situated between the anal and urinary pores. In the yellow perch (one fused ovary), the single ovarian duct appears to be a narrowing of the posterior ovary, which terminates in the abdominal pore. In contrast, ovarian ducts are incomplete in salmonids: In this case, ovulation involves the deterioration of intraovarian connective tissue, rupture of the ovarian wall, and release of the eggs into the ovarian central cavity, from which the eggs are expelled via muscular contractions through the urogenital papilla [3,7].

4.1.2 Anatomy of the Ovary

The ovary consists of oogonia, oocytes, and their surrounding follicle cells, supporting tissue or stroma, and vascular and nervous tissue (Figures 4.1 to 4.6). Most teleosts are cyclical breeders, and the ovary varies greatly in appearance at different times in the reproductive cycle. Ovaries have been classified in the broadest sense into three basic types, which also correspond to the respective spawning strategies [4,8,9]:

1. *Synchronous ovaries.* All oocytes develop and ovulate in unison (oocyte site distribution follows a single mode) and there is no replenishment from earlier stages. This is typical for fish that spawn once in their lifetime and then die (e.g., the anadromous *Oncorhynchus* species or the catadromous eels).
2. *Group synchronous ovaries.* At least two different size groups of oocytes are present at the same time, the larger group usually being more homogeneous (representing the eggs for the next spawning season). This is typical for fish with a single annual and short breeding period [e.g., brown trout (*Salmo trutta fario*)].
3. *Asynchronous ovaries.* Oocytes of all stages are present at some time. A variant of asynchronous ovaries can be found in the fractionally spawning fathead minnows, spawning approximately every 2 to 3 days [10], and medaka [11–13], spawning regularly almost every 24 to 48 hours throughout the short breeding period in the wild. The oocyte size frequency is thus continuous except in ripe ovaries, where a clear separation between ripe and yolked oocytes may occur [9]. In contrast, all oocyte stages are present at all times in zebrafish, who spawn approximately every second day when given the appropriate environmental stimuli (light, feed, temperature, etc.) and mating partners.

It must be kept in mind that the examples of synchronous and asynchronous ovulators described above simply represent the two extremes of a continuum of ovulation strategies found in teleost fish. Correspondingly, the physiology of the varying spawning strategies has direct implications for the morphology of the ovary [i.e., the cell types (somatic, germ cells) and relative numbers present at any one time].

In many species (e.g., channel catfish, *Ictalurus punctatus*), ridges of tissue growing from the proximal and distal margins of the ovary meet and fuse to form a cavity: the ovarian lumen or ovocoel. In zebrafish, carp (*Cyprinus carpio*), and other cyprinids (e.g., goldfish, *Carassius auratus*), putative ovaries become extended at their ventral edges, where stromal cells increase in number and spread laterally to the dorsal peritoneal wall. Concomitantly, small aggregations of cells appear on the peritoneal walls near the extended edges of the gonads and eventually fuse with these extensions to form a narrow ovarian cavity in the region between the gonads and the peritoneal wall [14–17]. In fathead minnows and carp, the distal ends of the developing ovary fuses laterally with the abdominal wall to form a dorsal cavity (see Figures 3.1, and 3.2, and 4.2). A secondary cavity also forms by the ventral fusion of a separate pair of tissue outgrowths. In contrast, in medaka and top-minnow, *Gobiocypris affinis*, the lateral edges of the fused ovary begin to elongate dorsally and the tips of the ovarian elongations fuse finally with the dorsal coelomic wall [17]. Similarly, in guppies *Poecilia reticulata*, a single ovary forms an ovocoel by fusion of tissue outgrowths with the peritoneal wall [18]. In other fish species the lumen is formed by invagination and subsequent fusion of the distal margin of the ovary [3]. The epithelium lining the lumen may have secretory functions in order to nourish the ovulated eggs. In viviparous species, the ovarian lumen may be further modified for the storage of sperm and for the physical and nutritional support of the developing embryos [3].

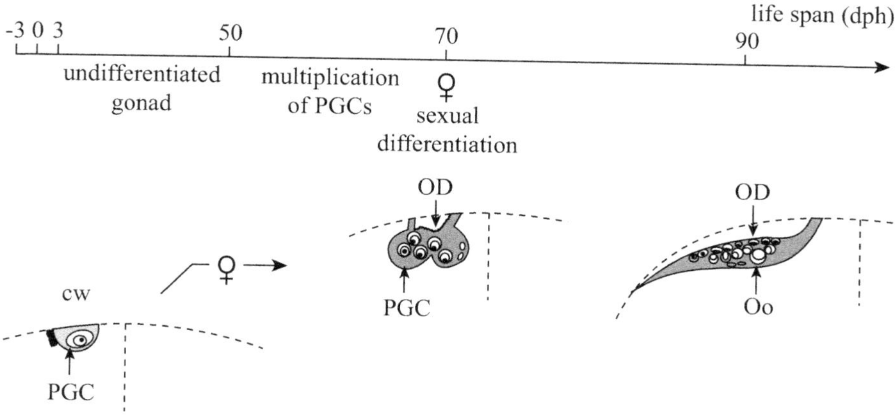

FIGURE 4.2 Process and timing of sexual differentiation in female carp. dph, days post-hatch; pgc, primordial germ cells; cw, coelomic wall; OD, oviduct; Oo, oogenesis. (From [19], with permission.)

4.2 HYPOTHALAMIC–PITUITARY–OVARIAN AXIS

Although the endocrinology of asynchronous spawners (fathead minnow, zebrafish, and medaka) are not as well known as is the case for synchronous spawners (salmonids: e.g., trout and salmon), the principal systems appear to have been highly conserved throughout the phylogenetic development of teleosts [10]. Indeed, the hypothalamus–pituitary (adenohypophysis)–gonadal axis (HPG axis) appears to be quite similar for many oviparous vertebrates [10,20–22]. In teleosts, gonadotropin-releasing hormone (GnRH), produced by the hypothalamus, does not reach the pituitary via a vascular portal system but rather, through direct innervation of the pituitary with hypothalamic neurons. The HPG axis is a dynamic system (see Figure 4.3) with positive and negative feedback loops, allowing for optimal control of developmental processes (*organizational level*) as well as of reproductive processes in the adult (*functional level*). GnRH, a decapeptide, stimulates the pituitary to produce and release two glycoprotein hormones analogous to the mammalian luteinizing hormone (LH) and follicle-stimulating hormone (FSH): namely, LH-like (GtH-II) and FSH-like (GtH-I), which in turn are involved in induction of oogenesis (FSH/GtH-I) and oocyte maturation as well as ovulation (LH/GtH-II; see Sections 4.3.1, 4.3.2, and 4.3.2.2). In turn, GnRH release is inhibited by dopamine, which is regulated by steroid levels [23].

4.3 CELLULAR STRUCTURE OF THE OVARY

4.3.1 Germ Cells (Oogenesis)

After the formation of primordial germ cells (PGCs) and transformation of the PGCs into oogonia (see also Sections 3.1 to 3.4), oogenesis can be described in four major steps:

1. Transformation of oogonia into oocytes (onset of meiosis)
2. Growth of the oocytes while under meiotic arrest

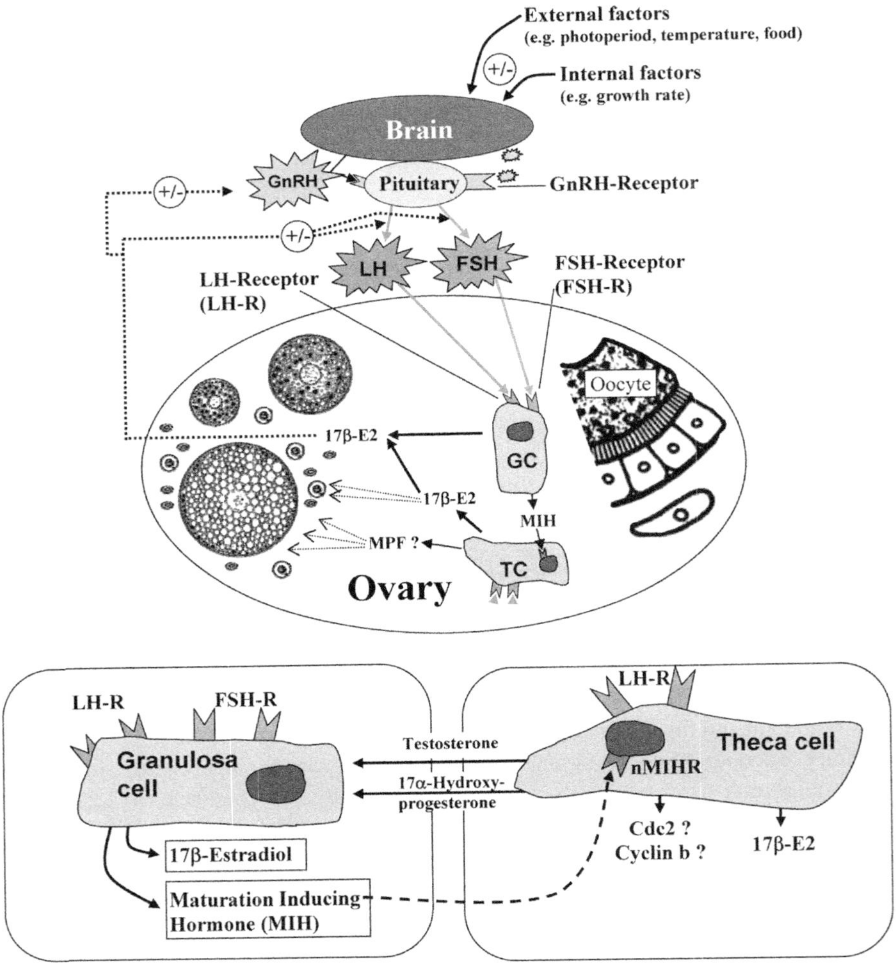

FIGURE 4.3 Schematic of the fish hypothalamic–pituitary–gonadal axis, demonstrating the positive and negative feedback loops as well as the somatic cells in the ovary, which are stimulated directly.

3. Resumption of meiosis (maturation)
4. Expulsion of the ovum from its follicle (ovulation)

One of the major effectors of oogenesis is FSH (Figure 4.3), which induces granulosa cells to express increased amounts of aromatase, which allows conversion of testosterone produced by theca cells to be converted into 17β-estradiol (E_2), which in turn promotes oocyte development as well as stimulating the liver to express and secrete vitellogenin (Vtg) (see Sections 4.3.2.1 and 4.3.2.2). Vitellogenin secreted by the liver into the bloodstream reaches the ovary and is then translocated into the developing oocyte via the granulosa cells (see below).

In the following section we describe some of the histologically (light microscope) observable changes that accompany these steps. For enhanced understanding and completion, some of the biological processes in the development of female germ cells are also highlighted. The different stages of oocyte maturation in an ovary section of a female medaka as well as for the three main experimental fish species (i.e., zebrafish, fathead minnows, and medaka) are shown in Figures 4.4 and 4.5, respectively.

Oogonia Shortly after gonadal sex differentiation or later in the adult life at the beginning of each reproductive cycle, or continuously in nonseasonal breeders, a fraction of the oogonia present in the ovaries undergo a series of mitotic divisions (clonal expansion) followed by their transformation into oocytes (onset of meiosis) [24]. The various stages of oocyte development and maturation have been described in detail morphologically as well as functionally in several reviews [4,8,9,17,24–26] as well as in a book dedicated to fish oocytes [6]. The terminology used for the description of most prominent stages of oocyte development (Figures 4.4 and 4.5) and follicular growth follows that of Yamamoto [27] and is based on conventional histological processing and staining with H&E (see Chapter 7 and the Appendix) as well as light-microscopic inspection.

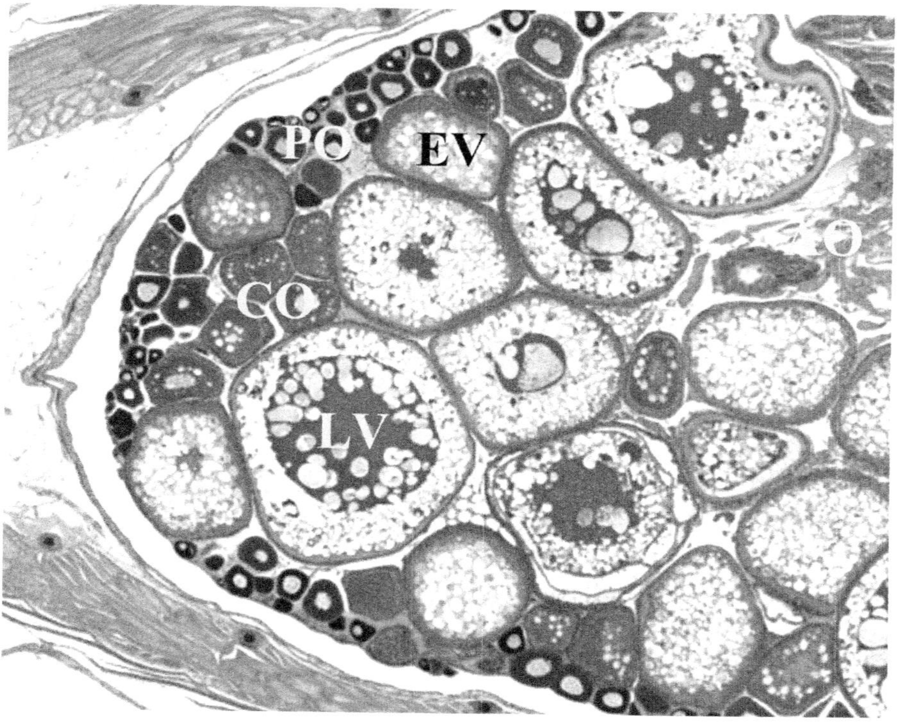

FIGURE 4.4 Various stages in the developmental/differentiation progression of the ovarian follicles of medaka. PO, perinucleolar oocyte; CO, cortical–alveolar oocyte; EV, early vitellogenic occyte; LV, late vitellogenic oocyte; AO, atretic oocyte. (Courtesy of R. D. Johnson, U.S. EPA, Duluth, Minnesota.) (See insert for color representation.)

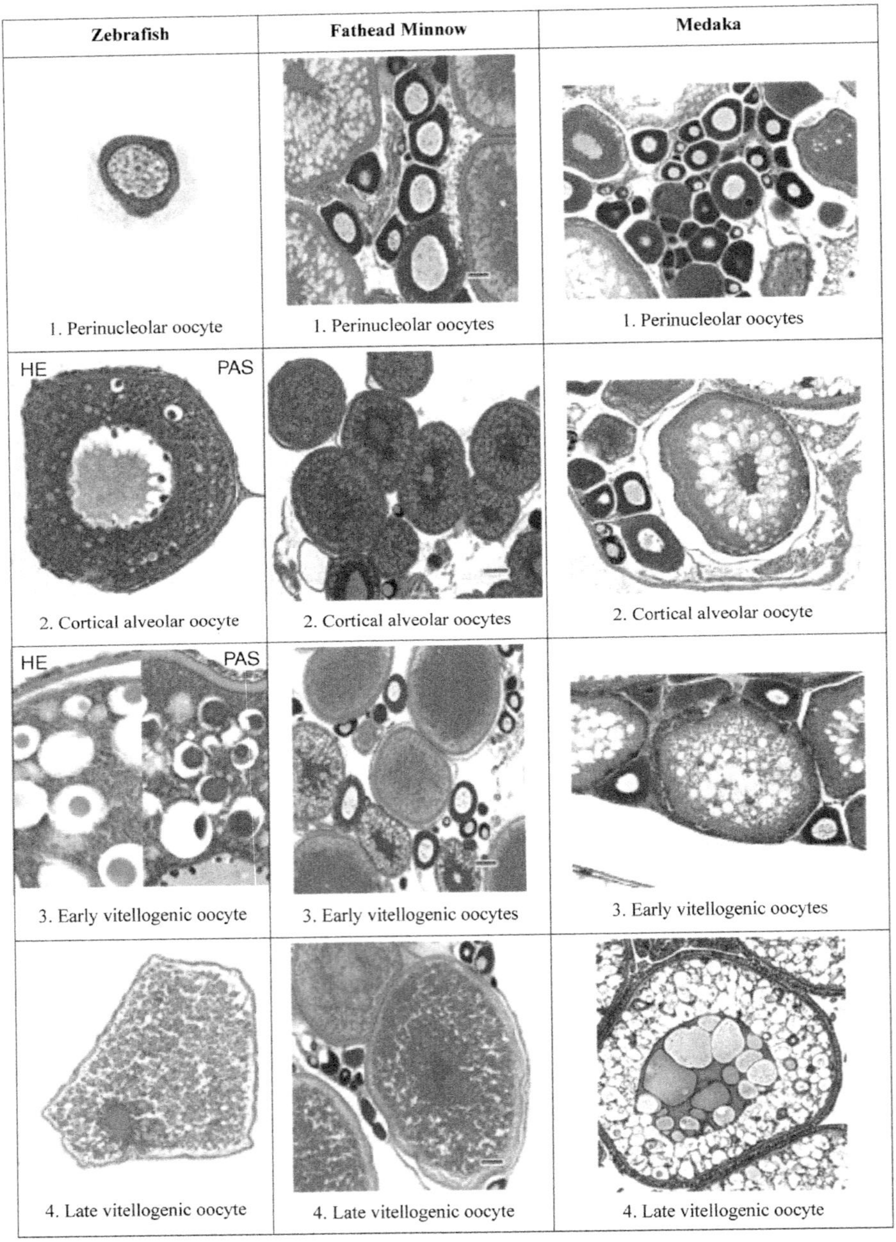

FIGURE 4.5 Oogenesis: various germ cell stages. (Zebrafish images courtesy of L. van der Ven and P. Wester, RIVM, Bilthoven, The Netherlands; fathead minnow images courtesy of J. Wolf, EPL Inc., Sterling, Virginia; medaka images courtesy of J. W. Fournie, U.S. EPA, Gulf Breeze, Florida.) (See insert for color representation.)

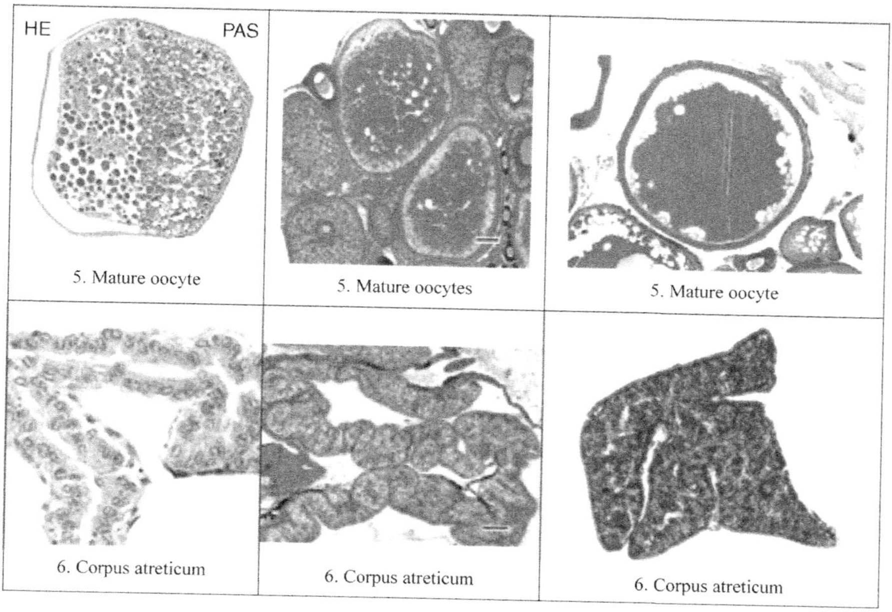

FIGURE 4.5 (*Continued*)

Chromatin Nucleolar Stage Teleost oocytes arise initially within the ovarian luminal epithelium. In this stage the chromosomes become arrested at the diplotene stage of the first meiotic prophase and the initial follicle is established. Pre-follicle cells (presumptive granulosa cells) and their associated basement membrane surround each oocyte and the resulting complex buds from the germinal nest as a primordial follicle [24]. This is the chromatin nucleolar stage of primary oocyte growth. The oocyte has a large nucleus surrounded by a thin layer of cytoplasm, whereas the nucleus contains a single large nucleolus [26].

Perinucleolar Stage Concomitant with oocyte growth, the nucleus (germinal vesicle) increases in size and multiple nucleoli appear, generally at the periphery. The cytoplasm stains uniformly, although late perinucleolus oocytes may have vacuoles in the cytoplasm, the presence of which usually characterizes the cortical alveolar stage (see below). At the end of the perinucleolar stage, the oocyte surface is extended to numerous microvilli, around which the chorion precursor material begins to accumulate in patches.

Cortical Alveolar Stage Yolk Vesicle (Cortical Alveoli) Formation This stage is characterized by the appearance of yolk vesicles (alveoli) in the cytoplasm. Using conventional H&E staining, these spherical structures appear empty but generally stain well with dyes such as Alcian blue, toluidine blue (metachromatic), and the periodic acid–Schiff (PAS) reagents (see Chapter 7 and the Appendix). This is due primarily to the fact that the alveoli contain large amounts of polysialoglycoproteins which are synthesized by the oocyte during middle-to-late previtellogenic growth [28]. The yolk vesicles increase in size and number to form several peripheral rows and give rise to cortical alveoli, which release their contents into the perivitelline space (between the oocyte and the overlying acellular vitelline envelope, sometimes also referred to as *chorion* or *zona radiata* [24]) and inside the egg membranes during fertilization. The contents (polysialoglycoproteins) released by the cortical alveoli are not yolk in a strict sense, since these contents do not serve as a nutrient source for the embryo. *Yolk* is often used synonymously with *vitellogenin* (Vtg) (see below). The term *yolk vesicle* [11,12] is considered by Selman et al. [29] and Bromage and Cumaranatunga [30] to be a misnomer, and Selman and Wallace [31] recommended that it be replaced by the term *cortical alveoli.* Oil droplets (i.e., fat vesicles, vacuoles, or globules; fatty or lipid droplets) begin to accumulate in the cytoplasm of oocytes in some teleosts, particularly marine species, at approximately the same time as cortical alveoli precursors. Fat vesicles are involved in the formation of oil or lipid droplets in fully developed eggs. As the contents of these lipid droplets shrink or are dissolved during histological processing of ovarian tissue (dehydration with alcohols) and appear empty with conventional H&E staining, Mayer et al. [32] stained the lipid droplets black by first stabilizing the unsaturated lipids with osmium tetroxide. The chorion commonly appears at the cortical alveolar stage, and some authors use the co-occurrence of the *zona radiata* and vacuolated cytoplasm to distinguish between cortical alveolar and perinucleolar oocytes [33]. However, the appearance of the *zona radiata* can vary among species, being reported at the late perinucleolar stage by some authors [6,28,34] and at the end of the cortical alveolar stage in blue fin tuna by Baglin [35].

Vitellogenic (Yolk) Stage The appearance of yolk proteins in fluid-filled spheres (yolk spheres, granules, or globules) is characteristic of vitellogenic or yolk-stage oocytes. The yolk granules are very small when they first appear, 2 to 3 μm, and may be difficult to detect with a light microscope. Yolk is made up primarily of proteolytic cleavage products (phosvitin and lipovitelline) of vitellogenin (Vtg), a glycophospholipoprotein (molecular weight 250 to 600 kDa), synthesized by the liver under hormonal (E_2) control [9,36,37]. Uptake of Vtg appears to be mediated via a specific vitellogenin receptor (VtgR) and subsequent internalization of the Vtg–VtgR complex through pinocytosis [24,29,31,33,38,39] of the oocyte membrane (oolemma) enhanced by "clathrin-coated pits." Uptake of Vtg appears to be stimulated by the FSH - like GtH-I (i.e., gonadotropin I) in rainbow trout (*Oncorhynchus mykiss*) and goldfish (*Carassius auratus*) [21,22,40–42], although this has to be confirmed for other teleosts. The yolk spheres may maintain their integrity throughout oocyte growth or fuse to eventually form a continuous mass of fluid yolk (yolk mass), which gives the eggs their characteristic transparency when inspected grossly. The fusion of yolk granules may begin soon after their initial formation or as late as final maturation. Although in most of the teleost species studied the cortical alveoli appear before the lipid droplets and yolk granules, there are exceptions [6]. Generally, an early vitellogenic follicle can be distinguished from a late vitellogenic follicle by the mere visual appraisal of yolk granules and thus aid in distinguishing developmental stages of the oocytes within the vitellogenic stage [34].

Ripe (Mature) Stage Following binding of pituitary-luteinizing hormone (LH/GtH-I) to the gonadotropin receptor in granulosa cells (see Figure 4.3) and the ensuing stimulation of events [e.g., production of maturation-inducing hormone (MIH)] by follicular cells, the acquisition of follicular maturational competence (FMC), and the MIH-dependent activation of maturation-promoting factor (MPF: cdc2 and cyclin B [41,43–48]) in the oocyte cytoplasm [44] and thus resumption of meiosis and cytoplasmic maturation, the final stage of oocyte development (ovulation), culminating in the release of oocytes from the follicle into the ovarian lumen, begins [49]. Acquisition of FMC in trout is characterized by an increase in MIH circulating levels and a concomitant drop in E_2 levels as well as by an increased FSH receptor and insulin-like growth factor 2 (IGF-2) mRNA levels [50]. Thus, FMC is not only influenced by the HPG-axis-dependent MIH via increased expression of MIH, but also by IGF-1 and IGF-2 in an HPG-axis-independent fashion [51–55]. The importance of IGF-1 and IG-2 for oocyte maturation appears largely teleost species–dependent; however, both IGF-1 and IGF-2, as well as the IGF receptor 1b mRNA stocks within the oocyte, appear crucial for later embryonic survival [48].

The start of the mature stage is indicated by the peripheral migration of the nucleus (germinal vesicle) and the dissolution of its membrane (germinal vesicle breakdown). Upon completion of the first meiotic division and the regression and expulsion of the first polar body (the polar body being the result of the regression of the second, yolk-poor oocyte following the first meiotic division), meiosis in the matured oocyte (egg) is arrested *de novo* in metaphase II [24,56]. Following the binding of MIH (in teleosts 17α,20β-dihydroxy-4-pregnen-3-one [41,43,57,58]) to an oocyte membrane receptor (mMIH-r) [43,57], oocyte meiosis is resumed. Concomitant with resumption of oocyte meiosis, oocytes undergo cytoplasmic maturation, which involves yolk proteolysis and hydration. The production of MIH by ovarian follicles [51] is also necessary to induce the last step of oogenesis, ovulation. Ovulation occurs when oocytes are shed from within their perifollicular cell sheath and are expelled into the ovarian lumen in the arrested phase of the second meiotic division [24]. However, unlike under meiotic resumption, ovulation requires MIH-dependent transcriptional activation that is regulated by a nuclear MIH receptor (nMIH-r) rather than a membrane-bound MIH receptor (mMIH-r). Although the location of the nMIHR is presently unclear in teleosts, analogy with rats would suggest that the nMIHR is located in granulosa and theca cells [5,24]. At this point the oocytes (eggs) are located within the ovarian lumen, ready to be spawned and fertilized. Fertilization induces not only the release of cortical alveolar contents into the perivitelline space, thus promoting hydration and hardening of the vitelline envelope, but also induces completion of the second meiotic division and expulsion of the second polar body (the result of the regression of the second, yolk-poor oocyte following the second meiotic division upon fertilization). Polar bodies, which are rarely observed, should not be confused histologically with the micropylar cell, a specialized granulosa cell, that is associated with the micropylar region, which covers the micropyle (a minute perforation in the chorion), the future entrance of sperm [12,59].

Although the individual steps involved in the maturation stage have been described in detail above, it must be clear that these individual steps are not necessarily specifically observable in a microscopic preparation and certainly not in a histological section of an ovary. This is due primarily to the fact that the time frame in which the individual steps reside can vary from prolonged to transient, meaning that some steps take place continuously and thus can rarely be observed. However, due to the multiple interactions of numerous factors in FMC, germinal vesicle breakdown (GVBD), and finally, maturation, it is not surprising to find that some of the EDCs can induce (diethylstilbestrol and tamoxifen) or inhibit (pentachlorophenol, methoxychlor) oocyte maturation, as demonstrated in zebrafish [44,60,61]. The mechanism by which diethylstilbestrol promotes oocyte maturation is governed by its capability to bind noncompetitively, yet with affinity similar to that of 17α,20β-dihydroxy-4-pregnen-3-one to the goldfish oocyte membrane progestin receptor α(mPRα) [62].

Post-ovulatory Follicles (Corpora Lutea) Ovulation results in ruptured, empty, or post-ovulatory (oocyte-free) follicles. Newly generated post-ovulatory follicles, sometimes referred to as Graafian follicles [63–66], are readily identifiable; however, they degenerate rapidly and thus may not be observed readily in histological sections. However, the term *Graafian follicle* is misplaced for fish histology, especially as the term stems from mammalian histology. Indeed, in mammalian histology, Graafian follicles are defined differently: The Graafian follicle is characterized by a large, fluid-filled antrum and an eccentric oocyte. The granulosa cells in Graafian follicles can be divided into two groups: The *zona granulosa* is a thin layer along the periphery of the follicle, and the *corona radiata* surrounds the oocyte. Consequently, for fish histology the term *postovulatory follicle* appears more appropriate.

Six stages of postovulatory follicles have been distinguished in the carp; the fifth and sixth stages are distinguished from one another by the size of the post-ovulatory follicle (for a review, see Saidapur [67]). The various stages distinguished as being post-ovulatory-specific do not differ substantially from the stages identified as being specific for follicular atresia (see below), although specifically at early stages, the major histological difference between these two is the absence or presence of oocyte debris, of which the membrane, particularly, is relatively persistent. Generally, a highly vascularized thecal cell layer characterizes post-ovulatory follicles. Whether or not hypertrophied granulosa cells can be observed appears largely a species-specific issue [67]. At moderate temperatures (13 to 19°C), post-ovulatory follicles are difficult to distinguish from intermediate stages of atretic oocytes. At higher temperatures, the resorption of the post-ovulatory follicles may be accelerated. In contrast to atretic follicles, post-ovulatory follicles in some species (e.g., medaka and zebrafish) may continue to express enzymes involved in steroidogenesis (3β-HSD and G-6-PDH) and to produce steroids (progestogens and testosterone) [8,67].

Atretic Follicles (Corpora Atretica) Follicular atresia is a common occurrence in the teleost ovary. Stress and other environmental insults (e.g., poor nutrition, physical factors, overpopulation or stressors that could decrease gonadotropin levels, xenobiotics) appear to increase the incidence of atresia [5,9,68]. However, this phenomenon is also observed in healthy female fish [4,5,9,69,70]. In the latter case, although probably minor in importance, atresia may serve to regulate derecruitment and/or resorption of malfunctioning younger-stage oocytes in order to preserve normal ovarian physiology during the breeding season. A more important role is attributed to atresia at the end of the breeding season, especially in fish with asynchronous ovaries (i.e., asynchronous spawning) and defined seasonal breeding patterns (e.g., in medaka and fathead minnows), where atresia could be seen as a "cleaning-up" process [9,71]. Indeed, atresia appears carried out via an apoptotic and not a necrotic process, thus clearly enabling the organism to remove excess or damaged follicles via a controlled process. However, the point in which and the extent to which apoptosis is involved in follicular atresia are currently a matter of dispute. Whereas Wood and van der Kraak [72] and Drummond et al. [73] maintain that apoptosis is a feature of normal follicular growth and post-ovulatory regression in teleost ovaries not associated with follicular atresia [72], studies in nonexposed and exposed feral white sucker populations [70] revealed that apoptosis is a part of the atretic process.

Also of importance in this context is the observation that previtellogenic, vitellogenic, and maturing follicle apoptosis in rainbow trout is suppressed by E_2, epidermal growth factor, and partially purified Chinook salmon (*Oncorhynchus tshawytscha*) gonadotropin [74], possibly explaining why reduced gonadotropin levels or xenobiotics may increase the incidence in atretic follicles [5,9,68]. Keeping the latter in mind and the fact that fish with asynchronous ovaries (e.g., medaka and fathead minnows) do not reach or end the spawning status at the very same time within a breeding season (inhomogeneity of reproductive ripeness within a cohort), a normal but highly variable incidence of atresia may be prevalent in feral fish, depending largely on the point within the breeding season and thus on the inherent gonadotropin and steroid levels. This may be important to acknowledge, especially when laboratory findings (with low levels of atresia) are compared or extrapolated to field situations or if the effects of chemical or chemical mixtures (e.g., bleached pulp mill effluents) on follicular atresia are evaluated [69]. Whether caused by physiologic stimuli and/or environmental factors, follicular atresia entails the degeneration of follicles at any stage of follicular development [8], although atretic follicles are often most noticeable in the later developmental stages [4]. Follicular atresia can be divided by light-microscopic description into four consecutive stages: 1(α), 2(β), 3(γ), and 4(δ) [8,75]. A division into five stages (α, β, γ, δ, and ε) of atresia was proposed by Khoo [76] after extensive investigation in the goldfish. Khoo fourth stage is characterized by the complete reabsorption of all remnants of oocytes by hypertrophied perifollicular granulosa cells, and the fifth stage by the collapse of granulosa cells to form an irregular cellular mass (Figure 4.5).

Generally, the four stages assembled by Saidapur [75] appear most applicable to the histopathological assessment of atresia in small fish, especially as this staging provides for the description of the most prominent features of atresia:

Stage 1. Breakdown of the ovum and its resorption and phagocytosis by hypertrophied granulosa cells (see Section 4.3.2.2)

Stage 2. Increased hypertrophy of the granulosa cells, invasion of follicular space by granulosa cells, further phagocytosis of the yolk, possibly in-growth of theca cells (see Section 4.3.2.1), and increased vascularization

Stage 3. Appearance of an intensely orange pigment in the granulosa cells and theca cells

Stage 4. Disintegration of the granulosa cell structures to form an irregular heap of cells

4.3.2 Female Somatic Supportive Tissue

The basic structure of the ovarian follicle is established when pre-follicle cells (presumptive granulosa cells) and their associated basement membrane envelope the oocyte completely (at late pachytene or early diplotene stages of meiosis; i.e., the chromatin–nucleolar stage of the oocyte). The granulosa cells (see Section 4.3.2.2) form a monolayer and surround the oocyte, with the microvilli of the oocyte making contact with the granulosa cells via heterologous gap junctions, first recognizable in cortical alveolar follicles. Homologous gap junctions also exist between granulosa cells as well as among theca cells, another type of supportive cells in the ovary (see Section 4.3.2.1). A heterogeneous layer of theca cells is added externally over an acellular basement membrane, which separates granulosa and theca cells (Figure 4.6). This arrangement remains essentially unchanged throughout follicular growth, but at the onset of ovulation, the heterologous gap junctions between the granulosa cells and the oocyte are lost. Thus, ovulated eggs of teleost fishes are devoid of surrounding granulosa cells [5,8].

Granulosa and thecal cells (Figures 4.6 and 4.7) are involved in the production and conversion of steroids in the ovary of fish (Figure 4.8). In general, theca cells are the site of precursor synthesis, and the granulosa cells are the site of conversion of the precursors to final products [21]. The transport mechanisms between the two cell layers are still unclear. The metabolites [e.g., the E_2 precursor testosterone (Figure 4.8)] may diffuse directly through the basement membrane or may be actively transported [77].

Considering the direct involvement of thecal and granulosa cells in the hormonal regulation of ovarian development in fish, and some important functions in oocyte maturation (via production of maturation-inducing hormone (17α,20β-dihydroxy-4-pregnen-3-one) and ovulation (see below), it appears self-explanatory that any xenobiotic-induced adverse effects on these cells (cytotoxicity, inhibition of enzymes, etc.) may result in an observable endocrine-disruptive effect without the direct involvement of a steroid or steroid analog. However, as described by Tokumoto and co-workers in zebrafish and goldfish [44,61,62], inhibition of oocyte maturation is observed with nonsteroidal EDC (PCP, methoxychlor), whereas an induction of oocyte maturation was observed with steroid analogs (e.g., diethylstilbestrol) and cyp19 (aromatase) inhibitors (e.g., tamoxifen).

4.3.2.1 Theca Cells Theca cells are not only the major site of testosterone and 17α-progesterone production in the ovary of teleosts, but have an additional function, for example, in mechanical processes during ovulation: In theca cells of medaka follicles, 5 hours after hydrocortisone stimulation, which induces ovulation, microfilaments organize into bands, and after 10 hours into bundles

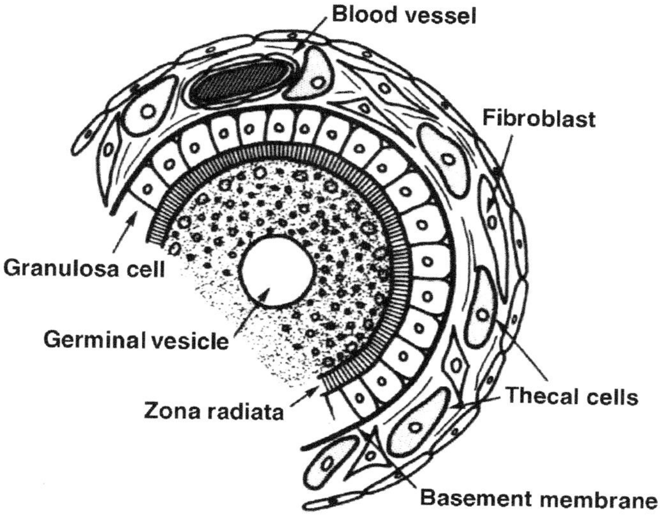

FIGURE 4.6 Diagrammatic representation of the follicle layer surrounding an early vitellogenic oocyte of salmonid fishes. The granulosa layer is separated from the thecal layer by a distinct basement membrane. The thecal layer is composed of thecal cells, fibroblasts, blood vessels, and so on. (Modified from [8].)

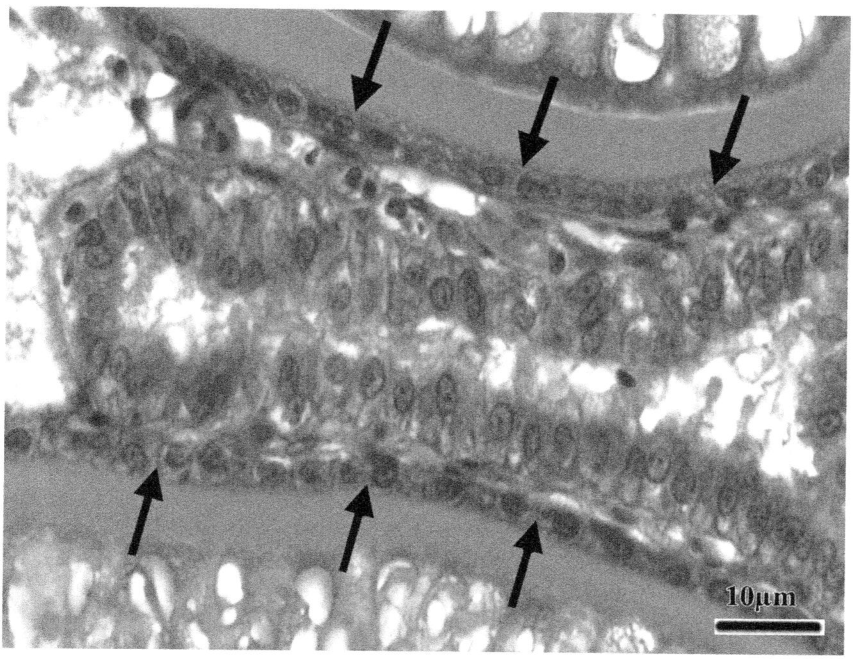

FIGURE 4.7 Fathead minnow theca and granulosa cell layer (arrows); enlarged and vacuolated theca and granulosa cells of the atretic oocyte in the middle. Scale bar = 10 μm. (Courtesy of J. Wolf, EPL Inc., Sterling, Virginia.) (See insert for color representation.)

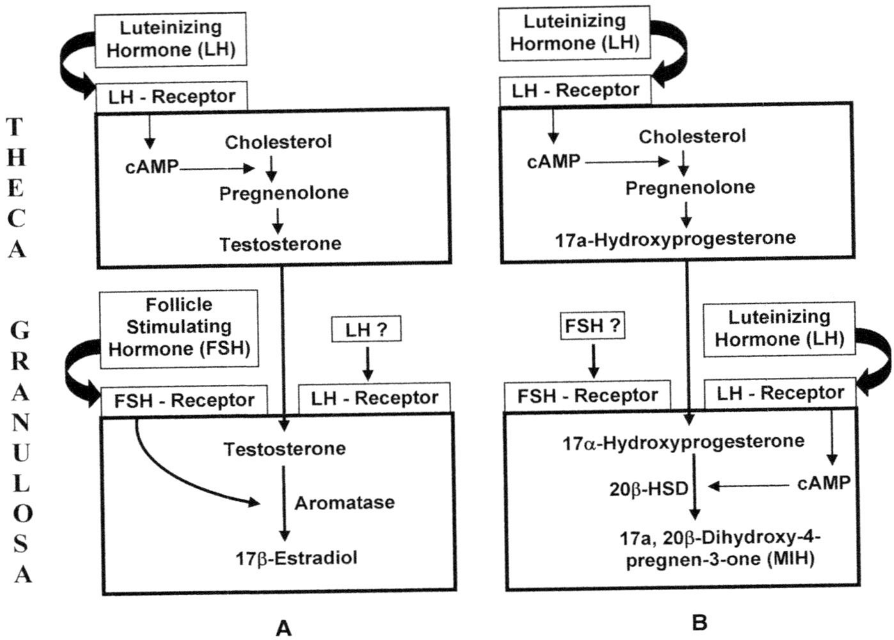

FIGURE 4.8 Proposed two-cell model for the production of (A) E_2 and (B) 17α,20β-dihydroxy-4-pregnen-3-one (MIH) in the ovarian follicle of salmonids. (Modified from [21] according to recent literature [5,24,56,82].)

in vitro. Pendergrass and Schröder suggest that these filaments are responsible for cell movement in medaka theca cells. Addition of cytochalasin B (inhibitor of microfilamentous contractile systems) prevented hydrocortisone-induced ovulation in medaka [78,79]. The presence of microfilaments in theca cells near the time of ovulation was also observed in a number of other teleost fish species [78,80,81], suggesting an active role of theca cells in the ovulation of teleosts.

Another important function of theca cells is the production of hormone precursors in response to gonadotropins, specifically testosterone (which is converted to E_2 by granulosa cells; Figure 4.8A) and 17α-hydroxyprogesterone (which is converted to the maturation-inducing hormone (MIH) by granulosa cells, Figure 4.8B), which results finally in a local increase of E_2 or MIH in the direct environment of the follicle. In isolated thecal layers (at various stages of oocyte maturation) of coho salmon (*Oncorhynchus kisutch*) MIH production was stimulated by both luteinizing hormone (LH, also termed gonadotropin II or GtH-II) and follicle-stimulating hormone (FSH, also termed gonadotropin I or GtH-I) [33] of coho salmon; LH was shown to be more potent [82]. The production of MIH in the theca layer may be due to a 10% contamination of the theca layer preparation with granulosa cells, since Nagahama found theca cells to express only receptors for LH [8]. LH has been suggested to induce testosterone formation in theca cells but to have no apparent effect on aromatase activity. Indeed, Planas et al. [82] suggest LH to be the determining factor in MIH production and induction of the germinal vesicle breakdown.

4.3.2.2 Granulosa Cells In the vertebrate ovarian follicle, cytoplasmatic connections (heterologous gap junctions) are established between the microvilli of an oocyte and their surrounding granulosa cells early in oocyte development. These gap junctions may serve as channels for the transport of nutrients from the granulosa cell into the growing oocyte [83]. In medaka, heterocellular gap junctions are first recognized in cortical alveoli follicles just prior to the onset of vitellogenesis [84]. In addition, granulosa cells form gap junctions with each other, resulting in homologous gap junctions [83].

Granulosa cells appear to be the major site of E_2 and MIH production. 3β-Hydroxysteroiddehydrogenase (3β-HSD) activity as well as the presence of 17α-hydroxyprogesterone and progesterone was demonstrated in the granulosa cells of various fish species, including medaka and zebrafish [8,85]. In medaka, granulosa cells appear to be active only during the spawning period or after an artificially induced ovulation or stimulation by gonadotropins [85,86]. Aromatase (converting testosterone to E_2) activity in granulosa cells, at least in the amago salmon (*Oncorhynchus rhodurus*), increases during vitellogenesis, reaching a peak in late vitellogenesis; thereafter, the activity declines rapidly [21].

MIH production in granulosa cells (at various stages of oocyte maturation) appears to be induced through binding of LH and FSH to their respective receptors [5,24,56]; see also Section 4.3. However, some species restrictions may apply, as Planas et al. [82] found stimulation of MIH production by LH but not by FSH in coho salmon granulosa cells.

Another function of granulosa cells is their role in the regulation of oocyte hydration during oocyte maturation The presence of granulosa cells was found to be strictly necessary for oocyte hydration, a process thought to be mediated by K^+ influx through the heterologous gap junctions between granulosa cells and the oocyte [87], a thesis supported by the observation that denuded oocytes of mummichog (*Fundulus heteroclitus*) do not hydrate when they undergo maturation [88].

REFERENCES

1. Yamamoto, T., Induced intersexes, in *Medaka (Killifish): Biology and Strains*. Keigaku Publishing Co., Tokyo, 1975, pp. 214–222.
2. Gray, M.A. and C.D. Metcalfe, Induction of testis–ova in Japanese medaka *(Oryzias latipes)* exposed to *p*-nonylphenol. *Environmental Toxicology and Chemistry*, 1997, **16**(5):1082–1086.
3. Redding, J.M. and R. Patino, Reproductive systems: Gross functional anatomy, in *The Handbook of Experimental Animals: The Laboratory Fish*, G.K. Ostrander, Editor. Academic Press, San Diego, CA, 2000, pp. 261–272.

4. Tyler, C.R. and J.P. Sumpter, Oocyte growth and development in teleosts. *Reviews in Fish Biology and Fisheries*, 1996, **6**(3):287–318.
5. Patino, R. and J.M. Redding, Reproductive systems, in *The Handbook of Experimental Animals: The Laboratory Fish*, G.K. Ostrander, Editor. Academic Press, San Diego, CA, 2000, pp. 489–500.
6. Babin, P.J., J. Cerdà, and E. Lubzens, in *The Fish Oocyte: From Basic Studies to Biotechnological Applications*, P.J. Babin, J. Cerdà, and E. Lubzens, Editors. Springer, New York, 2007.
7. Harder, W., *Anatomy of Fishes*, Vol. 2. E. Schweizerbart'sche Verlagsbuchhandlung, Stuttgart, Germany, 1975.
8. Nagahama, Y., The functional morphology of teleost gonads, in *Fish Physiology*, W.S. Hoar, D.J. Randall, and E.M. Donaldson, Editors. Academic Press, San Diego, CA, 1983, pp. 223–275.
9. Wallace, R.A. and K. Selman, Cellular dynamic aspects of oocyte growth in teleosts. *American Journal of Zoology*, 1981, **21**:325–343.
10. Ankley, G.T. and R.D. Johnson, Small fish models for identifying and assessing the effects of endocrine-disrupting chemicals. *ILAR Journal*, 2004, **45**(4):469–483.
11. Yamamoto, K. and H. Yoshioka, Rhythm of the development in the oocyte of the medaka, *Oryzias latipes*. *Bulletin of the Faculty of Fisheries, Hokkaido University*, 1964, **15**(1):5–19.
12. Yamamoto, T., Morphological and cytochemical studies on the oogenesis of the fresh water fish, medaka *(Oryzias latipes)*. *Japanese Journal of Ichthyology*, 1955, **4**:170–181.
13. Hirai, T., et al., Glycoprotein hormone receptor family in medaka *(Oryzias latipes)*. *Fish Physiology and Biochemistry*, 2003, **28**(4):345–346.
14. Strüssmann, C.A. and M. Nakamura, Morphology, endocrinology, and environmental modulation of gonadal sex differentiation in teleost fishes. *Fish Physiology and Biochemistry*, 2002, **26**(1):13–29.
15. Takahashi, H., Juvenile hermaphroditism in the zebrafish, *Brachydanio rerio*. *Bulletin of the Faculty of Fisheries, Hokkaido University*, 1977, **28**:57–65.
16. Uchida, D., et al., Oocyte apoptosis during the transition from ovary-like tissue to testes during sex differentiation of juvenile zebrafish. *Journal of Experimental Biology*, 2002, **205**:711–718.
17. Nakamura, M., et al., Gonadal sex differentiation in teleost fish. *Journal of Experimental Zoology*, 1998, **281**(5):362–372.
18. Takahashi, H., Modification of the development of female reproductive organs in the guppy, *Poecilia reticulata*, following androgen treatment in their juvenile period. *Bulletin of the Faculty of Fisheries, Hokkaido University*, 1974, **25**:174–199.
19. Gimeno, S. and C.T. Bowmer, Ecological risk assessment of substances with suspected estrogenic activity using standard laboratory fish tests, in *Environmental Toxicology and Risk Assessment: Standardization of Biomarkers for Endocrine Disruption and Environmental Assessment*, ASTM STP 1364, D.S. Henshel, M.C. Black, and M.C. Harrass, Editors. American Society for Testing and Materials, West Conshohocken, PA, 1999, pp. 182–206.
20. Adkins-Regan, E., Hormones and sexual differentiation, in *Hormones and Reproduction in Fishes, Amphibians, and Reptiles*, D.O. Norris and R.E. Jones, Editors. Plenum Press, New York, 1987, pp. 1–27.

21. Nagahama, Y., Gonadotropin action on gametogenesis and steroidogenesis in teleost gonads. *Zoological Science*, 1987, **4**(2):209–222.

22. Nagahama, Y., Gonadal steroid hormones: Major regulators of gonadal sex differentiation and gametogenesis in fish. *Sixth International Symposium on the Reproductive Physiology of Fish*, Institute for Marine Research, Bergen, Norway, 1999.

23. Peter, R.E. and K.L. Yu, Neuroendocrine regulation of ovulation in fishes: Basic and applied aspects. *Reviews in Fish Biology and Fisheries*, 1997, **7**:173–197.

24. Patino, R. and C.V. Sullivan, Ovarian follicle growth, maturation, and ovulation in the teleost fish. *Fish Physiology and Biochemistry*, 2002, **26**(1):57–70.

25. Yonkos, L.T., R. Reimschuessel, and A.S. Kane, Digital atlas of fathead minnow normal histology, http://aquaticpath.umd.edu/fhm/, 2003.

26. West, G., Methods of assessing ovarian development in fishes: A review. *Australian Journal of Marine and Freshwater Research*, 1990, **41**:199–222.

27. Yamamoto, K., Studies on the formation of fish eggs: Annual cycle in the development of the ovarian eggs in the flounder, *Liopsetta obscura*. *Journal of the Faculty of Science, Hokkaido University, Series VI, Zoology*, 1956, **12**:362–373.

28. Blazer, V.S., Histopathological assessment of gonadal tissue in wild fishes. *Fish Physiology and Biochemistry*, 2002, **26**(1):85–101.

29. Selman, K., V. Bair, and R.A. Wallace, The relationship of yolk vesicles and cortical alveoli in teleost oocytes and eggs. *International Symposium on Reproductive Physiology of Fish*, D.R. Idler, L.W. Crim, and J.M. Walsh, Editors. Memorial University of Newfoundland, St. Johns, Newfoundland, 1987, p. 216.

30. Bromage, N.R. and P.R.T. Cumaranatunga, Egg production in rainbow trout. *Recent Advances in Aquaculture*, Vol. 3. J.F. Muir and R.J. Roberts, Editors. Croom Helm, London, 1988, pp. 65–138.

31. Selman, K. and R.A. Wallace, Cellular aspects of oocyte growth in teleosts. *Zoological Science*, 1989, **6**:211–231.

32. Mayer, I., S.E. Shackley, and J.S. Ryland, Aspects of the reproductive biology of the bass, *Dicentrarchus labrax* L.: 1. An histological and histochemical study of oocyte development. *Journal of Fish Biology*, 1988, **33**(4):609–622.

33. Arukwe, A. and A. Goksoyr, Eggshell and egg yolk proteins in fish: hepatic proteins for the next generation: oogenetic, population, and evolutionary implications of endocrine disruption. *Comparative Hepatology*, 2003, **2**(1):4.

34. McDonald, K.K., et al., Reproductive indicators, in *Biomonitoring of Environmental Status and Trends (BEST) Program: Selected Methods for Monitoring Chemical Contaminants and Their Effects in Aquatic Ecosystems*. U.S. Department of the Interior, U.S. Geological Survey, Biological Resources Division, Washington, DC, 2000, pp. 37–42.

35. Baglin, R.E.J., Reproductive biology of western Atlantic bluefin tuna. *U.S. National Marine Fisheries Service Fishery Bulletin*, 1982, **80**(1):121–134.

36. Wester, P.W. and J.H. Canton, Histopathological study of *Oryzias latipes* (medaka) after long-term β-hexachlorocyclohexane exposure. *Aquatic Toxicology*, 1986, **9**(1):21–45.

37. Selman, K. and R.A. Wallace, Oocyte growth in the sheepshead minnow: Uptake of exogenous proteins by vitellogenic oocytes. *Tissue and Cell*, 1982, **14**(3): 555–571.

38. Mommsen, P.T. and P.J. Walsh, Vitellogenesis and oocyte assembly, in *Fish Physiology*, W. S. Hoar, D.J. Randall, and E.M. Donaldson, Editors. Academic Press, San Diego, CA, 1988, pp. 347–406.
39. Arukwe, A., Celluar and molecular responses to endocrine modulators and the impact on fish reproduction. *Marine Pollution Bulletin*, 2001, **42**:643–655.
40. Nagahama, Y., 17Alpha, 20beta-dihydroxy-4-pregnen-3-one, a maturation-inducing hormone in fish oocytes: mechanisms of synthesis and action. *Steroids*, 1997, **62**(1):190
41. Nagahama, Y., et al., Regulation of oocyte growth and maturation in fish. *Current Topics in Developmental Biology*, 1995, **30**:103–145.
42. Tyler, C.R., et al., An in vivo testing system for endocrine disruptors in fish early life stages using induction of vitellogenin. *Environmental Toxicology and Chemistry*, 1999, **18**(2):337–347.
43. Bhattacharya, S., et al., Molecular mechanism of oocyte maturation. *Journal of the Society for Reproduction and Fertility*, 2007, **63**(Suppl):45–55.
44. Tokumoto, T., M. Tokumoto, and Y. Nagahama, Induction and inhibition of oocyte maturation by EDCs in zebrafish. *Reproductive Biology and Endocrinology*, 2005,**3**:69.
45. Chattoraj, A., et al., Melatonin accelerates maturation inducing hormone (MIH): Induced oocyte maturation in carps. *General and Comparative Endocrinology*, 2005, **140**(3):145–155.
46. Basu, D., et al., Cdc2-cyclin B-induced G2 to M transition in perch oocyte is dependent on Cdc25. *Biology of Reproduction*, 2004, **71**(3):894–900.
47. Katsu, Y., M. Yamashita, and Y. Nagahama, Translational regulation of cyclin B mRNA by 17alpha,20beta-dihydroxy-4-pregnen-3-one (maturation-inducing hormone) during oocyte maturation in a teleost fish, the goldfish *(Carassius auratus)*. *Molecular and Cellular Endocrinology*, 1999, **158**(1–2):79–85.
48. Aegerter, S., B. Jalabert, and J. Bobe, Messenger RNA stockpile of cyclin B, insulin-like growth factor I, insulin-like growth factor II, insulin-like growth factor receptor Ib, and p53 in the rainbow trout oocyte in relation with developmental competence. *Molecular Reproduction and Development*, 2004, **67**(2):127–135.
49. Yamamoto, Y., et al., Differential expression and localization of four connexins in the ovary of the ayu *(Plecoglossus altivelis)*. *Molecular Reproduction and Development*, 2007, **74**(9):1113–1123.
50. Bobe, J., et al., Rainbow trout follicular maturational competence acquisition is associated with an increased expression of follicle stimulating hormone receptor and insulin-like growth factor 2 messenger RNAs. *Molecular Reproduction and Development*, 2003, **66**(1):46–53.
51. Mukherjee, D., et al., In vitro effects of insulin-like growth factors and insulin on oocyte maturation and maturation-inducing steroid production in ovarian follicles of common carp, *Cyprinus carpio*. *Comparative Biochemistry and Physiology, Part A: Molecular and Integrative Physiology*, 2006, **144**(1):63–77.
52. Weber, G.M., A.B. Moore, and C.V. Sullivan, In vitro actions of insulin-like growth factor-I on ovarian follicle maturation in white perch *(Morone americana)*. *General and Comparative Endocrinology*, 2007, **151**(2):180–187.
53. Weber, G.M. and C.V. Sullivan, Effects of insulin-like growth factor-I on in vivo final oocyte maturation and ovarian steroidogenesis in striped bass, *Morone saxatilis*. *Biology of Reproduction*, 2000, **63**(4):1049–1057.

54. Weber, G.M. and C.V. Sullivan, In vitro hormone induction of final oocyte maturation in striped bass *(Morone saxatilis)* follicles is inhibited by blockers of phosphatidylinositol 3-kinase activity. *Comparative Biochemistry and Physiology, Part B: Biochemistry and Molecular Biology*, 2001, **129B**(2–3):467–473.

55. Weber, G.M. and C.V. Sullivan, Insulin-like growth factor-I induces oocyte maturational competence but not meiotic resumption in white bass *(Morone chrysops)* follicles in vitro: Evidence for rapid evolution of insulin-like growth factor action. *Biology of Reproduction*, 2005, **72**(5):1177–1186.

56. Patino, R., et al., Gonadotropic control of ovarian follicle maturation: the two-stage concept and its mechanisms. *Comparative Biochemistry and Physiology Part B: Biochemistry and Molecular Biology*, 2001, **129**(2–3):427–439.

57. Rahman, M., et al., Characterization of ovarian membrane receptor for 17,20beta-dihydroxy-4-pregnen-3-one, a maturation-inducing hormone in yellowtail, *Seriola quinqueradiata. General and Comparative Endocrinology*, 2002, **127**(1):71–79.

58. Sorensen, P.W., et al., Evidence that 4-pregnen-17,20beta,21-triol-3-one functions as a maturation-inducing hormone and pheromonal precursor in the percid fish, *Gymnocephalus cernuus. General and Comparative Endocrinology*, 2004, **139**(1):1–11.

59. Amanze, D. and A. Iyengar, The micropyle: A sperm guidance system in teleost fertilization. *Development*, 1990, **109**(2):495–500.

60. Tabata, A., et al., Estrogenic influences of estradiol-17 beta, p-nonylphenol and bis-phenol-A on Japanese medaka (*Oryzias latipes*) at detected environmental concentrations. *Water Science and Technology*, 2001, **43**(2):109–116.

61. Tokumoto, T., et al., Diethylstilbestrol induces fish oocyte maturation. *Proceedings of the National Academy of Sciences USA*, 2004, **101**(10):3686–3690.

62. Tokumoto, T., M. Tokumoto, and P. Thomas, Interactions of diethylstilbestrol (DES) and DES analogs, with membrane progestin receptor-alpha and the correlation with their nongenomic progestin activities. *Endocrinology*, 2007, **148**(7):3459–3467.

63. Miles-Richardson, S.R., et al., Effects of waterborne exposure to 4-nonylphenol and nonylphenol ethoxylate on secondary sex characteristics and gonads of fathead minnows (*Pimephales promelas*). *Environmental Research, Section A*, 1999, **80**(2): S122–S137.

64. Miles-Richardson, S.R., et al., Effects of waterborne exposure of 17 β-estradiol on secondary sex characteristics and gonads of fathead minnows *(Pimephales promelas). Aquatic Toxicology*, 1999, **47**(2):129–145.

65. Miles-Richardson, S.R. *Effects of Estrogenic Compounds on the Fathead Minnow* (*Pimephales promelas*). PhD thesis, Michigan State University, East Lansing, MI, 1997.

66. Nichols, K.M., et al., Effects of exposure to municipal wastewater in situ on the reproductive physiology of the fathead minnow *(Pimephales promelas). Environmental Toxicology and Chemistry*, 1999, **18**(9):2001–2012.

67. Saidapur, S.K., Structure and function of postovulatory follicles (corpora lutea) in the ovary of nonmammalian vertebrates. *International Review of Cytology*, 1982, **75**:243–283.

68. Papoulias, D.M., D.B. Noltie, and D.E. Tillitt, An in vivo model fish system to test chemical effects on sexual differentiation and development: Exposure to ethinyl estradiol. *Aquatic Toxicology*, 1999, **48**(1):37–50.

69. Janz, D.M., et al., Elevated ovarian follicular apoptosis and heat shock protein-70 expression in white sucker exposed to bleachad kraft pulp mill effluent. *Toxicology and Applied Pharmacology*, 1997, **147**(2):391–398.

70. Janz, D.M., et al., Recovery of ovary size, follicle cell apoptosis, and HSP70 expression in fish exposed to bleached pulp mill effluent. *Canadian Journal of Fisheries and Aquatic Sciences*, 2001, **58**(3):620–625.
71. Janz, D.M. and G.J. van der Kraak, Suppression of apoptosis by gonadotropin, 17β-estradiol, and epidermal growth factor in rainbow trout preovulatory ovarian follicles. *General and Comparative Endocrinology*, 1997, **105**(2):186–193.
72. Wood, A.W. and G. van, der Kraak, Apoptosis and ovarian function: Novel perspectives from the teleosts. *Biology of Reproduction*, 2001, **64**:264–271.
73. Drummond, C.D., et al., Postovulatory follicle: A model for experimental studies of programmed cell death or apoptosis in teleosts. *Journal of Experimental Zoology*, 2000, **287**(2):176–182.
74. Wood, A.W. and G. van der Kraak, Inhibition of apoptosis in vitellogenic ovarian follicles of rainbow trout (*Oncorynchus mykiss*) by salmon gonadotropin, epidermal growth factor, and 17β-estradiol. *Molecular Reproduction and Development*, 2002, **61**(4):511–518.
75. Saidapur, S.K., Follicular atresia in the ovaries of nonmammalian vertebrates. *International Review of Cytology*, 1978, **54**:225–244.
76. Khoo, K.H., The corpus luteum of goldfish *(Carassius auratus* L.) and its functions. *Canadian Journal of Zoology*, 1975, **53**:1306–1323.
77. Kagawa, H., et al., Estradiol 17-β production in Amago salmon *(Oncorhynchus rhodurus)* ovarian follicles: Role of thecal and granulosa cells. *General and Comparative Endocrinology*, 1982, **47**(4):440–448.
78. Pendergrass, P. and P. Schroeder, The ultrastructure of the thecal cell of the teleost, *Oryzias latipes,* during ovulation in vitro. *Journal of Reproduction and Fertility*, 1976, **47**(2):229–233.
79. Schroeder, P.C. and P. Pendergrass, The inhibition of in vitro ovulation from follicles of the teleost, *Oryzias latipes*, by cytochalasin b. *Journal of Reproduction and Fertility*, 1976, **48**(2):327–330.
80. Nagahama, Y., Histochemistry and ultrastructure of pre- and post-ovulatory follicles in the ovary of the goldfish, *Carassius auratus*. *Canadian Journal of Zoology*, 1976, **54**:1128–1139.
81. Szöllösi, D. and B. Jalabert, La theque du follicle ovarian de la Truite. *Journal de Microscopie (Paris)*, 1974, **20**(92a).
82. Planas, J.V., et al., Regulation of ovarian steroidogenesis in vitro by follicle-stimulating hormone and luteinizing hormone during sexual maturation in salmonid fish. *Biology of Reproduction*, 2000, **62**(5):1262–1269.
83. York, W.S., R. Patino, and P. Thomas, Ultrastructural changes in follicle cell-oocyte associations during development and maturation of the ovarian follicle in Atlantic croaker. *General and Comparative Endocrinology*, 1993, **92**:402–418.
84. Iwamatsu, T., et al., Oogenesis in the medaka *Oryzias latipes*: Stages of oocyte development. *Zoological Science*, 1988, **5**:353–373.
85. Onitake, K. and T. Iwamatsu, Immunocytochemical demonstration of steroid hormones in the granulosa cells of the medaka, *Oryzias latipes*. *Journal of Experimental Zoology*, 1986, **239**(1):97–103.
86. Iwasaki, Y., Histochemical detection of d5-3b-hydroxasteroid dehydrogenase in the ovary of medaka, *Oryzias latipes,* during annual reproductive cycle. *Bulletin of the Faculty of Fisheries, Hokkaido University*, 1973, **23**:177–184.

87. Cerdá, J.L., T.R. Petrino, and R.A. Wallace, Functional heterologous gap junctions in *Fundulus heteroclitus* ovarian follicles maintain meiotic arrest and permit hydration during oocyte maturation. *Developmental Biology*, 1993, **160**(1):228–235.

88. Wallace, R.A., M.S. Greeley, Jr., and R. McPherson, Analytical and experimental studies on the relationship between Na^+, K^+, and water uptake during volume increases associated with *Fundulus* oocyte maturation in vitro. *Journal of Comparative Physiology, Part B: Biochemical, Systemic, and Environmental Physiology*, 1992, **162**(3):241–248.

CHAPTER 5

Male Gonad Anatomy and Morphology

5.1 GONADOGENESIS: TESTES

5.1.1 Location and Gross Organization

A mature testes filled with sperm may account for about 10% of a fish's body weight. In contrast, in nonspawning conditions of seasonal fish, for examples, the testes may regress to an insignificant and easily overlooked organ in many fish species [1]. The testes are usually slightly flattened with a whitish appearance; to the inexperienced eye, it may therefore be difficult when dissecting the fish to distinguish the testes from the whitish fat tissue around the intestine and ceca.

Testes of teleosts are typically paired structures (Figure 5.1), bilaterally adjacent to the swimbladder and extending almost the full length of the peritoneal cavity. The paired testes may fuse completely or along their posterior segments only (e.g., in poecilids). A thin layer of connective tissue, termed *mesorchium*, surrounds and suspends the testes. The mesorchium arises from the dorsal peritoneum and adheres to the swimbladder.

The efferent ducts (*vasa efferentia*) of each spermatogenic lobule (see Section 3.4) carry mature sperm toward the primary duct. Unlike most other vertebrates, the sperm ducts of teleosts are anatomically separate from the nephric or Wolffian duct [1,3]. The sperm ducts may be internalized within the testes or evident along the dorsal posterior region of each testis. The two sperm ducts (*vasa deferentia*) fuse posterior to the main sperm duct; lobular outgrowths from the sperm duct (probably having additional secretory functions) may be present. Alternatively, as in goldfish (*Carassius auratus*), the duct is modified to provide an accessory sperm storage organ, analogous to seminal vesicles in other vertebrate species. The main sperm duct generally terminates at the genital pore (Figure 5.2), which exits the body independently between the more anterior anal pore and the more posterior urinary pore. The genital and urinary pores are often associated with an externally prominent urogenital papilla [1].

Histological Analysis of Endocrine-Disruptive Effects in Small Laboratory Fish
By Daniel R. Dietrich and Heiko O. Krieger

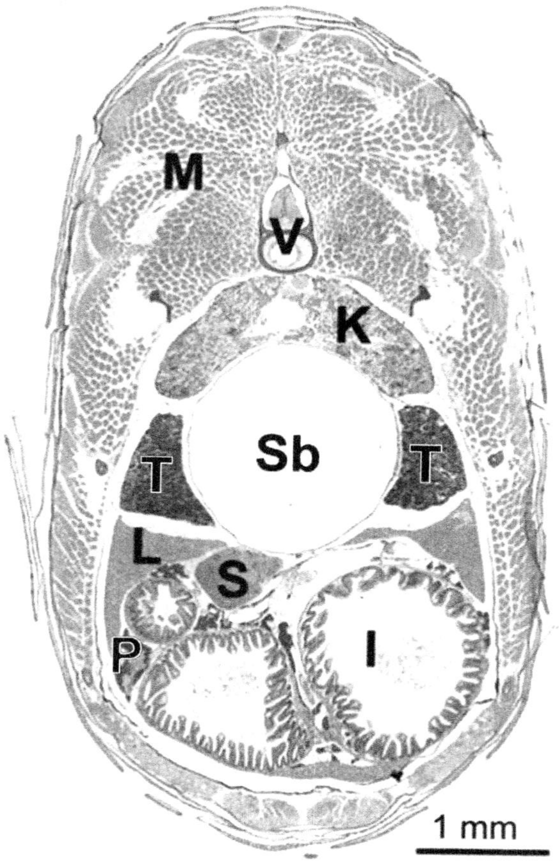

FIGURE 5.1 Relative size and location of organs in a male adult zebrafish. M, muscle; V, vertebra; K, kidney; Sb, swimbladder; T, testis; L, liver; S, spleen; P, pancreas, I, intestine. (Courtesy of L. van der Ven and P. Wester, RIVM, Bilthoven, The Netherlands.) (See insert for color representation.)

5.1.2 Anatomy of the Testes

Spermatogenesis within the testes takes place in the germinal epithelia, which are organized into lobular divisions composed of blind-ended tubes or sacs (Figure 5.3). A basement membrane separates the germinal epithelium from the interstitium. Distinction of testis types can be made on two different levels: on a gross anatomical level (gross structure: tubular vs. lobular; Figure 5.3), and on the intratubular or intralobular distribution of spermatogonia (fine structure: restricted vs. unrestricted). A distinction is made between the anastomosing tubular and the lobular testis types in fish. Figure 5.3A shows an anastomosing tubular type of testis structure, where the tubules loop at the testis periphery and form a continuously anastomosing tubular system, such as that found in the coelacantheiform (latimeria, *Latimeria*

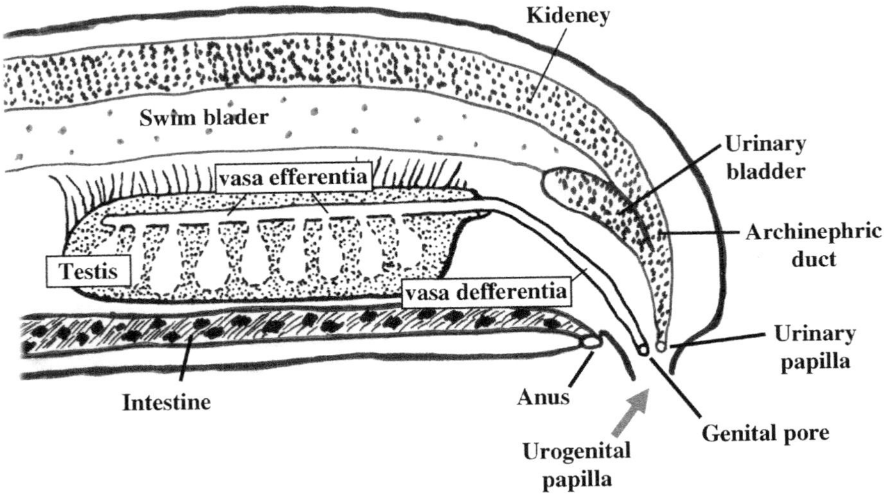

FIGURE 5.2 Schematic drawing of a male fish depicting the *vasa efferentia*, *vasa defferentia*, the genital pore, and the urogenital papilla.

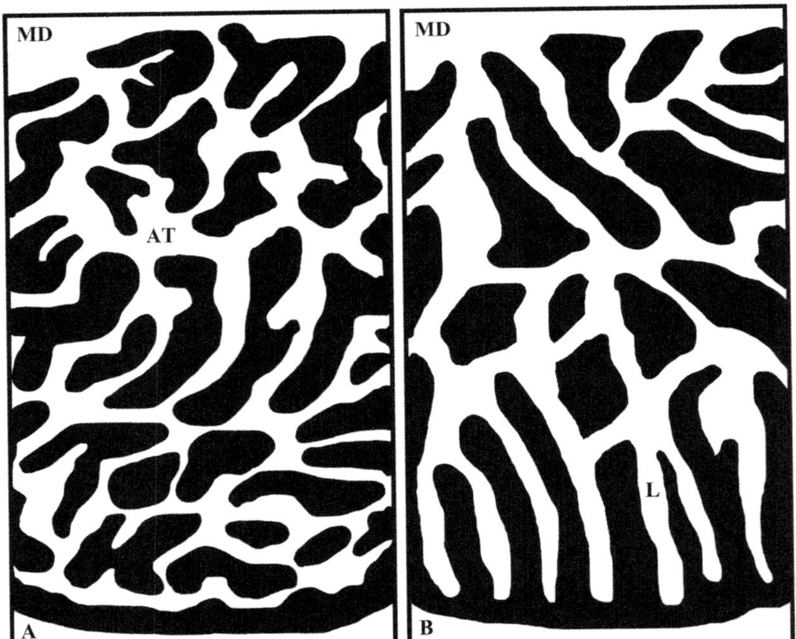

FIGURE 5.3 Gross structure of different testis types: (A) anastomosing tubular testis, hypothesized to be typical for lower fish species; (B) lobular testis, typical for perciform, atheriniform, beloniform, and possibly other, higher fish taxa. MD, main, longitudinal testis duct; AT, anastomosing tubular system; L, lobule. (Redrawn and modified from [5].)

chalumnae), acipenseriform (paddlefish, *Polyodon spathula*), elopiform (tarpon, *Megalops atlanticus*), salmoniform (rainbow trout, *Oncorhynchus mykiss*), siluriform (middle Tennessee bullheads, *Ictalurus natalis*), characiform (black tetra, *Gymnocorymbus ternetzi*), clupeiform (thread herring, *Opisthonema oglinum*), lepisoeiform (gar, *Lepisosteus platyrhyncus*), esociform (pike, *Esox lucius*), and cyprinidiform (zebrafish) fish (see Parenti and Grier [4] for an extensive review). A lobular testis structure (Figure 5.3B) means that the lobules terminate at the testis periphery and drain to a main duct. The lobular testis type is typical for perciform, cypriniform, atheriniform, beloniform, and possibly many other higher fish taxa. In the midportion of the testis, the lobules may form an anastomosing system [5].

Sperm production normally proceeds throughout the lobule. However, in atheriniformes (inland silverside, *Menidia beryllina*), cyprinodontiformes (killifish, *Nothobranchius rachovii*; swordfish, *Xiphophorus helleri*; guppy, *Poecilia reticulata*), and beloniformes (medaka), spermatogenesis is restricted to the terminal regions of the sacs (restricted lobular testis), farthest away from the efferent ducts (Figures 5.3B, 5.4, and 5.5A) [1,5,6]. However, in medaka, spermatogenesis not only takes

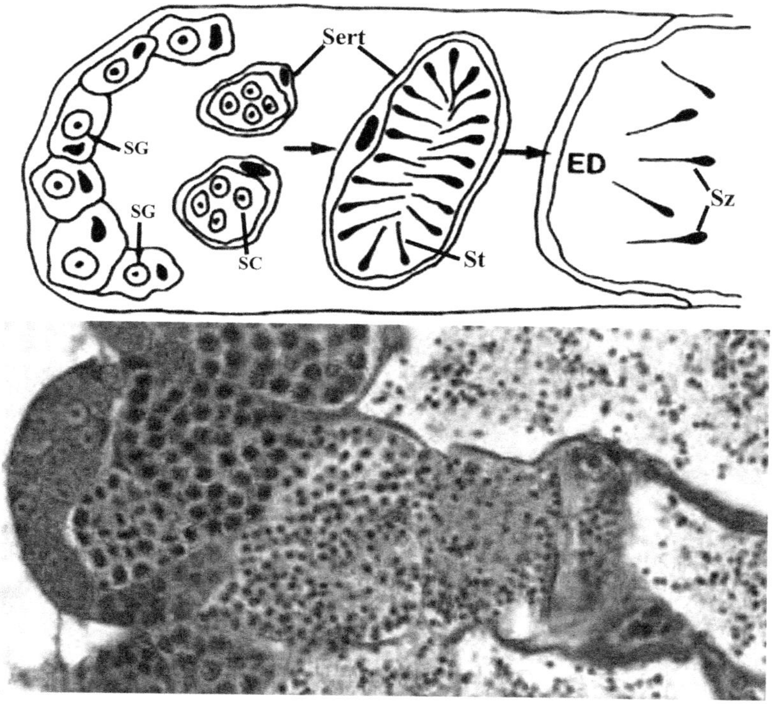

FIGURE 5.4 Schematic of directional spermatogenesis in medaka (A) and the corresponding histological section (B) with identification of the major cell types and structures present. SG, spermatogonia; SC, spermatocytes, St, spermatids; Sz, spermatozoa; Sert, Sertoli cells; ED, efferent duct. (A from [8] with permission, B courtesy of J.W. Fournie, U.S. EPA, Gulf Breeze, Florida.) (See insert for color representation.)

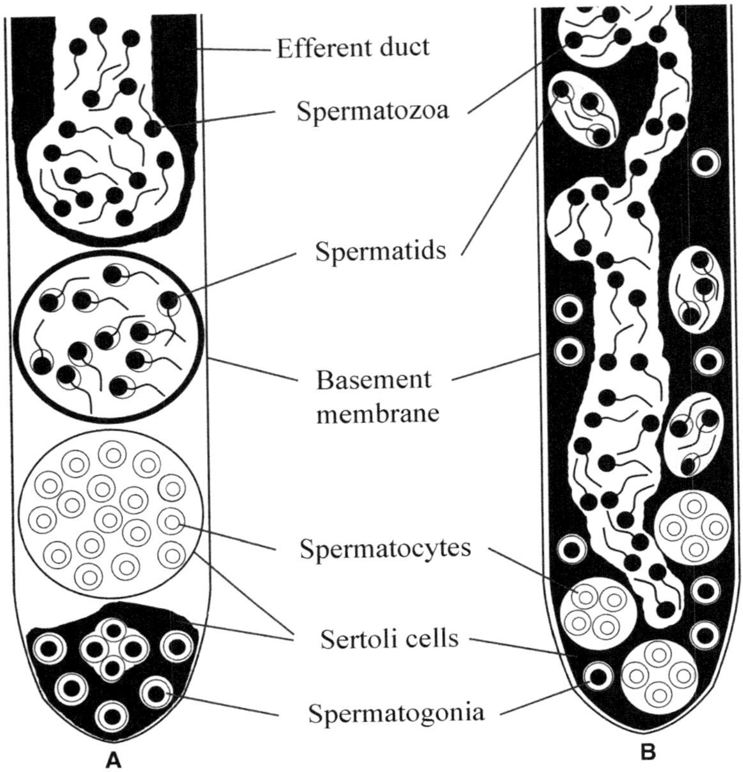

FIGURE 5.5 Diagrammatic representation of two tubular testicular types observed in teleosts. The restricted spermatogonial testis (A) is typical for *Atheriniformes* and *Beloniformes.* Cysts form at the distal end of the tubule but make contact with the efferent ducts prior to spermiation. In the unrestricted spermatogonial testis (B) typical for *Salmoniformes*, *Perciformes*, and *Cypriniformes*, spermatogonia are located along the length of the lobule. Cysts containing mature sperm are located adjacent to the lobular lumen. During spermiation, the cyst and lobular lumen become continuous. (Modified from [6], with permission.)

place in the head–tail (anterior–posterior) but also from the periphery–central lumen (cortical–medullary) direction (Figures 5.4 and 5.7). In many teleosts, however, the lobular testes present with a central lumen (sperm reservoir) where the cysts develop along the length of the lobules and remain stationary during spermatogenesis (Figure 5.5B). Mature spermatozoa are finally released into the central cavity [5–7].

Both testis types (anastomosing and lobular) can be defined further based on the distribution of spermatogonia within the tubules and lobules. An accurate and comprehensive description for the different types of teleost testes, namely, the restricted and unrestricted spermatogonial testicular type, can be found in reviews by Grier, Parenti and Grier, and Nagahama [3–6].

In a *restricted spermatogonial testis*, so far reported only for lobular testes of atherinomorph teleosts, among which are the beloniform medaka [4,5], spermatogonia are confined to small, peripheral spermatocysts (clusters of spermatogenic cells lined by peripheral Sertoli cells) within the tubules (Figure 5.5), which subsequently enlarge and extend toward the tubule lumen as spermatogenesis (the part of spermatogenesis that incorporates meiosis and ends in the production of sperm) proceeds. Spermatogenesis proceeds along the anterior-to-posterior axis in the testes, meaning that the rostral testis contains early germ cells (spermatogonia), and the caudal testis mature germ cells (Figure 5.4 and 5.5A). Due to the anterior-to-posterior as well as cortex-to-medullary gradient of development in medaka, more juvenile germ cell stages are found in the anterior and cortical portion of the gonads (Figures 5.4 and 5.7).

In the more common *unrestricted spermatogonial testis*, spermatogonia are distributed along the entire length of the tubule and lobule, respectively (Figure 5.5B), even during the various stages of recrudescence (Figure 5.6), as shown, for example, for the lobular testis of red drum (*Sciaenops ocellatus*).

5.1.3 Germinal Epithelium

Generally, the germinal epithelium of all chordates consists of the association between the gonadal basement membrane, the germ cells, and the Sertoli cells. The term *germ cell* includes spermatogonia, spermatocytes, spermatids, and sperm (spermatozoa) [4,5]. For example, at recrudescence a continuous germinal spermatogonial epithelium occurs along the entire length of the testicular lobules (Figure 5.6A) in *unrestricted* teleosts. Germ cells, with the exception of *stem-cell-like spermatogonia* (primary spermatogonia or spermatogonia A), which are separated from each other by the cytoplasm of Sertoli cells, reside within spermatocysts, whereas Sertoli cells are restricted entirely to the periphery of the spermatocysts [9]. In conjunction with the progression of recrudescence, germ cells within the spermatocysts divide and mature (see Section 5.1.4), whereby the tripartite structure of the germinal epithelium is lost and a spermatocyst having a Sertoli cell epithelium (i.e., one produced by Sertoli cell processes) evolves (Figure 5.6C). As the lobule now consists of spermatocysts with a tripartite structure and spermatocysts with a bipartite (germ cell, Sertoli cell) structure, the germinal epithelium is no longer *continuous* but is then defined as *discontinuous* (Figure 5.6C); the *continuous germinal epithelium* is observed only at the distal end of the lobule.

Spermatocyst enlargement and thus lobule elongation appears to be achieved via the continuous division of Sertoli cells and division and maturation of germ cells, eventually leading to the release of mature proximal germ cells into the testicular lumen (Figures 5.5B and 5.6D and E). The mechanism of lobule elongation (Figure 5.6B) in seasonally breeding fish appears to be lost toward the end of the breeding season, when primary spermatogonia (see Section 5.1.4) are absent from the distal ends of the lobules. Cyclicity (i.e., seasonality) has a major influence on the morphology of the testis and the presence of specific germ cell types, especially in

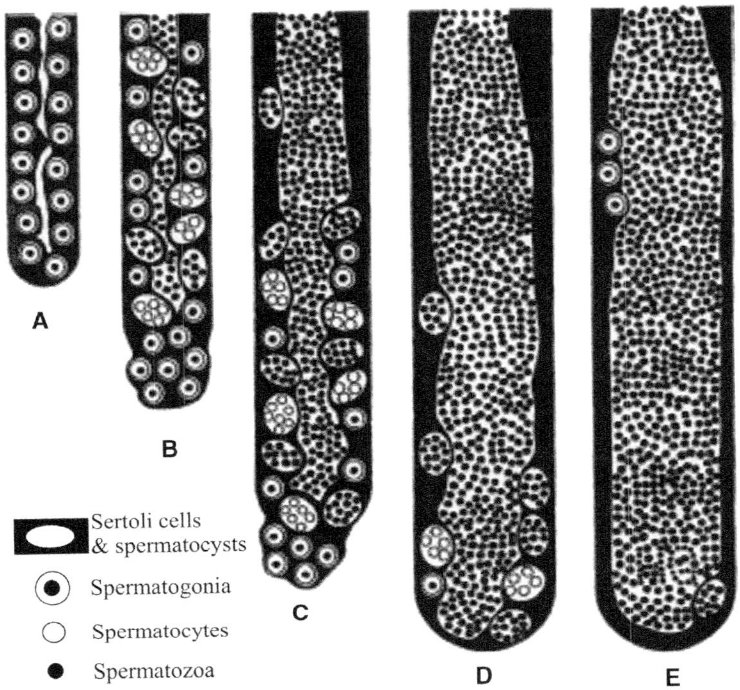

FIGURE 5.6 Scheme for testicular recrudescence in the lobular testis of the red drum (*Sciaenops ocellatus*, Perciformes). Changes in the germinal epithelium are represented without illustration or discussion of the basement membranes. (A) Gonadally regressed fish: testicular lobes have a germinal epithelium consisting of Sertoli cells and primary spermatogonia. (B) In early recrudescence: lobules elongate due to mitotic divisions of primary spermatogonia and Sertoli cells at their distal ends. A *continuous germinal epithelium*, wherein spermatocysts and spermatogonia are juxtaposed, extends for their entire lengths. (C) Mid-recrudescence: as distal, primary spermatogonia, and Sertoli cells continue to divide, producing lobule elongation and new spermatocysts, proximal germ cells are depleted from the epithelium as sperm form and are released into the lumen. Therefore, the tripartite structure of the germinal epithelium is lost, resulting in a Sertoli cell epithelium. During mid-recrudescence, spermatocysts become widely separated by the Sertoli cell epithelium, producing a *discontinuous germinal epithelium*. A *continuous germinal epithelium* remains at the distal end of the lobule. (D) Late recrudescence: a discontinuous germinal epithelium extends to the distal end of the lobule. (E) Termination of recrudescence: the testis becomes a storage organ for sperm. Few, if any, spermatocysts are detectable, and few inactive primary spermatogonia remain along the lobule wall. (From [5], with permission.)

fish caught in the wild [10]. Thus, an understanding of normal biological recrudescence as well as environmentally or xenobiotically induced alterations of recrudescence in seasonally breeding fish species (e.g., wild fathead minnows, wild medaka) is a key to any interpretation or extrapolation of histopathological findings in the gonads of feral fish in temperate zones. This pertains even more strongly to fish exposed to xenobiotics, such as, wild fathead minnows exposed to 6 ng/L 17α-ethinylestradiol (EE_2) for 120 days in a natural lake during the May–October breeding season [11].

Although laboratory and field studies with fathead minnows exposed to EE_2 may have suggested impairment of spermatogenesis or testis–ova [11–13], or in the case of the laboratory study by Laenge et al. [14], testis–ova as a specific endocrine-related mechanism, a comparison of histopathological findings demonstrates that in a four month exposure field study [11], the major gonadal effects reported were those of interstitial fibrosis and inhibited spermatogenesis and testicular development, but not testis–ova, with simultaneously loss of glycogen in the liver and renal inter- and intratubular proteinaceous fluid (edema and hyaline droplets). The continued dosing of Lake 206 with EE_2 at 5 to 6 ng/L for up to seven years resulted in the production of vitellogenin mRNA and protein, impacts on gonadal development as evidenced by intersex in male fathead minnows and altered oogenesis in females, and ultimately, near-extinction of fathead minnows from the lake [13]. Contrary to the authors' interpretation, the intersex in males and the altered oogenesis shown in the publication do not appear sufficiently intensive or vast to support convincingly causality for the demise of the fathead minnow population in Lake 206. Indeed, the major increase in vitellogenin levels in male and female fish (3000-to 30,000-fold and 8-to 80-fold, respectively) reported by Palace et al. [11] would have translated into major physiological stress to the fish, as also presented by the overt edema and hyaline droplets detected in the kidneys and the reduced glycogen content in the liver of male fish, and therefore must also have led to an increased rate of mortality among adult fish.

Similar pathological findings as well as increased mortality were observed by Wolf et al. [15] following 14 days exposure of fathead minnows to extreme concentrations (1860 ng/L) of 17β-estradiol (E_2), corroborating the latter interpretation. Whether or not continued exposure to EE_2 in Lake 206 also led to a feminization of juvenile fathead minnows could not be determined from the data presented by Kidd et al. [13] However, the current knowledge base on effects of EE_2 during fish development and adulthood [14,16,17] do suggest that the demise of the population in Lake 206 could have been the result of increased mortality of adult fish due to physiological stress, increased interstitial fibrosis of the gonads [18], and reduced recruitment of fecund and fertile progenies. The foregoing findings highlight the caution required when extrapolating histopathological findings of an experimental cohort to the field, and thus to an entire population, or vice versa, when a single finding or even a limited number of histopathological findings in wild-caught fish (e.g., testis–ova resulting from an exposure to endocrine active compounds) are interpreted as being nearly solely causative for declining fish populations [14,16].

5.1.4 Male Germ Cells (Spermatogenesis)

Following the establishment of an association between a spermatogonium and a Sertoli cell, spermatogenesis progresses through a series of mitotic and meiotic divisions that produce sequentially smaller cells and a haploid chromosome

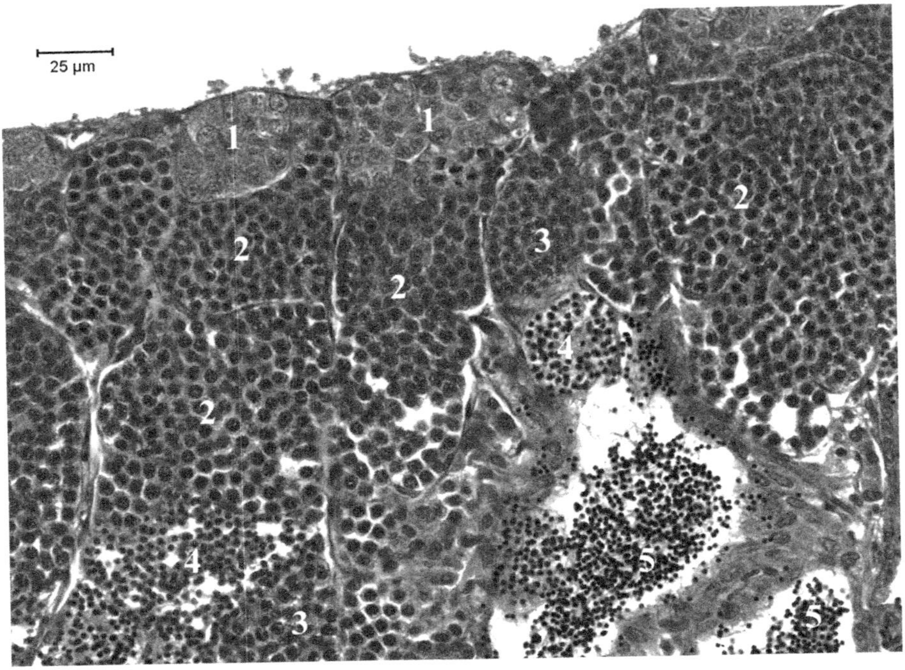

FIGURE 5.7 Reproductively active medaka testes: 1, spermatogonia; 2, early primary spermatocyte; 3, late primary spermatocytes (dividing diplotene) or secondary spermatocyte; 4, spermatids; 5, spermatozoa. (Courtesy of J.W. Fournie, U.S.EPA, Gulf Breeze, Florida.) (See insert for color representation.)

configuration [3,8,19,20]. Germ cells progress through six distinct cytological stages during spermatogenesis: primary spermatogonia (spermatogonia A), secondary spermatogonia (spermatogonia B), primary spermatocytes, secondary spermatocytes, spermatids, and spermatozoa (Figures 5.7 to 5.9) [3,8,19,21,22].

Primary spermatogonia are individually surrounded by Sertoli cell processes, but primary spermatogonia are not within the spermatocysts (Figures 5.5 to 5.9). Each primary spermatogonium undergoes a series of mitotic divisions (see also Figure 5.16) to produce a cluster of secondary spermatogonia encapsulated within a spermatocyst [8,24]. The spermatocyst arises from Sertoli cells associated with the original germ cell (primary spermatogonium), thus restricting the Sertoli cells and their processes to the periphery of the spermatocysts. Clusters of cells at the same stage of development (resulting from divisions of the original germ cell, i.e., the secondary spermatogonia), and all subsequent maturational stages of germ cell development, occur within spermatocysts [4,5]. Secondary spermatogonia (spermatogonia B) are smaller than primary spermatogonia, with large, lightly basophilic nuclei and little cytoplasm and are the result of clonal expansion (mitotic division). Primary spermatocytes (1SCs), the result of another round of mitotic divisions, are smaller still (approximately 25% smaller than secondary spermatogonia), with increasingly

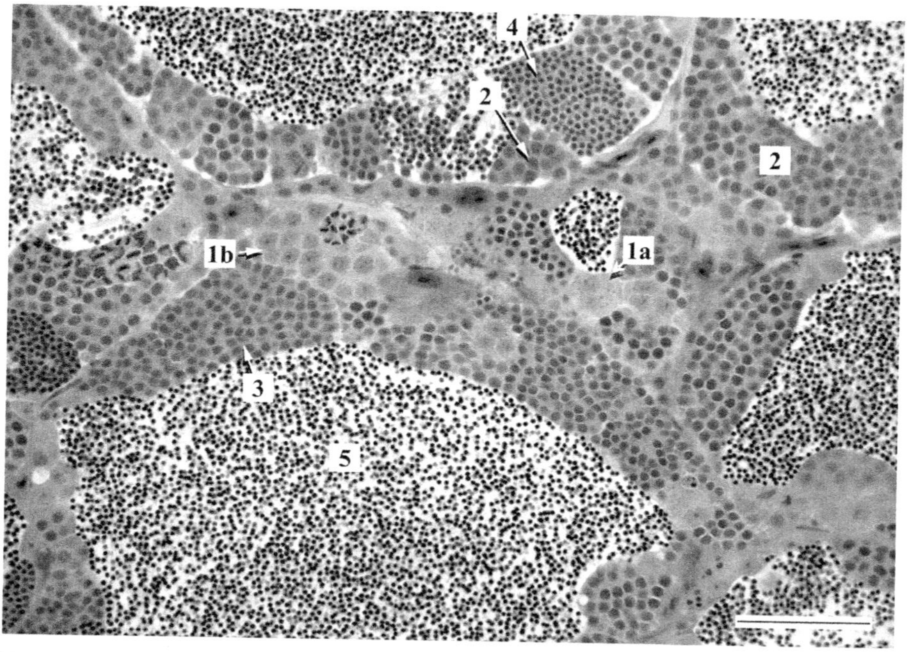

FIGURE 5.8 Reproductively active fathead minnow testes: 1, putative spermatogonia A; 1, spermatogonia B; 2, putative early primary spermatocyte (1SC); 3, putative late primary spermatocytes (dividing diplotene) or secondary spermatocyte (2SC); 4, spermatids; 5, spermatozoa. (Modified from [23] and nomenclature adapted from [4].) Scale bar = 50 μm.

basophilic nuclei (Figure 5.9) [2]. Primary spermatocytes, which are actually in meiosis I (an extended process with several distinctive stages continuing over one or more days), undergo the first meiotic division to produce secondary spermatocytes (2SCs, Figures 5.10 and 5.16). Still enveloped and thus contained within the spermatocyst, these cells are again smaller and have increasingly dense basophilic nuclei. In contrast with the "long-lived" and thus readily observed primary spermatocytes, secondary spermatocytes present with a very short transition time [25,26]. Secondary spermatocytes undergo a second meiotic division. The resulting spermatids have condensed, intensely basophilic nuclei and very little cytoplasm. At this stage the cyst opens up, releasing the spermatids into the lumen, where final maturation takes place (see Figure 5.16). Spermatids undergo maturation events that include nuclear shape changes, development of a flagellum, loss of cytoplasmic volume, and phagocytosis of cytoplasmic remnants (residual bodies) prior to their release into the lobular lumen as spermatozoa (spermiation) [3,8,24]. To complete the formation of mature spermatozoa in a zebrafish, it takes approximately 10 days in total. Following maturation, sperm is stored in tubule lumina and within the ductus deferens.

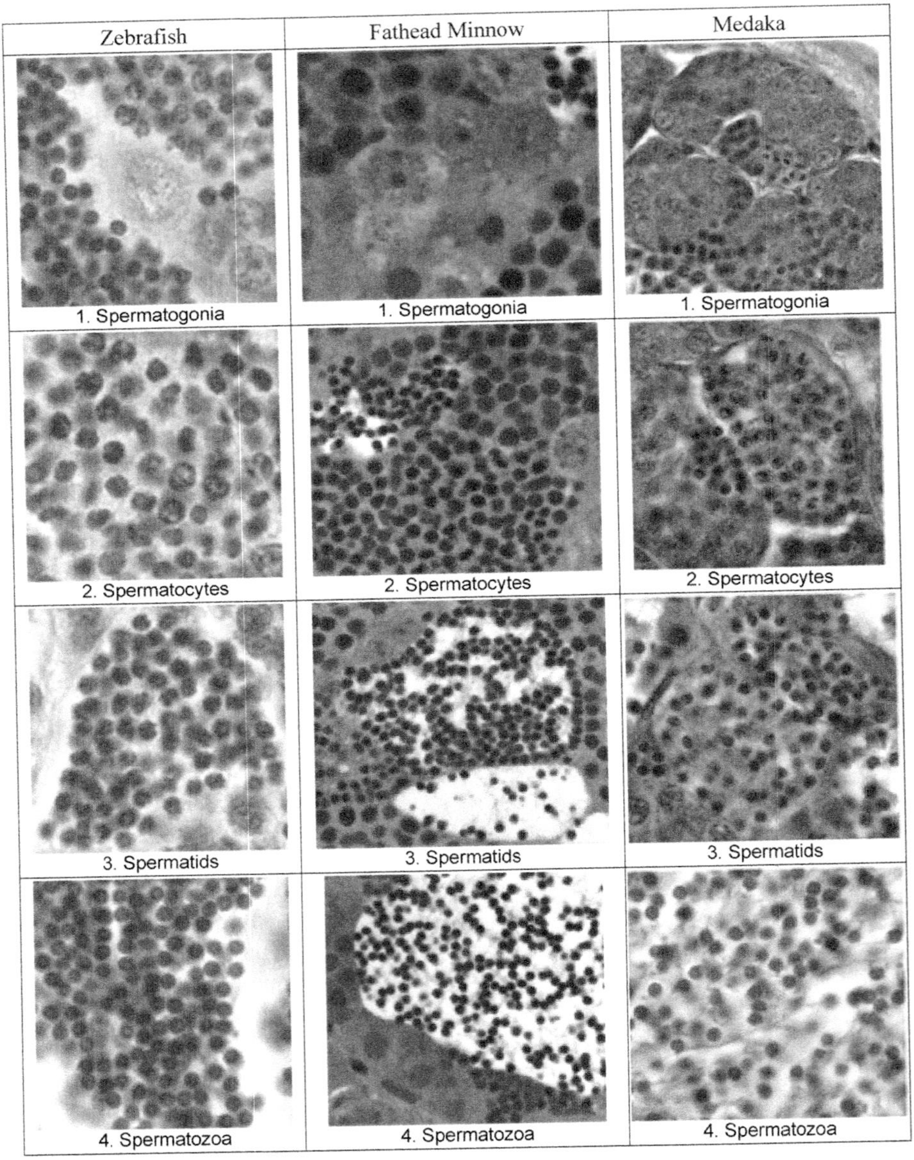

FIGURE 5.9 Spermatogenesis, different sperm cell stages. (Zebrafish images courtesy of L. van der Ven and P. Wester, RIVM, Bilthoven, The Netherlands; fathead minnow images courtesy of J. Wolf, EPL Inc., Sterling, Virginia, USA; medaka images courtesy of J. W. Fournie, U.S. EPA, Gulf Breeze, Florida.) (See insert for color representation.)

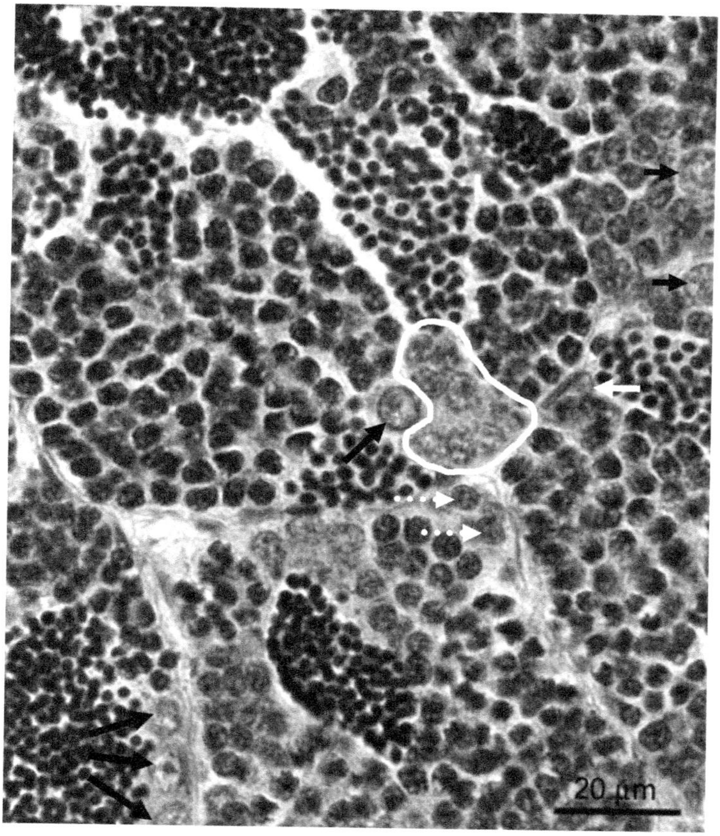

FIGURE 5.10 Detail of normal testis in zebrafish, showing all stages of spermatogenesis and the location of interstitial Leydig (white dotted arrows) and Sertoli cells (white arrow). Large solitary spermatogonium A cells (arrows) can be discerned. After division, these spermatogonia give rise to a generation of spermatogonia B (white lining), which occur in clusters of four or more cells. (Courtesy of L. van der Ven and P. Wester, RIVM, Bilthoven, The Netherlands.) (See insert for color representation.)

5.1.5 Male Somatic Supportive Tissue

5.1.5.1 Sertoli Cells Although the basic functions of Sertoli cells appear comparable in mammals and fish [7], fish demonstrate much larger species diversity than mammals in functional and morphological specialization [5]. For example, in teleost fish testes, contrary to the situation in both agnathans and elasmobranches, Sertoli cells (Figures 5.10 to 5.12) envelop spermatocysts that are located within anastomosing tubules or lobules [5], as described previously (see Section 5.1.2 to 5.1.4). Between spermatocyst formation and maturation of spermatozoa, the morphology and function of the Sertoli cells appears to remain unchanged. At spermiation, spermatocysts and thus Sertoli cells break down to release

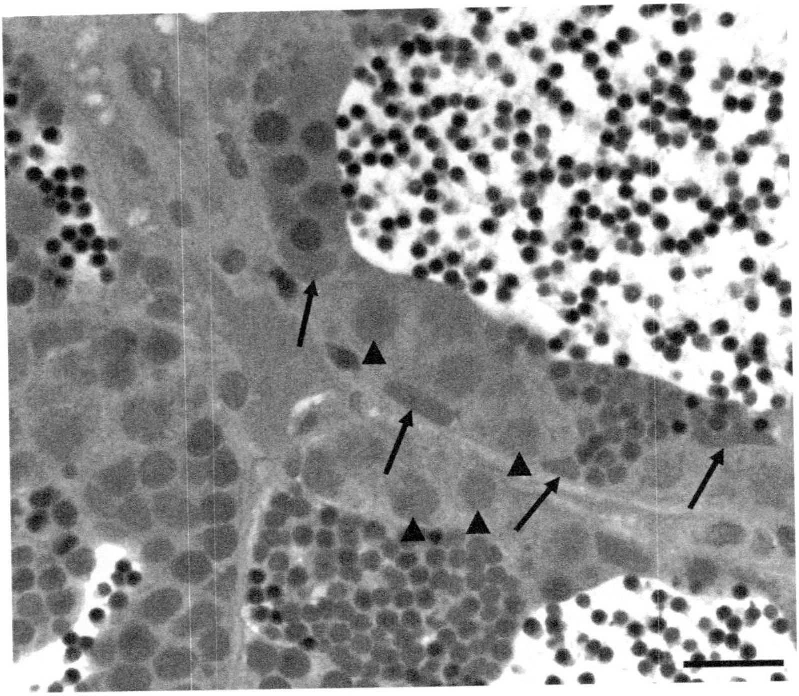

FIGURE 5.11 Sertoli cells (arrows) in fathead minnow testis. Note the morphological diverse spermatogonia (arrowheads). Scale bar = 10 μm. (Courtesy of J. Wolf, EPL Inc., Sterling, Virginia.) (See insert for color representation.)

spermatozoa and then degenerate [7]. In contrast, in restricted atherinomorph fish (e.g., the medaka), Sertoli cells differentiate into a secretory columnar form, thereby transforming into efferent duct cells after spermiation [5]. Due to the intimate relationship between Sertoli cells and spermatogenesis it is not surprising to find that the number of Sertoli cells per testis is one of the most important factors determining the quantity of sperm produced. Indeed, germ survival and development apparently are critically dependent on a germ cell's contact with Sertoli cells. Sertoli cell proliferation, at least in the Nile tilapia (*Oreochromis niloticus*) and the African catfish (*Clarias gariepinus*) [27], was most apparent during spermatogonial proliferation [i.e., specifically from primary spermatogonia (spermatogonia A) to spermatocyte cysts], allowing the cyst-forming Sertoli cells to provide sufficient space for the growing germ cell clones. Once spermatocytes have proceeded to undergo meiotic divisions (as of the primary spermatocyte stage; see also Section 5.1.5.4 and Figure 5.16), the number of Sertoli cells per cyst appears to remain stable.

In addition to their function within the tripartite (basal membrane, Sertoli cell, spermatogonia A) and then bipartite structure (spermatocyst: Sertoli cells and germinal cells) of the germinal epithelium (see Figure 5.14), Sertoli cells have been

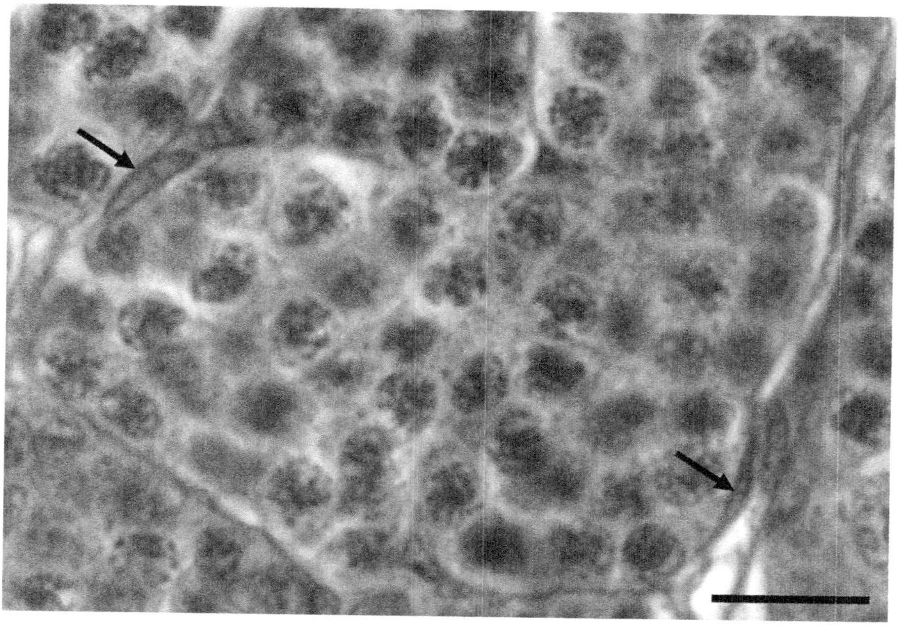

FIGURE 5.12 Sertoli cells (arrows) in medaka. Scale bar = 10 μm. (Courtesy of J. W. Fournie, U.S. EPA, Gulf Breeze, Florida.) (See insert for color representation.)

demonstrated to be able to phagocytose the residual bodies that are cast off by maturing spermatids [7,15,28]. Residual body phagocytosis has been observed in fish with the restricted testicular type as well as the unrestricted type [28]. In addition, as a common function across all chordate taxa, teleost Sertoli cells also phagocytose nondischarged or degenerating germ cells and, especially in seasonal breeders, old spermatozoa that were not released at the end of the breeding season [5]. This phagocytic activity usually results in Sertoli cell cytoplasm replete with large lysosomes and digestive vacuoles [29–31].

When appraised under the light microscope and especially when the testicular tissue has undergone changes subsequent to exposure to a xenobiotic (e.g., E_2 [30]), the identification of Sertoli cells is difficult. For example, in fathead minnows, mere localization of Sertoli-like cells at the periphery of the spermatocysts or the close association with the primary spermatogonia and the basement membrane (Figures 5.10 to 5.12), coupled with a strongly basophilic appearance [15] and an "apoptotic-like" feature [32], usually does not suffice for identification. Similarly, van der Ven and Wester [2] have investigated the appearance and morphological alteration of Sertoli cells extensively in male zebrafish following exposure to a wide range of endocrine-modulating compounds. These authors observed Sertoli cell hypertrophy in zebrafish following exposure to the anti-androgen flutamide (Figure 5.13). As flutamide binds to the androgen receptor (see Section 5.1.5.4 and Figure 5.16) without the downstream transcriptional activation

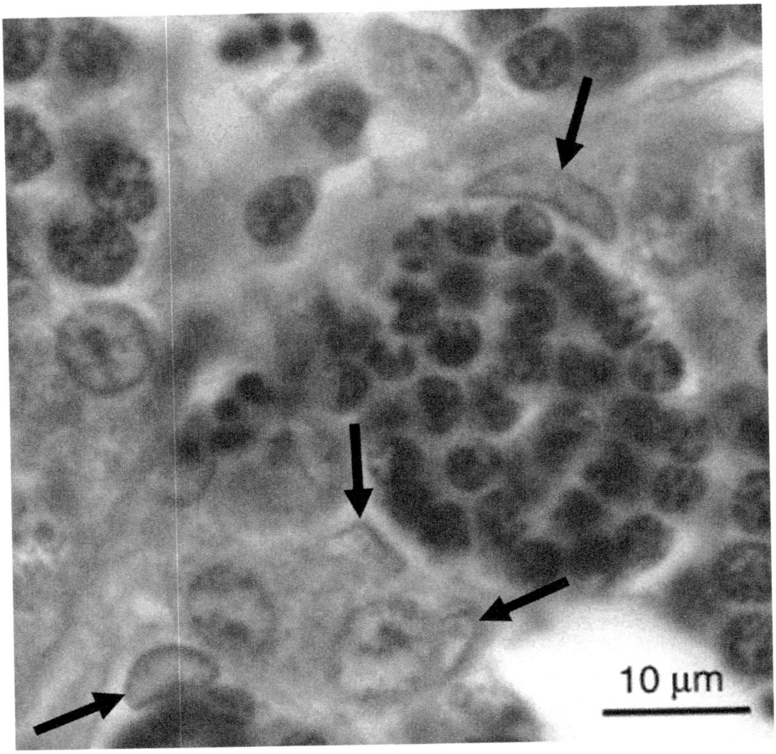

FIGURE 5.13 Sertoli cell hypertrophy (arrows) in male zebrafish following exposure to flutamide (anti-androgen). (Courtesy of L. van der Ven and P. Wester, RIVM, Bilthoven, The Netherlands.) (See insert for color representation.)

of the androgen- responsive element DNA in the nucleus [33], compensatory pathways may have been activated in Sertoli cells, finally resulting in the hypertrophy observed. Hypertrophied Sertoli cells (Figure 5.13) in male small fish testes have been observed to undergo morphological alteration, with some studies describing hypertrophied Sertoli cells presenting with a more rounded nuclei. Consequently, positive identification of Sertoli cells is difficult to achieve using merely light microscopy. However, additional confirmation can be gained by the use of electron microscopy.

In another study using high-power magnification [34], the same authors noted that Sertoli cell hypertrophy and hyperplasia were the most predominant alterations in the zebrafish testis following exposure to nonaromatizable 17-methyldihydrotestosterone (MDHT). Whether MDHT treatment induced a greater rate of spermatogonial proliferation and rapid advancement of spermatogenesis, thus providing a growth and proliferation stimulus to Sertoli cells via positive feedback from the expanding numbers of germ cells, as suggested by Kurokawa [35], cannot be established from the present data but certainly warrants further investigation.

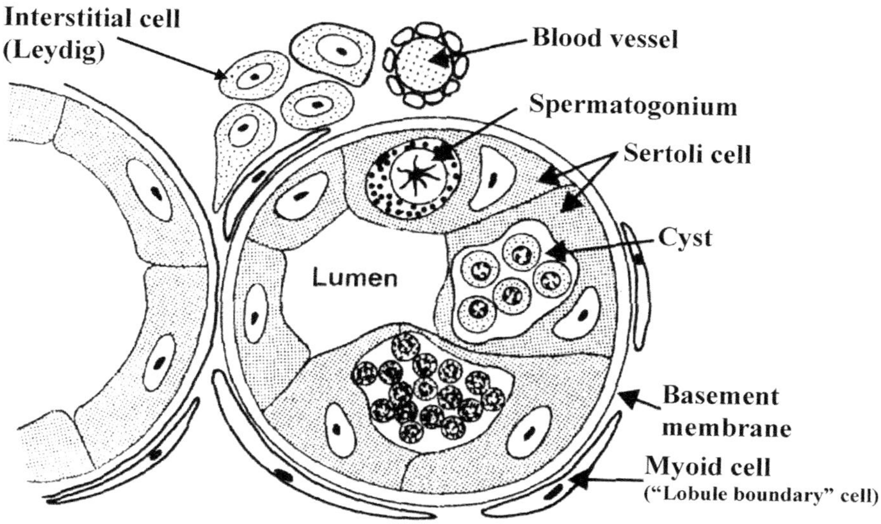

FIGURE 5.14 Diagrammatic representation of a cross section of a teleost lobule showing the localization of Sertoli and Leydig cells and gross intralobular organization. The term *lobule boundary cell* [6] (Figure 5.5) is an antiquated descriptive term for indiscriminant identification of Leydig and myoid as well as possibly Sertoli cells. (Modified from [37].)

5.1.5.2 Leydig Cells Leydig cells, the abundance of which varies within teleost species, are usually distributed singly or in small groups in the interstices between the lobules of the teleost testis (Figures 5.14 and 5.15) [3,36]. The testicular interstitial cells of teleosts have ultrastructural features typical of steroid-producing cells (see Section 5.1.5.4 and Figure 5.16). They are large polygonal cells characterized by extensive agranular endoplasmic reticulum and mitochondria with tubular cristae. Commensurate with the function as a steroidogenic cell [36,38–44], 3β-HSD (3β-hydroxysteroid dehydrogenase), an enzyme involved in steroid hormone synthesis, has been demonstrated histochemically in the interstitial cells of the testis of numerous teleosts [3]. Exposure of male fathead minnows to ketoconazole, known to inhibit the cyp P450-mediated step in ergosterol biosynthesis, caused a marked proliferation of interstitial (Leydig) cells [45]. In addition, mRNA transcripts for two key steroidogenic enzymes, cytochrome P450 side-chain cleavage (CYP11A) and cytochrome P450 c17α hydroxylase/17,20 lyase (CYP17), were elevated. Both the changes in transcript levels and the proliferation of gonad tissue represent potential adaptive or compensatory responses to impaired steroidogenic capacity. In contrast, medaka exposed from hatch until sexual maturity to 10 ng/L EE_2 [46] presented with apoptotic cells, namely Leydig and Sertoli cells, spermatocytes, and spermatids, suggesting that the high concentration of estrogenic EE_2 resulted from repressed androgen production in Leydig cells and therefore abrogated maintenance of the sufficient by high levels of 11-KT required for spermatogenesis.

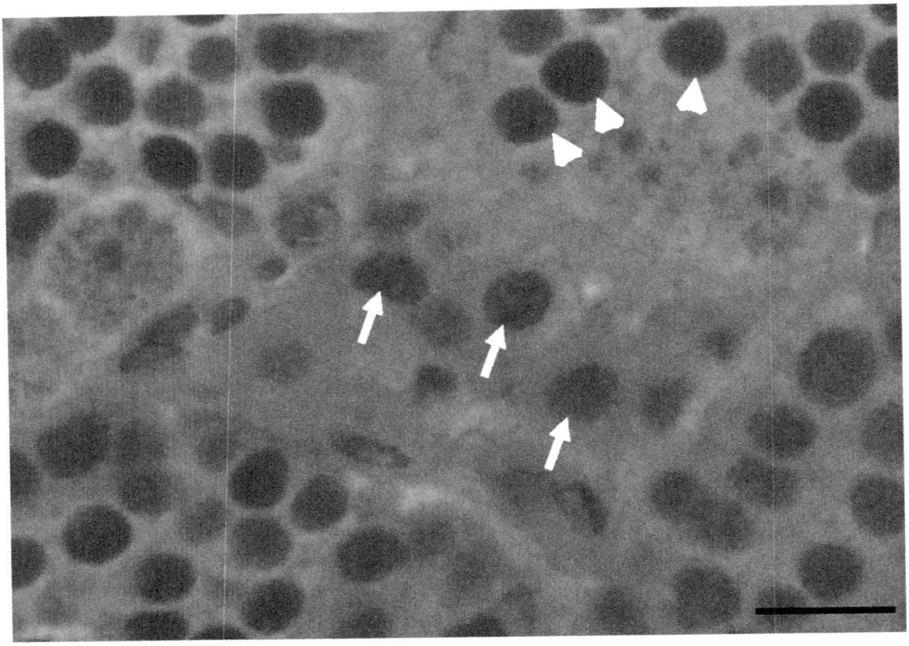

FIGURE 5.15 Leydig cells (arrows) in the fathead minnow testis. Morphologically, they resemble spermatocytes (arrowheads) but are present only within interstitial areas. Scale bar = 10 μm. (Courtesy of J. Wolf, EPL Inc., Sterling, Virginia.) (See insert for color representation.)

5.1.5.3 Lobule Boundary Cells The term *lobule boundary cells* (Figure 5.14) has been expanded to describe a Leydig cell homolog in fish and amphibians [5], despite the fact that Leydig cells have never been shown to form part of the boundary wall at any phase of testis differentiation. Similarly, the term *lobule boundary cells* was used previously to describe what are now known as Sertoli cells in medaka [9] or as Sertoli cell homologs in general [5]. After some more profound investigation, especially of the amphibian testicular ultrastructure, it became clear that the term *lobule boundary cells* [6] (Figure 5.5) probably best, albeit imprecisly describes myoid boundary cells with contractile features as well as Leydig cells, especially when used in conjunction with cells clearly located outside the testicular lobules [5]. Consequently, use of the stricter terms *myoid, Leydig,* and *Sertoli cells* is advised, and use of the term *lobule boundary cell*, in general, should be discontinued [7].

5.1.5.4 Sex Steroid Control of Spermatogenesis Although two distinct gonadotropic hormones (GTHs), GTH I (FSH) and GTH II (LH), have been identified and characterized in several fish species representing at least four teleost orders, their direct interaction with their respective receptors (e.g., FSH-R and LH-R) and downstream effects on hormone production, crosstalk, and signal transduction in Sertoli and Leydig cells, including the effect on spermatogenesis, is not yet well

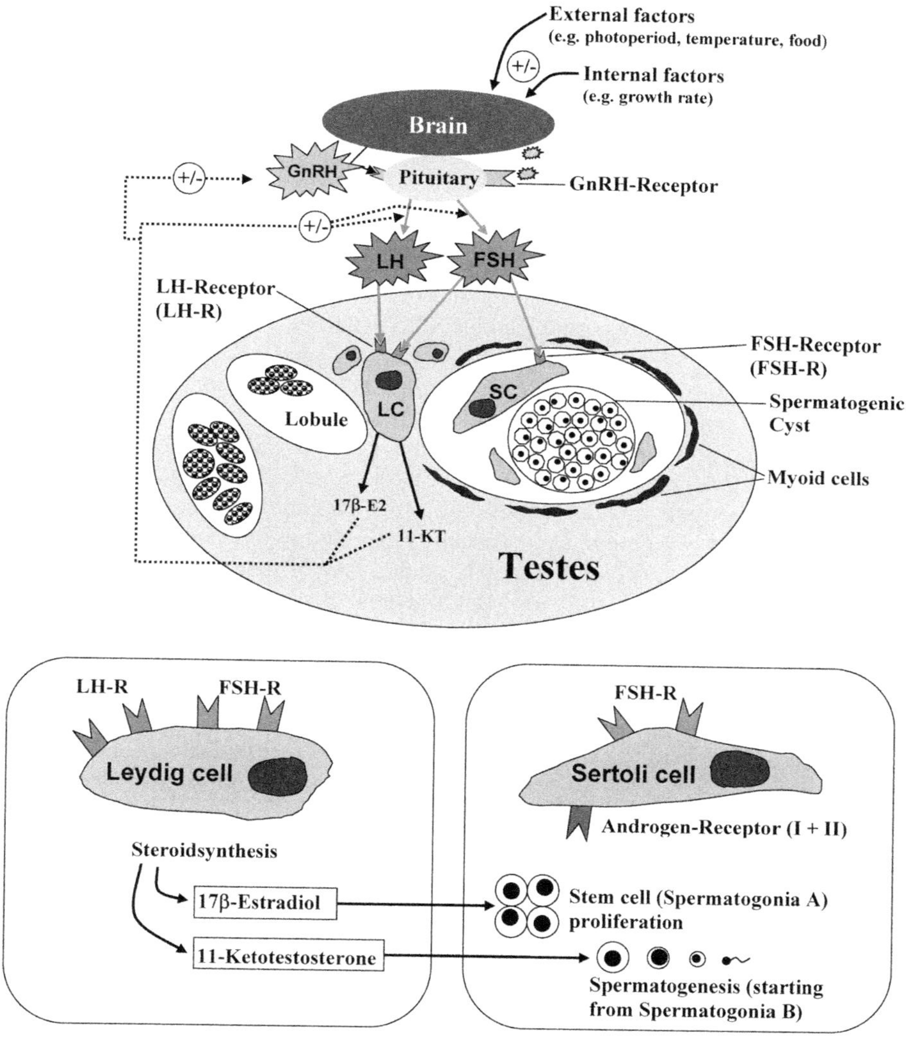

FIGURE 5.16 Schematic of the brain–pituitary–gonad axis, positive and negative feedback loops in conjunction with gonadotropic hormone–expressing gonadosomatic cells, steroidogenesis, and associated paracrine stimulation and sex steroid influence on spermatogenesis.

understood. Consequently, the schematic shown in Figure 5.16 is not directly applicable to zebrafish, medaka, and fathead minnows, but rather incorporates findings from several different fish species [27,28,36,38,40,42,43,47–60] to form a more general descriptor of the steroidal control of spermatogenesis. However, whenever possible, specific information pertaining to zebrafish, medaka, and fathead minnows is indicated.

Estrogen receptors as well as LH receptors are located in Leydig cells [61,62], which are the site of production of 11-ketotestosterone (11-KT) [63], the functional androgen in teleosts (see Figure 5.16). In conjunction with the findings above, establishment of the hypothalamo-pituitary axis in fish and the localization of gonadotropin hormone mRNA in the pituitary of most teleosts species [61] strongly suggest that Leydig cells, in conjunction with the FSH (GTH-I) receptor expressing Sertoli cells, influence germ cell development (i.e., spermatogenesis) [64]. Indeed, LH hCG (human chorionic gonadotropin) was demonstrated to induce Leydig cell activation in juvenile Japanese eel (*Anguilla japonica*) *in vivo* [47] and *in vitro* [40] and adult rainbow trout (*Oncorhynchus mykiss*) testicular tissue *in vitro* [65], followed by Sertoli cell activation via 11-ketotestosterone, which was subsequently followed by spermatogonial proliferation, meiosis, and spermatogenesis. This interpretation gains further support by the observation that in medaka, interstitial cells appear before differentiation of the spermatogonia [3], yet after sexual differentiation of the gonad into testes. However, contrary to intuitive expectation, mitotic divisions of spermatogonia (Figure 5.16), specifically spermatogonial stem cell renewal, is autoregulated by E_2 [39,66]. It is thus not surprising that estrogen receptors α and β were found to be variably expressed in the testis of various fish species [49,62,67–69], depending largely on the stage of spermatogenesis [49,70]. Moreover, very low levels of E_2 (pg/mL) have been detected in plasma samples of undifferentiated juvenile grey mullet (*Mugil cephalus*) [71] and adult male Atlantic croaker (*Micropogonias undulatus*) [70] as well as in the testes (>0.4 pg E_2/mg organ) of adult male medaka [35].

The situation above therefore suggests a sequential and variant appearance of several steroids during spermatogenesis. Indeed, the mitotic proliferation and meiotic differentiation of germinal cells during spermatogenesis are dependent on the presence of, and intimate contact with, the surrounding activated somatic Leydig and Sertoli cells. Steroid, mainly androgen, synthesis primarily occurs in Leydig cells under the control of LH, which, following binding to the LH receptor, promotes the release of downstream factors (e.g., steroidogenic acute regulatory protein), which facilitates the transport of cholesterol into the mitochondria, where it is converted by the P450s into pregnelonone (Figure 5.16) or the zebrafish [72] homolog of steroidogenic factor 1 (SF-1) [72], known for its regulation of multiple genes encoding steroid hydroxylases in steroidogenic cells [27,28,61]. In contrast, FSH receptor was found to be expressed in Sertoli cells surrounding type A/early type B spermatogonia and Leydig cells of male Japanese eels, suggesting the involvement of FSH in the induction of early steroidogenesis and androgenesis in Leydig cells [56].

The latter is supported by the finding that application of trilostane, a specific 3β-hydroxysteroid dehydrogenase (3β-HSD) inhibitor, prevented progression of spermatogonia from type A to type B and beyond, despite FSH activation of Leydig and Sertoli cells via the FSH receptor in Japanese eel testicular fragments. Thus, FSH and androgens [e.g., testosterone and 11-ketotestosterone (11-KT)] signal to germ cells [37,64] via the FSH receptor- and androgen receptor-expressing Sertoli cells. This may be seen as an early event in that FSH plasma peaks earlier than LH in the reproductive cycle of Japanese eels and salmonid species [61]. However,

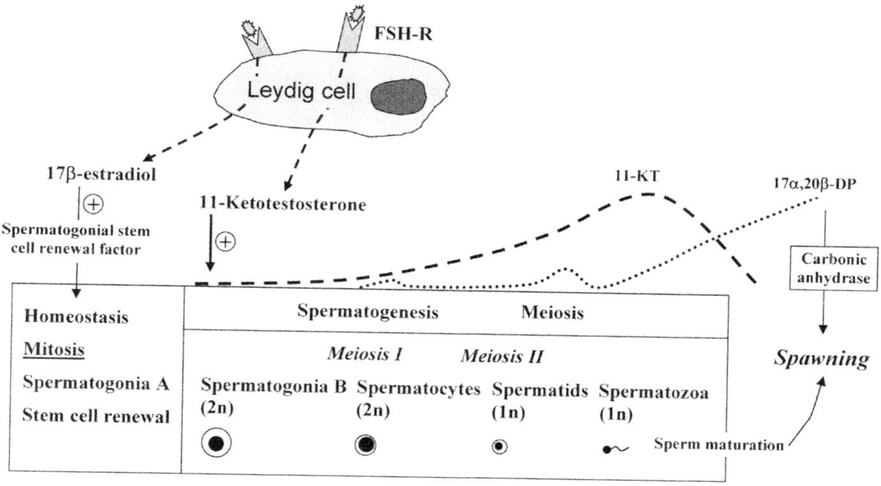

FIGURE 5.17 Variant influence of respective steroids on spermatogenesis.

the latter cannot be generalized, as cyprinid and perciform [73] species demonstrate concomitant appearance of FSH and LH mRNA levels during their reproductive cycle. As indicated earlier, E_2 is important for the expression of the spermatogonial stem cell renewal factor eSRS34 in male Japanese eels [38,58], allowing for clonal expansion of spermatogonia A in the absence of spermatogenesis. FSH is assumed to up-regulate anti-Müllerian hormone (AMH) expression through a nonclassical cAMP–PKA pathway involving transcription factors AP2 and NFkappaB in Sertoli cells [74] at low levels and thus to suppress the onset of spermatogenesis.

From the latter it appears that low levels of FSH and E_2 support spermatogonial stem cell homeostasis and renewal, while increasing levels of FSH at the beginning of the reproductive period promote the synthesis of androgens (Figure 5.17), thereby suppressing the expression of AMH. Indeed, 11-KT, produced in Leydig cells [35] following the conversion of testosterone to 11β-hydroxytestosterone (Figure 5.18), activates Sertoli cells via direct interaction with androgen receptors to down-regulate expression of the AMH homolog eSR21 (in Japanese eels [59]) and concomitantly, stimulate the production of activin B. cDNAs encoding for AMH homologs have recently been isolated from Atlantic salmon (*Salmo salar*) and several other fish species ([75] and references therein), demonstrating the more general representation of this regulation pathway across fish species.

Activin B binds to activin type I and II receptors on spermatogonia A to stimulate the *de novo* synthesis of G1/S cyclins and CDKs (cyclin-dependent kinases), leading to the initiation of mitosis and thus to the generation of spermatogonia B as the first step in spermatogenesis. As shown in Figures 5.16 and 5.17, 11-KT levels appear to increase from the onset of spermatogenesis, reaching the highest levels during meiosis II (i.e., the conversion of spermatids to spermatozoa). Conversely, another product of Leydig cell steroidogenesis, 17α, 20β-dihydroxy-4-pregnen-4-one

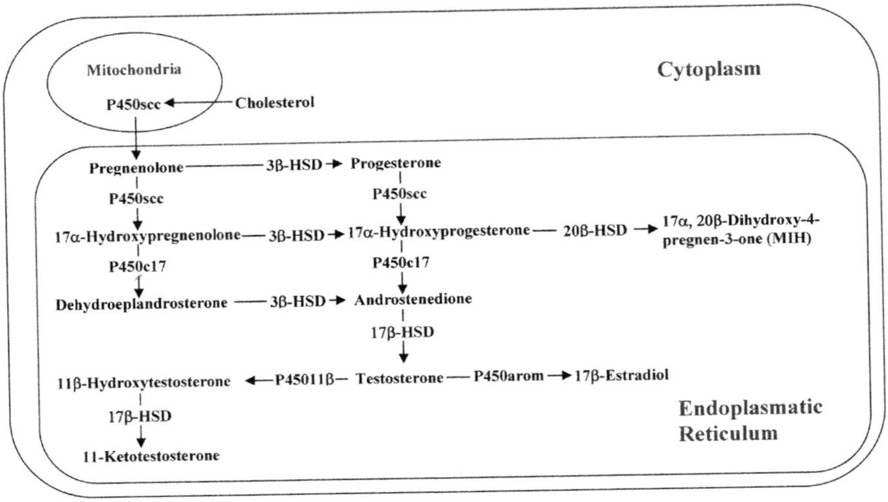

FIGURE 5.18 Steroidogenesis as observed in the Leydig cells of fish.

(17α,20β-DP), is involved in spermatogenesis only intermittently (Figures 5.16 to 5.18). Amer et al. [60], investigating the involvement of sex steroids in the early developmental stages of spermatogenesis in Japanese huchen (*Hucho perryi*), detected small peaks of 17α,20β-DP serum levels at the appearance of spermatocytes and spermatids (Figure 5.17), whereas serum levels increased dramatically at the time of spawning, reaching their maximum value at the onset of milt production.

The concurrent detection of low levels of E_2 throughout the entire reproductive cycle suggests that the importance of E_2 for spermatogonial stem cell replacement and mitosis, as found for the Japanese eel [38,58], may also hold true for the Japanese huchen. Moreover, the data above demonstrate succinctly that a delicate balance of varying individual levels of sex steroids, gonadotropins and even growth factors [e.g., the insulin growth factor (IGF-1)] [76–88], is a prerequisite for maintaining normal gametogenesis in male as well as in female fish [89]. Consequently, any disturbance of gonadotropin or sex steroid synthesis or their respective receptors and downstream effectors could potentially translate to an adverse effect on gametogenesis [46], and thus long-term survival of a population [13,17].

REFERENCES

1. Redding, J.M. and R. Patino, Reproductive systems: Gross functional anatomy, in *The Handbook of Experimental Animals: The Laboratory Fish*, G.K. Ostrander, Editor. Academic Press, San Diego, CA, 2000, pp. 261–272.

2. van der Ven, L.T.M. and P.W. Wester, Toxicological pathology atlas of small laboratory fish with special reference to normal histology and effects of endocrine disruption in the zebrafish (*Danio rerio*). http://www.rivm.nl/fishtoxpat, 2003.
3. Nagahama, Y., The functional morphology of teleost gonads, in *Fish Physiology*, W.S. Hoar, D.J. Randall, and E.M. Donaldson, Editors. Academic Press, San Diego, CA, 1983, pp. 223–275.
4. Parenti, L.R. and H.J. Grier, Evolution and phylogeny of gonad morphology in bony fishes. *Integrative and Comparative Biology*, 2004, **44**:333–348.
5. Grier, H.J., Comparative organization of Sertoli cells including the Sertoli cell barrier, in *The Sertoli Cell*, L.D. Russell and M.D. Griswold, Editors. Cache River Press, Clearwater, FL, 1993, pp. 704–739.
6. Grier, H.J., et al., Structural evidence for two different testicular types in teleost fishes. *American Journal of Anatomy*, 1980, **159**(3):331–345.
7. Russell, L.D. and M.D. Griswold, *The Sertoli Cell*. Cache River Press, Clearwater, FL, 1993, p. 747.
8. Pudney, J., Spermatogenesis in nonmammalian vertebrates. *Microscopy Research and Technique*, 1995, **32**(6):459–497.
9. Grier, H.J., Sperm development in the teleost *Oryzias latipes*. *Cell and Tissue Research*, 1976, **168**:419–431.
10. Smith, R.J.F., Seasonal changes in the histology of the gonads and dorsal skin of the fathead minnow, *Pimephales promelas*. *Canadian Journal of Zoology*, 1978, **56**(10):2103–2109.
11. Palace, V.P., et al., Induction of vitellogenin and histologic effects in wild fathead minnows from a lake experimentally treated with the synthetic estrogen, ethynylestradiol. *Water Quality Research Journal of Canada*, 2002, **37**(3):637–650.
12. van Aerle, R. et al., Windows of sensitivity for the estrogenic effects of ethinylestradiol in early life-stages of fathead minnow, *Pimephales promelas*. *Ecotoxicology*, 2002, **11**(6):423–434.
13. Kidd, K.A., et al., Collapse of a fish population after exposure to a synthetic estrogen. *Proceedings of the National Academy of Sciences of the USA*, 2007, **104**(21):8897–8901.
14. Laenge, R., et al., Effects of the synthetic estrogen 17α-ethynylestradiol on the life-cycle of the fathead minnow *(Pimephales promelas)*. *Environmental Toxicology and Chemistry*, 2001, **20**(6):1216–1227.
15. Wolf, J.C., et al., Qualitative and quantitative histomorphologic assessment of fathead minnow *Pimephales promelas* gonads as an endpoint for evaluating endocrine-active compounds: A pilot methodology study. *Toxicologic Pathology*, 2004, **32**(5):600–612.
16. Balch, G.C., C.A. Mackenzie, and C.D. Metcalfe, Alterations to gonadal development and reproductive success in Japanese medaka *(Oryzias latipes)* exposed to 17alpha-ethinylestradiol. *Environmental Toxicology and Chemistry*, 2004, **23**(3):782–791.
17. Brown, K.H., I.R. Schultz, and J.J. Nagler, Reduced embryonic survival in rainbow trout resulting from paternal exposure to the environmental estrogen 17alpha-ethynylestradiol during late sexual maturation. *Reproduction*, 2007, **134**(5):659–666.
18. Bouma, J., J.G. Cloud, and J.J. Nagler, In vitro effects of estradiol-17β on myoid and fibroblastic cell proliferation in the immature rainbow trout testis. *Fish Physiology and Biochemistry*, 2003, **28**(4):191–192.

19. Grier, H.J., Cellular organization of the testis and spermatogenesis in fishes. *American Zoologist*, 1981, **21**:345–357.
20. U.S. EPA, A short-term test method for assessing the reproductive toxicity of reproductive toxicity of endocrine-disrupting chemicals using the fathead minnow (*Pimephales promelas*). U.S. Environmental Protection Agency, Washington, DC, 2002, pp. 141–144.
21. Yonkos, L.T., R. Reimschuessel, and A.S. Kane, Digital atlas of fathead minnow normal histology. http://aquaticpath.umd.edu/fhm/, 2003.
22. Shibata, N. and S. Hamaguchi, Evidence for the sexual bipotentiality of spermatogonia in the fish, *Oryzias latipes*. *Journal of Experimental Biology*, 1988, **245**(1):71–77.
23. Jensen, K.M., et al., Aspects of basic reproductive biology and endocrinology in the fathead minnow (*Pimephales promelas*). *Comparative Biochemistry and Physiology Part C: Toxicology and Pharmacology*, 2001, **128**(1):127.
24. Patino R., and J.M. Redding, Reproductive systems, in *The Handbook of Experimental Animals: The Laboratory Fish*, G.K. Ostrander, Editor. Academic Press, San Diego, CA, 2000, pp. 489–500.
25. Alberts, B., et al., *Molecular Biology of the Cell*, 3rd ed., Garland Publishing, New York, 1994.
26. Sinha, G.M., et al., Chronology of meiosis and spermiogenesis in a fresh-water teleost fish, *Colisa fasciata* (Bloch and Schneider). *Zeitschrift für mikroskopisch-anatomische Forschung*, 1979, **93**(3):442–448.
27. Schulz, R.W., et al., Sertoli cell proliferation in the adult testis—evidence from two fish species belonging to different orders. *Biology of Reproduction*, 2005, **73**(5):891–898.
28. Schulz, R.W. and T. Miura, Spermatogenesis and its endocrine regulation. *Fish Physiology and Biochemistry*, 2002, **26**:43–56.
29. Loir, M., et al., Cell–cell interactions in the testis of teleosts and elasmobranchs. *Microscopy Research and Technique*, 1995, **32**(6):533–552.
30. Miles-Richardson, S.R., et al., Effects of waterborne exposure of 17 β-estradiol on secondary sex characteristics and gonads of fathead minnows *(Pimephales promelas)*. *Aquatic Toxicology*, 1999, **47**(2):129–145.
31. Miles-Richardson, S.R., et al., Effects of waterborne exposure to 4-nonylphenol and nonylphenol ethoxylate on secondary sex characteristics and gonads of fathead minnows (*Pimephales promelas*). *Environmental Research, Section A*, 1999, **80**(2): S122–S137
32. Dietrich, D.R., et al., Optimization of tissue preparation for the quantification of germinal cells in histologic sections of fathead minnow gonads. *SETAC Europe 13th Annual Meeting*, Hamburg, Germany, 2003.
33. Farla, P., et al., Antiandrogens prevent stable DNA-binding of the androgen receptor. *Journal of Cell Science*, 2005, **118**(1):4187–4198.
34. van der Ven, L.T.M., P.W. Wester, and J.G. Vos, Histopathology as a tool for the evaluation of endocrine disruption in zebrafish (*Danio rerio*). *Environmental Toxicology and Chemistry*, 2003, **22**(4):908–913.
35. Kurokawa, H., Germ cells are essential for sexual dimorphism in the medaka gonad. *Proceedings of the National Academy of Sciences of the USA*, 2007, **23**(43): 16958–16963.

36. Nobrega, R.H. and I. Quagio-Grassiotto, Morphofunctional changes in Leydig cells throughout the continuous spermatogenesis of the freshwater teleost fish, *Serrasalmus spilopleura* (Characiformes, Characidae): An ultrastructural and enzyme study. *Cell Tissue Research*, 2007, **329**(2):339–349.

37. Billard, R., et al., Endocrine control of spermatogenesis in teleost fish. *Canadian Journal of Fisheries and Aquatic Science*, 1982, **39**:65–79.

38. Miura, T., et al., Estradiol-17β stimulates the renewal of spermatogonial stem cells in males. *Biochemical and Biophysical Research Communications*, 1999, **264**:230–234.

39. Miura, T. and C.I. Miura, Molecular control mechanisms of fish spermatogenesis. *Fish Physiology and Biochemistry*, 2003, **28**(4):181–186.

40. Miura, T., et al., Hormonal induction of all stages of spermatogenesis in vitro in the male Japanese eel, *Anguilla japonica. Proceedings of the National Academy of Sciences of the USA*, 1991, **88**(13):5774–5778.

41. Miwa, S., L. Yan, and P. Swanson, Localization of two gonadotropin receptors in the salmon gonad by in vitro ligand autoradiography. *Biology of Reproduction*, 1994, **50**:629–642.

42. Nagahama, Y., Endocrine regulation of gametogenesis in fish. *International Journal of Developmental Biology*, 1994, **38**:217–229.

43. Nagahama, Y., Gonadal steroid hormones: Major regulators of gonadal sex differentiation and gametogenesis in fish. *Sixth International Symposium on the Reproductive Physiology of Fish*, Institute of Marine Research, Bergen, Norway, 1999.

44. Nagahama, Y., H. Kagawa, and G. Young, Cellular sources of sex steroids in teleost gonads. *Canadian Journal of Fisheries and Aquatic Sciences*, 1982, **39**:56–64.

45. Ankley, G.T., et al., Ketoconazole in the fathead minnow *(Pimephales promelas):* Reproductive toxicity and biological compensation. *Environmental Toxicology and Chemistry*, 2007, **26**(6):1214–1223.

46. Weber, L.P., et al., Increased kidney, liver, and testicular cell death after chronic exposure to 17alpha-ethinylestradiol in medaka *(Oryzias latipes). Environmental Toxicology and Chemistry*, 2004, **23**(3):792–797.

47. Miura, T., et al., Induction of spermatogenesis in male Japanese eel, *Anguilla japonica,* by a single injection of human chorionic gonadotropin. *Zoological Science*, 1991, **8**(1):63–73.

48. Liu, S., et al., Expression of cytochrome P45011β (11β-hydroxylase) gene during gonadal differentiation and spermatogenesis in rainbow trout, *Oncorhynchus mykiss. Journal of Steroid Biochemistry and Molecular Biology*, 2000, **75**(4–5):291–298.

49. Callard, G.V., et al., Stage-dependent changes in steroidonic enzymes and estrogen receptors during spermatogenesis in the testis of the dogfish, *Squalus acanthias. Endocrinology*, 1985, **117**:1328–1335.

50. Miura, C., T. Miura, and M. Yamashita, PCNA protein expression during spermatogenesis of the Japanese eel *(Anguilla japonica). Zoological Science*, 2002, **19**:87–91.

51. Miura, T., K. Yamauchi, and Y. Nagahama, Human chorionic gonadotropin induces all stages of spermatogenesis in vitro in the male Japanese eel (*Anguilla japonica*). *Developmental Biology*, 1991, **146**(1):258–262.

52. Chang, C.-F., E.-L. Lau, and B.-Y. Lin, Stimulation of spermatogenesis or of sex reversal according to the dose of exogenous estradiol-17β in juvenile males of protandrous black

porgy, *Acanthopagrus schlegeli. General and Comparative Endocrinology*, 1995, **100**(3):355.

53. van der Ven, L.T., et al., Effects of the estrogen agonist 17beta-estradiol and antagonist tamoxifen in a partial life-cycle assay with zebrafish (*Danio rerio*). *Environmental Toxicology and Chemistry*, 2007, **26**(1):92–99.
54. Schulz, R.W., et al., Pituitary gonadotrophs are strongly activated at the beginning of spermatogenesis in African catfish, *Clarias gariepinus. Biology of Reproduction*, 1997, **57**(1):139–147.
55. Kusakabe, M., et al., Changes in mRNAs encoding steroidogenic acute regulatory protein, steroidogenic enzymes and receptors for gonadotropins during spermatogenesis in rainbow trout testes. *Journal of Endocrinology*, 2006, **189**:541–554.
56. Ohta, T., et al., Follicle-stimulating hormone induces spermatogenesis mediated by androgen production in Japanese eel, *Anguilla japonica. Biology of Reproduction*, 2007, **77**:970–977.
57. Ozaki, Y., C. Miura, and T. Miura, Molecular cloning and gene expression of Spo11 during spermatogenesis in the Japanese eel, *Anguilla japonica. Comparative Biochemistry and Physiology, Part B: Biochemistry and Molecular Biology*, 2006, **143**(3):309–314.
58. Miura, T., et al., Complementary deoxyribonucleic acid cloning of spermatogonial stem cell renewal factor. *Endocrinology*, 2003, **144**(12):5504–5510.
59. Miura, T., et al., Spermatogenesis-preventing substance in Japanese eel. *Development*, 2002, **129**:2689–2697.
60. Amer, M.A., et al., Involvement of sex steroid hormones in the early stages of spermatogenesis in Japanese huchen *(Hucho perryi). Biology of Reproduction*, 2001, **65**:1057–1066.
61. Schulz, R.W., et al., Gonadotropins, their receptors, and the regulation of testicular functions in fish. *Comparative Biochemistry and Physiology Part B: Biochemistry and Molecular Biology*, 2001, **129**:407–417.
62. Bouma, J. and J.J. Nagler, Estrogen receptor-α protein localization in the testis of the rainbow trout (*Oncorhynchus mykiss*) during different stages of the reproductive cycle. *Biology of Reproduction*, 2001, **65**(1):60–65.
63. Le Gac, F. and M. Loir, Control of testis function in fish: In vitro studies of gonadotropic regulation in the trout (*Salmo gairdneri*). *Reproduction, Nutrition and Development*, 1988, **28**:1031–1046.
64. Baroiller, J.-F., Y. Guiguen, and A. Fostier, Endocrine environmental aspects of sex differentiation in fish. *Cellular and Molecular Life Sciences*, 1999, **55**: 910–931.
65. Loir, M., Spermatogonia of rainbow trout: II. In vitro study of the influence of pituitary hormones, growth factors and steroids on mitotic activity. *Molecular Reproduction and Development*, 1999, **53**(4):434–442.
66. Andreassen, T.K., et al., Molecular cloning, characterisation, and tissue distribution of oestrogen receptor alpha in eelpout *(Zoarces viviparus). General and Comparative Endocrinology*, 2003, **132**(3):356.
67. Halm, S., et al., Cloning, characterisation, and expression of three oestrogen receptors (ERα, ERβ1 and ERβ2) in the European sea bass, *Dicentrarchus labrax. Molecular and Cellular Endocrinology*, 2004, **223**(1–2):63–75.

68. Hornung, M.W., et al., Mechanistic basis for estrogenic effects in fathead minnow *(Pimephales promelas)* following exposure to the androgen 17alpha-methyltestosterone: conversion of 17alpha-methyltestosterone to 17alpha-methylestradiol. *Aquatic Toxicology*, 2004, **66**(1):15–23.

69. Katsu, Y., et al., Functional associations between two estrogen receptors, environmental estrogens, and sexual disruption in the roach *(Rutilus rutilus). Environmental Science and Technology*, 2007, **41**(9):3368–3374.

70. Loomis, A.K. and P. Thomas, Binding characteristics of estrogen receptor (ER) in Atlantic croaker *(Micropogonias undulatus)* testis: Different affinity for estrogens and xenobiotics from that of hepatic ER. *Biology of Reproduction*, 1999, **61**:51–60.

71. Chang C.-F., et al., The concentrations of plasma sex steroids and gonadal aromatase during controlled sex differentiation in grey mullet, *Mugil cephalus. Aquaculture*, 1999, **177**(1–4):37–45.

72. Parker, K.L., et al., Steroidogenic factor 1: An essential mediator of endocrine development. *Recent Progress in Hormone Research*, 2002, **57**(1):19–36.

73. Hassin, S., M.C.H. Holland, and Y. Zohar, Early maturity in the male sriped bass, morone saxatilis: Follicle-stimulating hormone and luteinizing hormone gene expression and their regulation by gonadotropin-releasing hormone analogue and testosterone. *Biology of Reproduction*, 2000, **63**:1691–1697.

74. Lasala, C., et al., Subcellular and molecular mechanisms regulating anti-Mullerian hormone gene expression in mammalian and nonmammalian species. *DNA and Cell Biology*, 2004, **23**(9):572–585.

75. Maugars, G., Endocrine regulation of early sexual maturation in male Atlantic salmon parr. University of Agricultural Sciences, Umeå, Sweden, 2007, p. 48.

76. Dubois, W. and G.V. Callard, Culture of intact Sertoli/germ cell units and isolated Sertoli cells from *Squalus* testis: II. Stimulatory effects of insulin and IGF-I on DNA synthesis in premeiotic stages. *Journal of Experimental Biology*, 1993, **267** (2):233–244.

77. Weber, G.M. and C.V. Sullivan, Effects of insulin-like growth factor-I on in vivo final oocyte maturation and ovarian steroidogenesis in striped bass, *Morone saxatilis. Biology of reproduction*, 2000, **63**(4):1049–1057.

78. Le Gac, F., et al., Insulin-like growth factor (IGF-I) mRNA and IGF-I receptor in trout testis and in isolated spermatogenic and Sertoli cells. *Molecular Reproduction and Development*, 1996, **44**(1):23–35.

79. Patino, R., et al., Gonadotropic control of ovarian follicle maturation: The two-stage concept and its mechanisms. *Comparative Biochemistry and Physiology, Part B: Biochemistry and Molecular Biology*, 2001, **129**(2–3):427–439.

80. Richards, R.G., et al., Estradiol stimulates tyrosine phosphorylation of the insuline-like growth factor-1 receptor and insulin receotor substrate-1 in the uterus. *Proceedings of the National Academy of Sciences of the USA*, 1996, **93**(21):12002– 12007.

81. Loir, M. and F. Le Gac, Insulin-like growth factor-I and -II binding and action on DNA synthesis in rainbow trout spermatogonia and spermatocytes. *Biology of Reproduction*, 1994, **51**(6):1154–1163.

82. Vera Cruz, E.M. and C.L. Brown, The influence of social status on the rate of growth, eye color pattern and insulin-like growth factor-I gene expression in Nile tilapia, *Oreochromis niloticus. Hormes and Behavior*, 2007, **51**(5):611–619.

83. Weber, G.M., A.B. Moore, and C.V. Sullivan, In vitro actions of insulin-like growth factor-I on ovarian follicle maturation in white perch *(Morone americana). General and Comparative Endocrinology*, 2007, **151**(2):180–187.

84. Mukherjee, D., et al., In vitro effects of insulin-like growth factors and insulin on oocyte maturation and maturation-inducing steroid production in ovarian follicles of common carp, *Cyprinus carpio. Comparative Biochemistry and Physiology, Part A: Molecular and Integrated Physiology*, 2006, **144**(1):63–77.

85. Weber, G.M. and C.V. Sullivan, Insulin-like growth factor-I induces oocyte maturational competence but not meiotic resumption in white bass *(Morone chrysops)* follicles in vitro: Evidence for rapid evolution of insulin-like growth factor action. *Biology of Reproduction*, 2005, **72**(5):1177–1186.

86. Kamangar, B.B., J.C. Gabillard, and J. Bobe, Insulin-like growth factor-binding protein (IGFBP)-1, -2, -3, -4, -5, and -6 and IGFBP-related protein 1 during rainbow trout postvitellogenesis and oocyte maturation: molecular characterization, expression profiles, and hormonal regulation. *Endocrinology*, 2006, **147**(5):2399–2410.

87. Aegerter, S., B. Jalabert, and J. Bobe, Messenger RNA stockpile of cyclin B, insulin-like growth factor I, insulin-like growth factor II, insulin-like growth factor receptor Ib, and p53 in the rainbow trout oocyte in relation with developmental competence. *Molecular Reproduction and Development*, 2004, **67**(2):127–135.

88. Bobe, J., et al., Rainbow trout follicular maturational competence acquisition is associated with an increased expression of follicle stimulating hormone receptor and insulin-like growth factor 2 messenger RNAs. *Molecular Reproduction and Development*, 2003, **66** (1):46–53.

89. Miura, C., T. Higashino, and T. Miura, A progestin and an estrogen regulate early stages of oogenesis in fish. *Biology of Reproduction*, 2007, **77**:822–828.

CHAPTER 6

Endocrine-Disrupting Compounds

Although many studies presenting abnormalities of spermatogenesis or oogenesis have been published (see Tables 6.1 to 6.12), it is important to note that at present (i.e., based on the extensive review of available literature), no studies have yet presented convincing results suggesting a relationship between histological lesions in the gonads of fish exposed to endocrine-disrupting chemicals (EDCs) and altered reproduction or reproductive success [1], with the exception of those experiments where very high doses of EDCs have been used and nearly 100% permanent sex reversal or histologically atrophic gonadal tissue was achieved. Indeed, when employing reproduction parameters (female fecundity and hatchability) concomitantly with histopathological evaluation of the gonads and vitellogenin induction in male fish, it is surprising to note that impaired reproductive success occurred consistently at much greater concentrations of the respective endocrine disrupter than the effects on vitellogenin induction or gonadal histology, irrespective of whether the fish species was considered an r- (zebrafish) or a K-strategist (fathead minnow, medaka) [2–6]. This may be due primarily to the fact that most population models are geared toward the detection of changes in female fecundity, hatchability and survival success of the hatchlings, and not on the presence or absence of specific gamete stages or pathological lesions in either the male or female gonad. On the other hand, population models (see Wester et al. [5] for a more detailed discussion) appear to predict that only major changes in sex ratio in a species will affect the continued survival of the population, irrespective of the reproductive strategy. K-strategist population dynamics also appear to be affected by the number of offspring and juvenile survival, whereas this does not appear to be the case for r-strategist populations. The latter prediction appears to be supported by recent reports of a completely annihilated fathead minnow population following 3 years of continuous exposure to 17α-ethinylestradiol (EE_2, 4.8 to 6.1 ng/L) in Lake 260 of the Canadian experimental lakes area [7]. Although the authors do report a major increase in vitellogenin concentration, a complete reproductive failure (lack of age 0 offspring) in the spring of the second and third years of EE_2 exposure, follicular atresia in females, and fibrosis-inhibited gametogenesis and primary-stage oocytes in the testis of males, the experiment did not look at EE_2-induced mortality, morbidity, and changes in fecundity and fertility of adults, nor for increased mortality, morbidity, or inhibited

Histological Analysis of Endocrine-Disruptive Effects in Small Laboratory Fish
By Daniel R. Dietrich and Heiko O. Krieger

survival success of offspring generated. As suggested in Chapter 5, the pathology as presented would not appear to suggest completely or even substantially inhibited gametogenesis, but rather, point to increased mortality of adult fish, as is also indicated by loss of the remaining age 1–4 in the second and third years of EE_2 application, and thus lack of enough reproducing adult fish for the maintenance of a healthy population. The latter explanation would also confirm the interpretation of Wester el al., [5], whereby the population strategy of a given fish species is irrelevant for the survival of individual populations once the number of breeding pairs is decreased to below the critical number.

In contrast to the view of Kidd et al. [7] and in consideration of a representative part of the literature found on steroid estrogen concentrations in sewage treatment plant (STP) effluents directly at the STP outflow, the actual concentrations of EE_2 in effluents and surface waters rarely exceeds 1.8 and 0.7 ng/L, respectively [8–20], demonstrating that the experimental lakes area experiment [7] represents a chronic high-exposure situation, apparently often observed in close proximity to Canadian STP environments [11,21]. In view of this, one could conclude that only massive overt changes in gonad histopathology or inhibition of gametogenesis in nearly all individuals of one or both sexes could potentially have a long–term effect on population dynamics and therefore lead to a decline in the affected population.

Although the latter hypothesis could be tested using continuous exposure over several generations of small laboratory fish (e.g., medaka, fathead minnows, zebrafish, guppies), the relevance of these findings for the description of population effects of endocrine-disrupting chemicals in the environment appears questionable. For example, Haubruge and colleagues [22] reported dose-dependent reduced (40 to 75%) sperm numbers following three-week semistatic exposure of male guppies (*Poecilia reticulata*) to high doses (274 and 549 μg/L) of bisphenol A (BPA) using spermatozeugmata numbers per male and mature sperm numbers evaluated in three spermatozeugmata per male. However, no significant treatment-related effects were reported with regard to the number of spermatozeugmata or sperm length, suggesting that BPA affected spermatogenesis rather than spermiation (see Chapter 5). In contrast, a 164-day flow-through study with male and female fathead minnows exposed to 1, 16, 64, 160, and 640 μg/L BPA did not demonstrate any effect on male or female germ cell distribution [23], with the exception of the highest dose (see also Section 6.1.1), despite employing the statistically most rigorous experimental design possible.

The latter study would therefore strongly support the interpretation that effects reported for high doses of EDCs are unlikely to reflect a real environmental risk and thus environmental relevance for indigenous fish populations. Indeed, an analysis of endocrine disruption in wild roach living in rivers that receive treated sewage effluents [24] demonstrated that despite continuous exposure to high levels of natural estrogens [25], no statistical differences were found between the proportions of male and female roach maturing in each year class at each site. Furthermore, subsequent histopathological analysis of the gonads revealed that intersex fish (male fish with testes containing female germ cells) were found at the reference as well as at the sites receiving sewage treatment effluents; and that gonadal growth appeared to be inhibited only in severely intersex fish, whereas progression of spermatogenesis was delayed in a large proportion of all intersex and exposed male fish. The fact that intersex fish—wild roach in the field study by Jobling et al. [24], or laboratory-exposed medaka [3]—are able to spawn and fertilize eggs and thus to secure the next generation highlights the difficulty of interpreting and evaluating the potential population impact of histopathological changes, including inhibited gametogenesis, in the gonads of fish exposed to moderate-to-high concentrations of estrogens and other EDCs.

6.1 INDIVIDUAL EFFECTS

6.1.1 Inhibition of Gametogenesis

The term *inhibition of gametogenesis* includes reduced gamete division and maturation and thus is applicable to germ cells, male as well as female, in general. It is interesting to note, however, that the predominant proportion of literature reporting on inhibition of gametogenesis relates to observations in male fish, where as substantially fewer reports are found for effects in female fish [26–28]. Whether this is a matter of "politically and historically motivated" scientific bias toward the expectation that "effects of EDC are primarily expected in males" [6,29–31] and therefore female fish had not been evaluated as often, or indeed, this is a true biological fact, is currently difficult to ascertain given the present scientific documentation (see Section 6.3.2). Inhibition of gametogenesis is also often reported in conjunction with stromal fibrosis (see Sections 6.1.6, 6.2.1.3, and 6.2.2.3) and thus may reflect germ cell degeneration (necrosis/apoptosis; see Sections 6.1.2, 6.2.1.4, and 6.2.2.4), inflammation (Sections 6.2.1.6 and 6.2.2.6), germ cell replacement, and tissue scarring (fibrosis) (e.g., as observed in guppy exposed to methyl mercury chloride) [32], medaka exposed to cyproterone acetate or 17β-estradiol [4,27], or in fathead minnows exposed to 17α-ethinylestradiol [33]. Most, although certainly not all, endocrine-disrupting compounds have been observed to induce inhibition of gametogenesis at the level of germ cell division and maturation (i.e., an increased proportion of more juvenile germ cells) [6], concurrent with a reduction in the number of mature germ cells (Figures 6.1 and 6.2), and simultaneously with other pathological effects [e.g., necrosis/atresia of germ cells, as well as necrosis, inflammation, and fibrosis (scarring) of interstitial tissues] (see Sections 6.2 and 6.2.2).

Few other compounds (e.g., bisphenol A (BPA) [3,6], 4-nonylphenol (4-NP) [34]) appeared to induce an inhibition of spermatogenesis, as the sole pathological observation, beyond the induction of testis–ova in male medaka [3,34]. Whether this is due to a lack of overall pathological evaluation of the testicular tissue or the absence of other pathology is difficult to determine from the literature. However, it is important to state that a dramatic shift from mature to more juvenile germ cells, as well as the presence of testis–ova in medaka, were observed for both compounds at very high concentrations only, whereas no effects of BPA were observed on spermatogenesis, oogenesis, or general gonad histopathology at concentrations up to 160 μg/L in fathead minnows [23]. Consequently, the question must be raised whether other compound-induced pathological effects in the gonads (e.g., germ cell degeneration, necrosis, atresia, apoptosis) and in other organs (degeneration, necrosis, apoptosis, inflammation, protein accumulation, hyperplasia, hypertrophy, fibrosis) may be of greater importance for the functionality of the gonadal tissue, and therefore for the reproductive success of an individual fish and the population, than the mere numerical change in the maturational stages of germ cells.

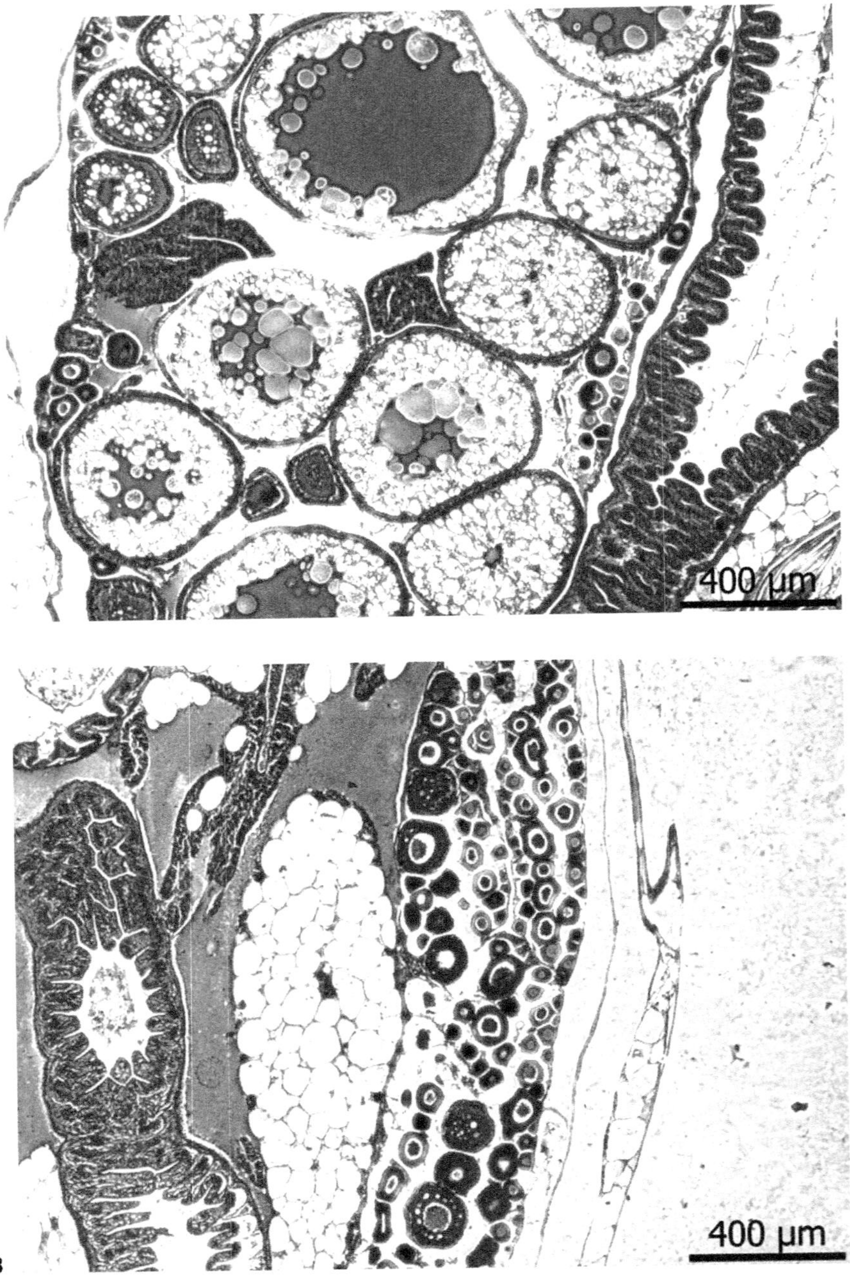

FIGURE 6.1 Ovaries from (A) a control (100 μg/L acetone) medaka (ca.103 days old) with oocytes in different stages of maturity and (B) a medaka exposed to 100 μg/L nonylphenol for 100 days containing only early stages of oocytes. (Courtesy of David Janz, University of Saskatchewan, Saskatoon, Canada [39].) (See insert for color representation.)

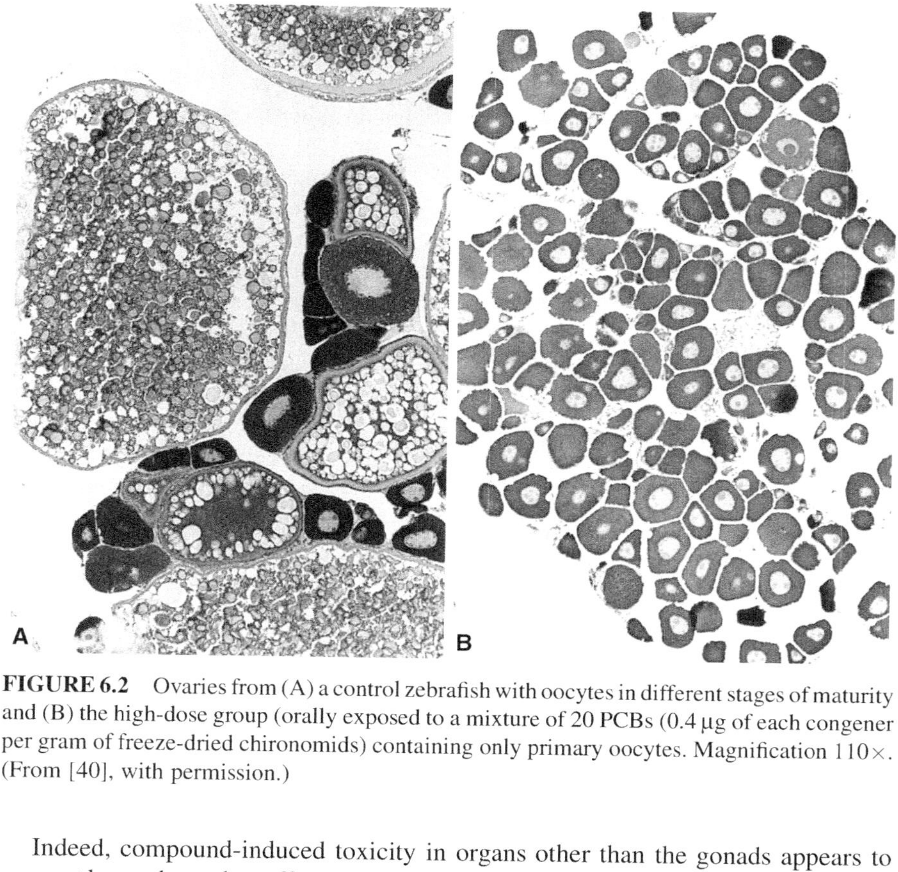

FIGURE 6.2 Ovaries from (A) a control zebrafish with oocytes in different stages of maturity and (B) the high-dose group (orally exposed to a mixture of 20 PCBs (0.4 μg of each congener per gram of freeze-dried chironomids) containing only primary oocytes. Magnification 110×. (From [40], with permission.)

Indeed, compound-induced toxicity in organs other than the gonads appears to occur at lower doses than effects on the gonads per se or on gametogenesis specifically [26,35], thus strongly advocating a thorough histopathological assessment of the major organs for a proper toxicological evaluation of a given compound [26,36,37]. Furthermore, the findings above raise the question of whether non- or extremely weak endocrine-modulating compounds can induce histopathological effects nearly indistinguishable from the effects induced by veritable EDCs [38] and therefore whether or not the exclusive focus on a specific morphological alteration in the gonads (e.g., gametogenesis) may be misleading rather than illuminating in the quest for mechanistic understanding and environmental risk assessment of a given compound. Indeed, despite severe craniofacial malformations, lack of a caudal fin, renal tubular necrosis and karyorrhexis, and overt ovarian atresia, PCB-104-exposed zebrafish were fertile and capable of producing healthy offspring [38].

6.1.2 Necrosis and Apoptosis (Gamete and Stromal Cells)

Necrosis of germ cells was reported in male fathead minnows exposed for 14 days to a high concentration of 17β-estradiol [41], 12 days to high concentrations of methyltestosterone [42], and 42 days to low concentrations of 4-nonylphenol (4-NP) and nonylphenol (NP) [43], in 1 week-old medaka exposed for 96 hours to high levels of 11-ketotestosterone or flutamide [44], as well as in guppy exposed to low-to-moderate levels of methyl mercury for three months [32]. Interestingly, no reports were found on sperm necrosis in zebrafish, although Weber et al. [26] report that NP and 17α-ethinylestradiol induced enhanced pyknosis in kidney epithelial and hematopoietic interstitial cells as well as in male germ cells of zebrafish exposed for up to 60 days post-hatch. Pyknosis is currently considered a step in the process of cell necrosis (oncotic apoptosis) [45], thus the results by Weber et al. [26] would constitute the description of a necrotic process. Furthermore, with the exception of the diazinone-induced findings in bluegill sunfish (*Lepomis macrochirus*) [46], no reports were available that mentioned necrosis of stromal cells or female germ cells of fish (i.e., including species beyond the three small fish focused on in this review; see Section 6.3.2). Whether the latter is because it was not specifically addressed (evaluated), because nondescriptive terms (e.g., severe damage of oocytes [47], stromal and follicular cell degeneration [48,49], degenerated spermatozoa and spermatocytes [4]), were used, or because necrosis was not observed is difficult to ascertain. However, differences in exposure protocol (length of exposure, compound employed, tissue processing) as well as in the pathological evaluation may explain some of these inconsistencies. If necrosis of germ and interstitial cells occurs following chemical exposure, one can assume that the intracellular enzymes released during the necrotic process will provide an inflammatory and regenerative/scarring response. Indeed, in many cases testicular inflammation and interstitial cell hyperplasia/fibrosis is reported following exposure of fish to endocrine-active compounds and other xenobiotics, although not always concurrently with cell necrosis [4,32,33,50]. Whether or not interstitial fibrosis is secondary only to a primary necrotic and/or inflammatory event or indeed also a primary xenobiotic-induced pathological lesion is discussed in more detail below.

Apoptosis of germ cells and epithelial cells in other organs is observed in male and female fish exposed to single xenobiotics or mixtures [51–53]. In male fish germ cell apoptosis is observed primarily in conjunction with germ cell phagocytosis by Sertoli cells and thus appears to be a normal physiological process (see Section 3.1.5.1) as well as a process that can be induced by xenobiotics [41,43,51,54,55]. However, the term *apoptosis* has never been clearly dissociated from *necrosis*, especially when used in conjunction with phagocytosed germ cells within Sertoli cells [43]. Of importance appears to be the fact that while necrosis is observed in germ cells of zebrafish exposed to 100 μg/L 4-NP from 2 to 60 dph [26], exposure of adult medaka to the same concentration of 4-NP for 42 days induced germ cell apoptosis and no necrosis [51].

As apoptosis was determined by TUNEL (terminal dideoxynucleotidyl transferase dUTP nick end labeling), which determines apoptotic cells but also recognizes fragmented DNA of necrotic cells [56], it is at present unclear whether this difference in cell death pattern constitutes an experimental artifact, or if biologically true (nonartifactual), the effect of the time of xenobiotic exposure of the fish (post-hatch vs. adult stage), or a species-specific effect. Weber et al. [51] also reported an increased apoptosis of Sertoli and Leydig cells in zebrafish testes following 4-NP exposure. However, TUNEL staining would be expected to stain Sertoli cells, as Sertoli cells by definition of their function as phagocytic cells would be expected to engulf apoptotic germ cells and thus would need to be TUNEL positive in the cytoplasm. As only TUNEL-positive cells were counted, it remained unclear whether Sertoli cell nuclei stained TUNEL-positive.

Quite different is the issue of apoptotic Leydig cells, as Leydig cells do not have a phagocytic function. TUNEL positivity indicates that these cells are undergoing apoptosis or necrosis as the result of 4-NP exposure and thus would suggest that zebrafish testis are encountering enhanced loss of steroidogenic and possibly also nurse cell capacity concomitantly with increased scarring and fibrosis. Substantial loss of supportive cells would, as a consequence, induce an inhibited germ cell proliferation and maturation and therefore testicular regression, as observed in medaka [1,27,51,57], zebrafish [51,54] and fathead minnows [6,33,41,43,55,58] exposed to endocrine-active compounds but also to compounds with no endocrine activity (e.g., methyl mercury) [32,35]. This would suggest a mechanistic pathway different from the previously described primary inhibition of spermatogenesis or juvenilization of small fish testes following exposure to endocrine-active compounds in the absence of degenerative, inflammatory, and fibrotic histopathological lesions [3,6,23,28,34,37].

In female fish, germ and supportive cell apoptosis is restricted to follicular cell apoptosis, which is used synonymously with the term *follicular cell atresia* and is discussed in more detail below.

6.1.3 Atretic Follicles

Whether caused by normal physiologic stimuli or environmental stress, atresia entails the degeneration of follicles at any stage of follicular development [59], although atretic follicles are often most noticeable in the later developmental stages of the ovary [60]. Stress and other environmental insults (e.g., poor nutrition, physical factors, overpopulation or stressors such as xenobiotics that could decrease gonadotropin levels) appear to increase the incidence of atresia [61–63]. However, this phenomenon is also observed in healthy female fish [52,53,60–62]. In the latter case, although probably minor in importance, atresia may serve to regulate de-recruitment and/or resorption of malfunctional younger-stage oocytes in order to preserve normal ovarian physiology during the breeding season. Exposure to estrogens or estrogen mimics [26–28,38,39,42,55,57,63] [e.g., nonylphenol (Figure 6.1) or PCB mixtures (Figure 6.2)], aromatase inhibitors (e.g., fadrozole) [64], as well as xenobiotic mixtures (e.g., bleached pulp mill effluent) [52,53], were demonstrated to induce increased follicular atresia, especially of vitellogenic follicles, resulting in an accumulation of previtellogenic follicles and consequently, decreased fecundity [1]. The mechanism(s) by which follicular atresia can be induced is not well understood, although some findings may lend themselves to provide some mechanistic insight. Atresia of vitellogenic follicles can be inhibited by low levels of 17β-estradiol (E_2), epidermal growth factor, or gonadotropin [65], thereby possibly explaining why reduced gonadotropin levels, androgenic hormones (e.g., methyltestosterone [42] and 11-ketotestosterone [44]), or xenobiotics (e.g, the aromatase inhibitor fadrozole [64] or the anti-androgen flutamide [44]), and in consequence the decreased endogenously available levels of E_2 and its precursors may increase the incidence in atretic follicles [61–63]. Indeed, support for this interpretation is provided by the more recent experiment in zebrafish chronically exposed to dietary levels of 2,3,7,8-TCDD [66]. The TCDD-mediated decreased serum E_2 concentrations and thus vitellogenesis was assumed to have led to inhibition of the transition from pre-vitellogenic-stage follicles to vitellogenic-stage follicles, as well as the induction of follicular atresia due to lacking estrogenic signaling within the vitellogenic follicles. The overall reduction of egg production by more than 50% and of spawning success by as much as 96% may be suggestive of a potential population effect. However, it must be kept in mind that daily dietary exposures of 0.08, 0.32, or 0.80 ng TCDD per female are extremely high exposure levels rarely found in the environment, unless one includes accidental release areas (e.g., Seveso in Italy) and thus represent a hypothetical high-dose situation.

The mechanisms by which specific antiandrogens (e.g., cyproterone acetate or vinclozolin) induce increased follicular atresia and consequently the maturational arrest of adult medaka [27] and fathead minnow [67] ovaries in a previtellogenic state is at present unknown, although it is suspected that these anti-androgens adversely affect Δ^5–3β-hydroxysteroid dehydrogenase levels in the ovaries and liver and thus lead to a regression and masculinization of the female fish.

Interestingly, the response of medaka and fathead minnows to anti-androgens is similar, although with lower intensity, to the response observed for the same fish including zebrafish, exposed to low concentrations of estrogens [1,28,55,63,68,69] (Figure 6.3) and high concentrations of estrogen mimics in zebrafish and medaka [26,38,57]. Indeed, Papoulias et al. [63] injected 0.5 to 2.5 ng 17α-ethinylestradiol (EE_2) per egg into medaka eggs and demonstrated that exposed XX females appeared to have a higher incidence of atretic follicles than control group females (Figure 6.3). There were fewer incidences of atretic oocytes in the lowest-dose treatment than in the higher treatment concentrations. The authors speculated that the high incidence of atresia observed in EE_2-exposed XX and XY females suggests that embryonic exposure to EE_2 may have disrupted the development of gonadotropin-producing cells or impaired the release or uptake of gonadotropins in some other, as yet unclear way. Thus, the effects of both low and high concentrations of true estrogens and estrogen mimics, respectively, could act through a negative feedback mechanisms on the pituitary–hypothalamic

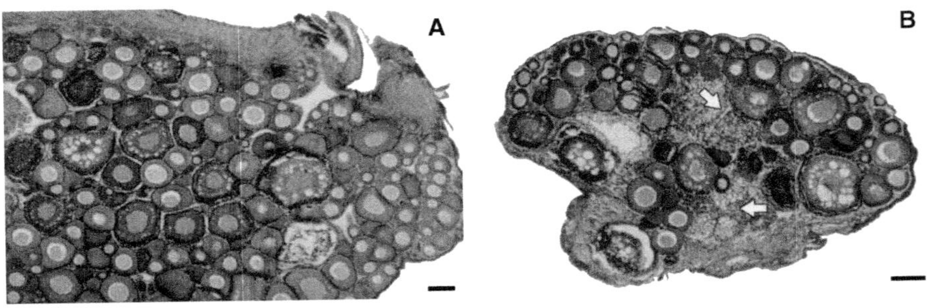

FIGURE 6.3 Ovaries of control (A) and 0.5 ng ethinylestradiol (B) medaka exposed from egg stage. Arrows indicate atretic follicles, Masson's trichrome stain. Scale bars 100 μm. (From [63], with permission.)

system, suppressing the release of gonadotropins [i.e., luteinizing hormone (LH) and follicle-stimulating hormone (FSH); see Section 3.2.4]. As FSH was shown to promote vitellogenin uptake into the oocytes [70], a low level of LH and FSH would explain the relatively increased number of perinucleolar oocytes and previtellogenic follicles and the decreased number of vitellogenic follicles in the ovaries of estrogen and estrogen mimics of exposed fish. Similarly, lack of an FSH signal and abrogated vitellogenesis could induce vitellogenic follicles to enter into atresia, if atresia is a hormone-mediated process, and thus explain the dose–response observed in the estrogen- and estrogen-mimics-induced follicular atresia. Of primary importance appears to be the fact that oocyte atresia as well as other effects are primarily a result of an imbalance of steroidal signaling and their up- and downstream effector signals, mediators (e.g., enzymes), and products generally observed at high concentrations of a given EDC.

6.1.4 Sertoli Cell Hypertrophy and Hyperplasia

To describe xenobiotic-mediated effects on male gonads [i.e., on specific testicular cell types (e.g., Sertoli cells)], it is assumed that the affected cell types can be readily identified. However, complications arise when one considers the morphological changes in some testicular cells following exposure to endocrine-active compounds. This obviously leads to confusion and a divergence in opinions as to which cell types are actually present (i.e., the possibility of misidentification of Sertoli cells). Indeed, abnormally enlarged Sertoli cells (hypertrophy), often in parallel with increased numbers (hyperplasia), have been associated with exposure to some estrogenically-active compounds [31,37,43,55,71–73]. In some studies, the same authors noted that Sertoli cell hypertrophy and hyperplasia were the most predominant, if not exclusive, alteration in the zebrafish testis following exposure to 17α-methyldihydrotestosterone (Figure 6.4), thus emphasizing the need to identify Sertoli cells correctly in order to evaluate the effects of endocrine disrupters at the histopathological level.

For example, a study conducted by Wolf et al. [41,74] with fathead minnows exposed to 17β-estradiol (E_2) for 10 days described a type of cell which was found to have the appearance of a hypertrophied primary spermatogonia but could not be distinguished clearly from what could also be described as a hypertrophied Sertoli cell. The authors suggested that the inability to define accurately which cell category the cells represented was due primarily to the co-localization of Sertoli cells and primary spermatogonia in fathead minnows as well as to potential morphological alterations that both cell types could potentially have undergone following E_2 treatment [75–78].

Although there is a consensus description among various sources in which the normal Sertoli cells of various osteichthyes are characterized by small pyramidal-shaped or elongated nuclei and minimal or attenuated cytoplasm [55,79–81], the light-microscopic appearance of altered Sertoli cells as represented in the scientific literature is more ambiguous. For example, Miles-Richardson et al. [43] reported hyperplasia and hypertrophy of Sertoli cells (Figure 3.6) as well as germ cell syncytia in fathead minnows following exposure to 1.1 or 3.4 μg/L nonylphenol or to 10 nM E_2 [43,55]. The hypertrophic Sertoli cells, under light-microscopic appraisal, were depicted as having a "vesiculate nucleus and a prominent nucleolus" [55]. Concurrent electron microscopical evaluation did indeed demonstrate that hypertrophic Sertoli

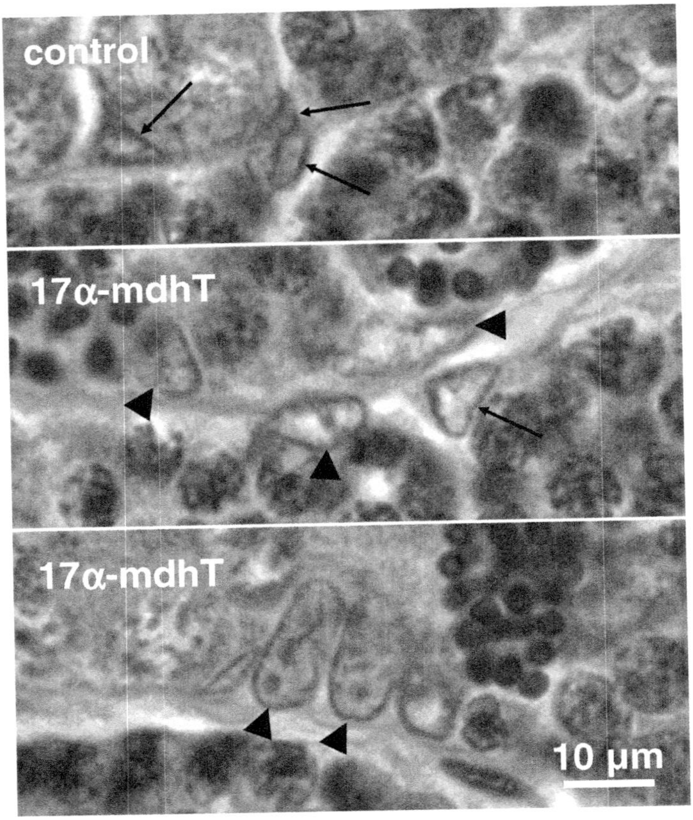

FIGURE 6.4 Sertoli cells in control and 17α-methyldihydrotestosterone-exposed zebrafish. Note the expected triangular morphology of Sertoli cells in control and treated zebrafish (arrows) in addition to the hypertrophied and elongated appearance of some "Sertoli cells" (arrowheads), denoted by the authors as Sertoli cell proliferation. Such altered Sertoli cells may prove very difficult to distinguish morphologically from primary spermatogonia in a cell-by-cell identification process during cell-type quantification. The clustered appearance of these cells in the bottom image suggests proliferation. (Courtesy of L. van der Ven and P. Wester, RIVM, Bilthoven, The Netherlands.) (See insert for color representation.)

cells distended with myelin figures and necrotic spermatozoa were present (Figure 6.5). However, later reassessment of the enlarged Sertoli cells, as identified under the light microscope, suggested that cells were probably spermatogonia rather than enlarged Sertoli cells (see Section 4.2.2.1).

In another study, "hypertrophic cuboidal to columnar Sertoli cells with spherical nuclei" were observed in the testes of control platyfish (*Xiphophorus maculatus*) and increased numbers of these cells were present in the testes of platyfish that had been exposed to nonylphenol [81]. Some of inconsistencies in the characterization of hypertrophic Sertoli cells are undoubtedly due to interspecies differences. It is also possible that the hypertrophic appearance of some Sertoli cells may be dependent on

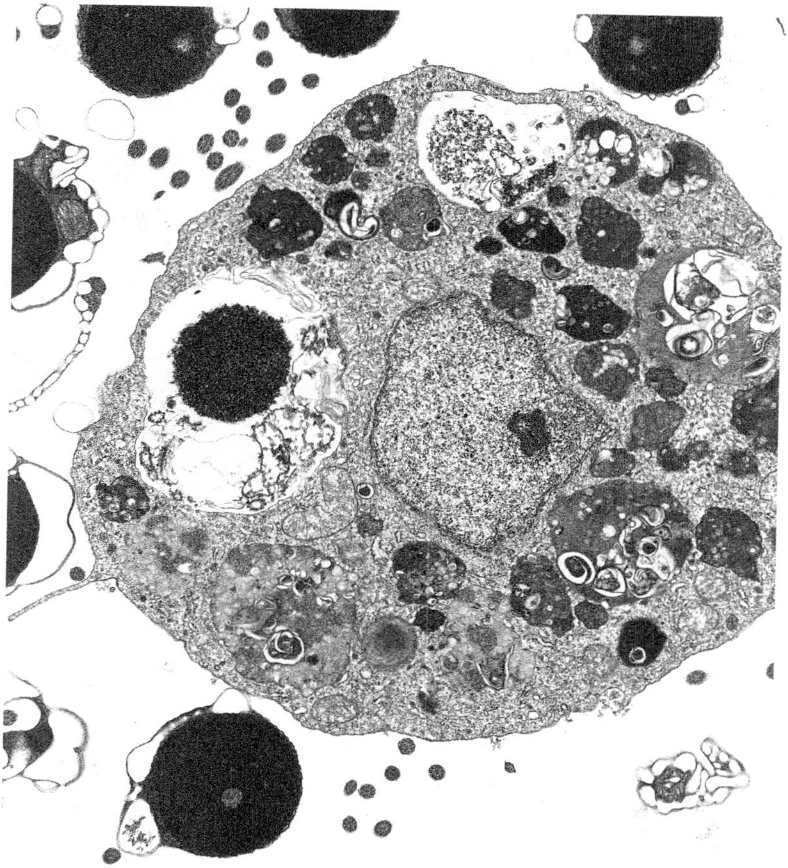

FIGURE 6.5 Transmission electron micrograph of testis from a fathead minnow exposed to 10 nM 17β-estradiol, week 3 post-exposure. Note the Sertoli cell distended with degenerate spermatozoa, myelin figures, and residual bodies. Magnification 17,092×. (From [55], with permission.)

the nature of the cytomegalic stimulus (i.e., the morphology of these cells may vary according to the physiological requirements of the testis during a given period of development or according to the type of exogenous chemical exposure). Hypertrophied Sertoli cells (Figure 6.6) in male small fish testes have been observed to undergo morphological alteration, with some studies describing hypertrophied Sertoli cells presenting with a more rounded nucleus (compare Figure 6.7D with E). This effect has been reported by various authors following exposure to estrogens, androgens, or to compounds with estrogen-androgen-like activity (e.g., in the platyfish [81] and in zebrafish [73]) (Figures 6.4 to 6.7). An additional consideration is the fact that piscine Sertoli cells have few defining light-microscopic features. Even at the ultrastructural level, justification for the homologous existence of Sertoli cells in fishes is primarily

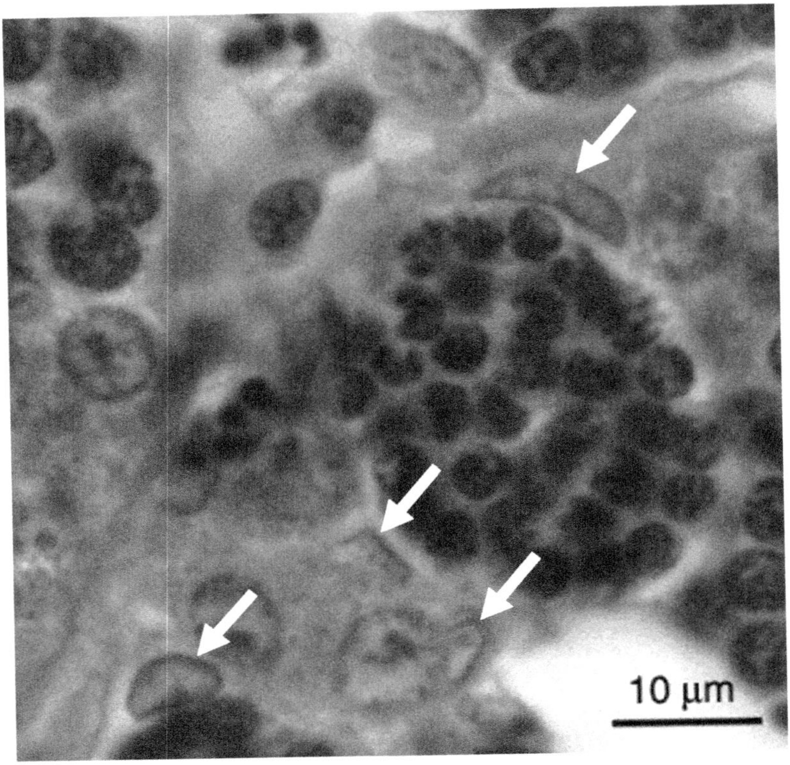

FIGURE 6.6 Sertoli cell hypertrophy (arrows) in male zebrafish following exposure to flutamide (anti-androgen). (Courtesy of L. van der Ven and P. Wester, RIVM, Bilthoven, The Netherlands.) (See insert for color representation.)

dependent on the finding of residual spermatogenic materials in the cytoplasm of such cells and the presence of junctional complexes that define the Sertoli cell barrier [75,76]. Thus, it is not surprising that fish Sertoli cells have been confused with other cell types, such as spermatogonia, when viewed by light microscopy [79].

Consequently, ultrastructural analysis appears to be the method of choice for the definitive identification of piscine Sertoli cells. Unfortunately, electron microscopy is an impractical tool for assessing cell proportions because the number of cells that can be viewed at any one time is very limited. Methods to identify Sertoli cells accurately are currently being sought intensively. At present, the enzyme γ-GT is thought to be specifically expressed by Sertoli cells [82]. If this proves to be true, this marker could present a valuable tool for the unequivocal identification of this cell type. However, no immuno- or histochemical stain for γ-GT allowing unequivocal identification of Sertoli cells in normal fish testes and in testes following exposure to xenobiotics with/without endocrine modulating activity has been published to date. Indeed, all efforts aimed at developing a histochemical stain for stain for γ-GT have failed in testes of control as well as E_2-exposed fathead minnows [41].

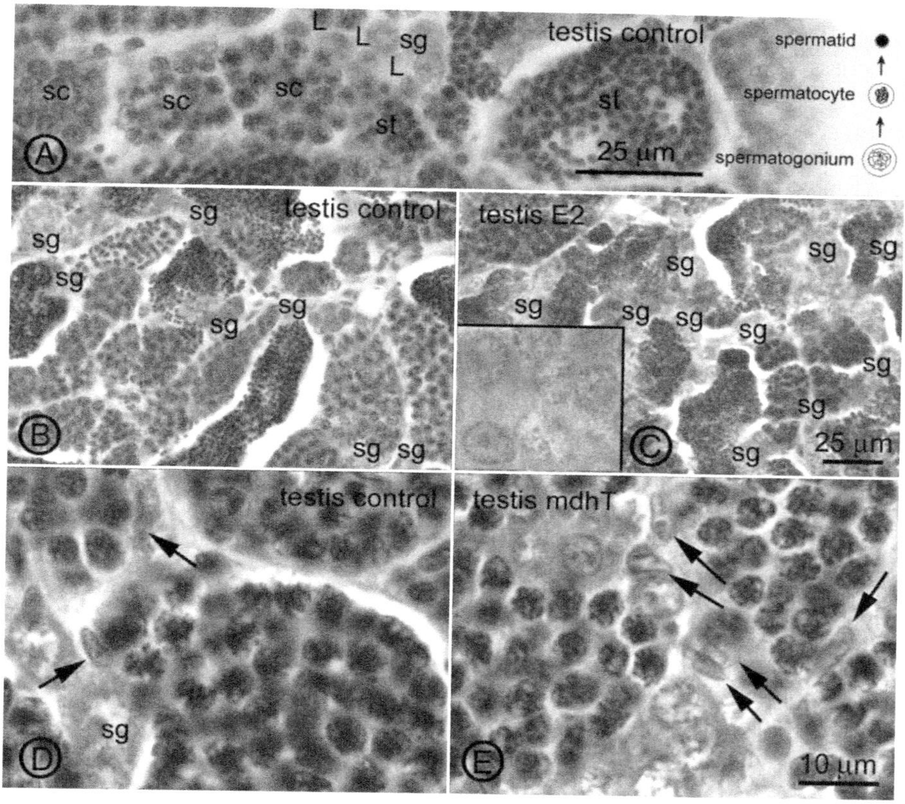

FIGURE 6.7 Medium and high-power magnification images, illustrating normal adult zebrafish testis histology (A, B, D) and detailed effects of endocrine disruption (C, E); H&E staining. Exposure regimes: 1 nM 17β-estradiol (E2; C), 1 μg/L 17α-methyldihydrotestosterone (mdhT; E), sg, spermatogonium; sc, spermatocyte cyst; st; spermatid cyst; L, interstitial Leydig cells. (A) Spermatogonia show increased density after exposure to E_2 (C) compared to control males (B); arrows in (D) and (E) indicate Sertoli cell nuclei, which are more abundant in the mdhT exposed animal (E) than in the control animal (D). Magnification in (B) as in (C), and in (D) and (C); insert as in (E). (From [37], with permission.) (See insert for color representation.)

Until further standards of recognition become available (e.g., the identification of Sertoli cells via routine immunohistochemical staining), the most rational approach may be to utilize descriptive terminology for those cells that cannot be associated with well-defined cell categories by light microscopy. In particular, two types of altered cells that were observed in the fathead minnow study by Wolf et al. [41,74] were termed *vacuolated cells* (VCs) and *apoptotic body cells* (ABCs). The cellular origins of VCs and ABCs were not always obvious during light-microscopic examination; consequently, to justly label such cells, descriptive appellations such as VCs and ABCs were created for cell-counting purposes. The general appearance of many VCs (those with one to several large, clear, cytoplasmic vacuoles that did not displace the nucleus) strongly suggested that they were vacuolated spermatogonia or spermatocytes that had undergone degenerative changes as a consequence of E_2 exposure. It is quite possible that other VCs (those with greatly expanded, pale, foamy cytoplasm) were derived from a different cell type (e.g., activated macrophages or Sertoli cells).

The presence of large vacuoles in Sertoli cells has been attributed to phagocytosis of residual bodies and other cellular elements [83]. Although it could be contended that VCs were artifacts of tissue handling or processing, this notion is difficult to justify, as VCs (and ABCs) were evident in all three fixation–embedding combinations employed in the study by Wolf et al. [41] and in each testis inspected to varying degrees. The existence of basophilic bodies within the cytoplasmic vacuoles of some cells (ABCs) supports the concept that phagocytic cells are present (and probably resident) within the testicular lobule. It is relevant that among chordata, a unifying function of Sertoli cells is the phagocytosis of effete spermatozoa and spermatogenesis byproducts [75,83,84], and that degenerated and presumptively apoptotic sperm have been observed ultrastructurally within the Sertoli cells of fathead minnows and other fishes [43,55,75]. The question, however, remains whether Sertoli cell hypertrophy and hyperplasia are lesions specifically associated with endocrine-disrupting compounds or whether these are secondary effects.

Some light may be shed on this question from the most recent long-term study, in which fathead minnows were exposed to bisphenol A (BPA) at concentrations ranging between 0 and 640 μg/L for 164 days [23]. This study quantitatively assessed the number of each spermatogenic cell type as well as Sertoli cells, Leydig cells, ABCs interstitial tissue and unknown cells (ITUCs) as well as VCs in the testes of control and BPA exposed male fathead minnows. While no effects of BPA were observed on Sertoli cells ABCs, VCs, and ITUCs, Leydig cell numbers were decreased in testes of fish exposed to 160 and 640 μg/L. Given that the ABCs and VCs observed in this study represent Sertoli cells (see above), BPA treatment affected neither the Sertoli cell spermatocyst structural or phagocytic function (see Section 5.1.5.1). In contrast, Sertoli hypertrophy and associated phagocytosis were observed in testes of guppy and other small fish, including goldfish (*Carassius auratus*), treated with methyl mercury or cadmium [32,85,86].

The increased number of phagocytosed sperm cell remnants in methyl mercury–treated guppy was interpreted as being the result of an abortive spermatogenesis [32]. Consequently, any compound, with or without endocrine-disruptive capacity (see Table 6.5) inducing an abrogated spermatogenesis (e.g., EE_2 in zebrafish [54] and guppy [87] or E_2 in fathead minnows [55], guppy, or platyfish [31,71]) could also provide for increased-severity grades of Sertoli cell hypertrophy and possibly hyperplasia, given that Sertoli cell hypertrophy can be causally related to increased Sertoli cell phagocytic activity. The Sertoli cell hyperplasia within this context could also be interpreted as a cell-proliferative response of the testes to counter enhanced sperm degeneration and inflammation within the testes with a higher phagocytic capacity. Sertoli cell proliferation, however, may also be seen as a direct proliferative effect of EDCs on Sertoli cells. Unfortunately, most of the observations above remain at the descriptive pathology level and therefore preclude more intimate understanding of the processes involved in Sertoli cell hypertrophy and hyperplasia.

FIGURE 2.1

FIGURE 2.3

FIGURE 2.5

FIGURE 2.6

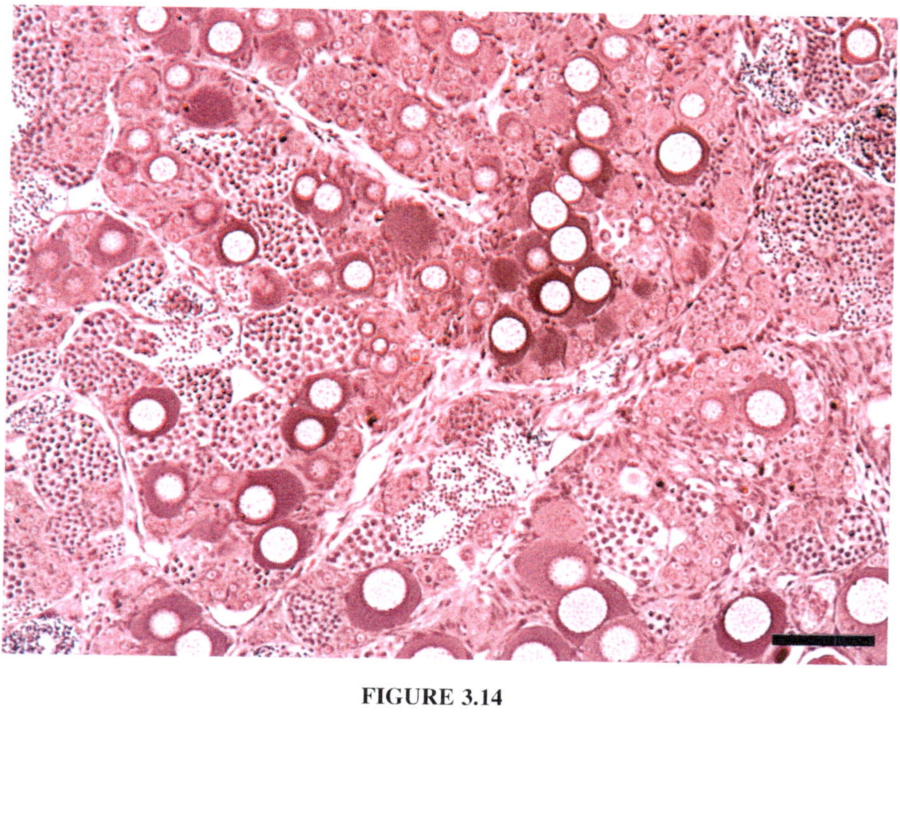

FIGURE 3.14

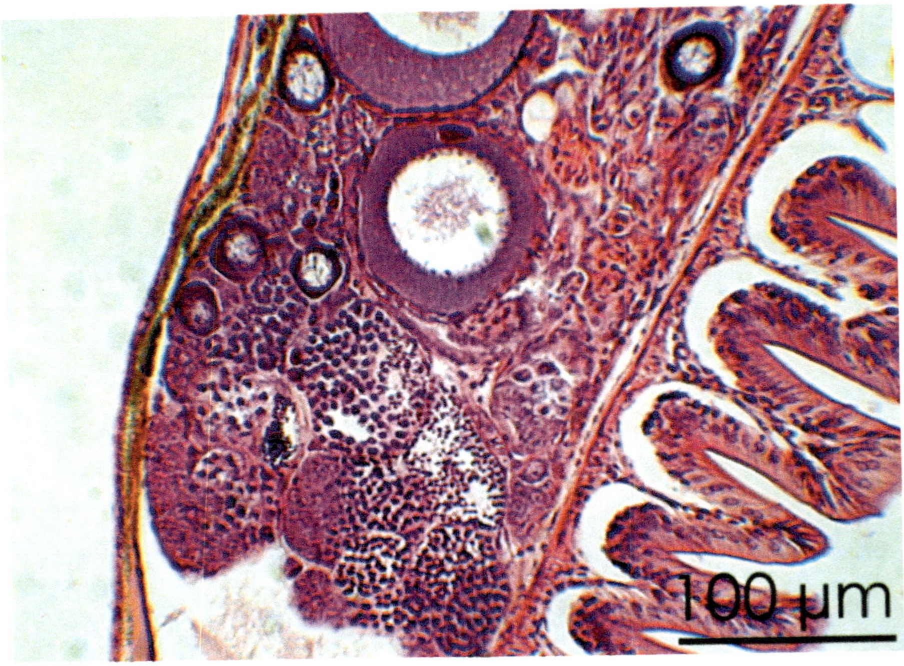

FIGURE 3.17

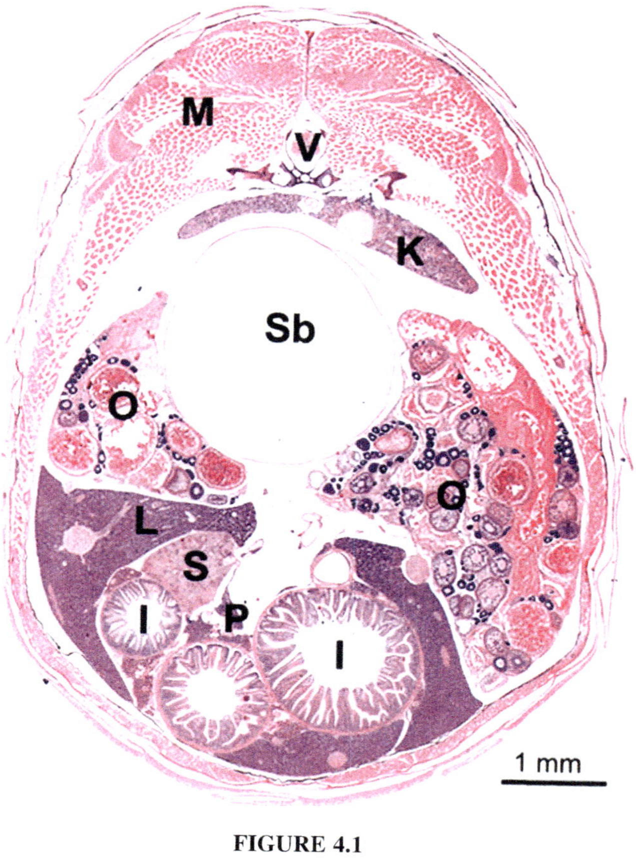

FIGURE 4.1

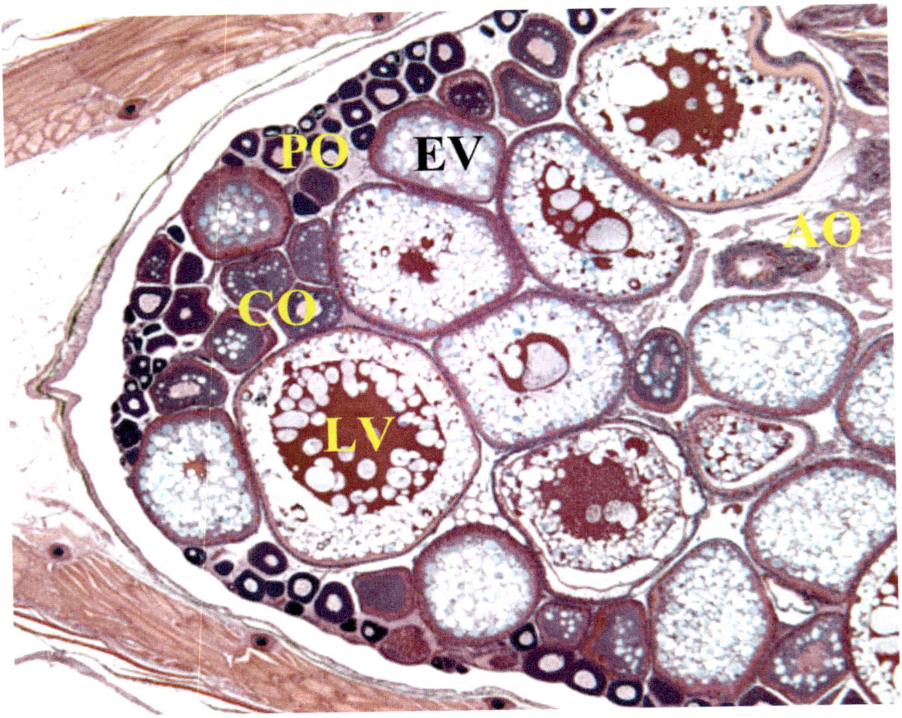

FIGURE 4.4

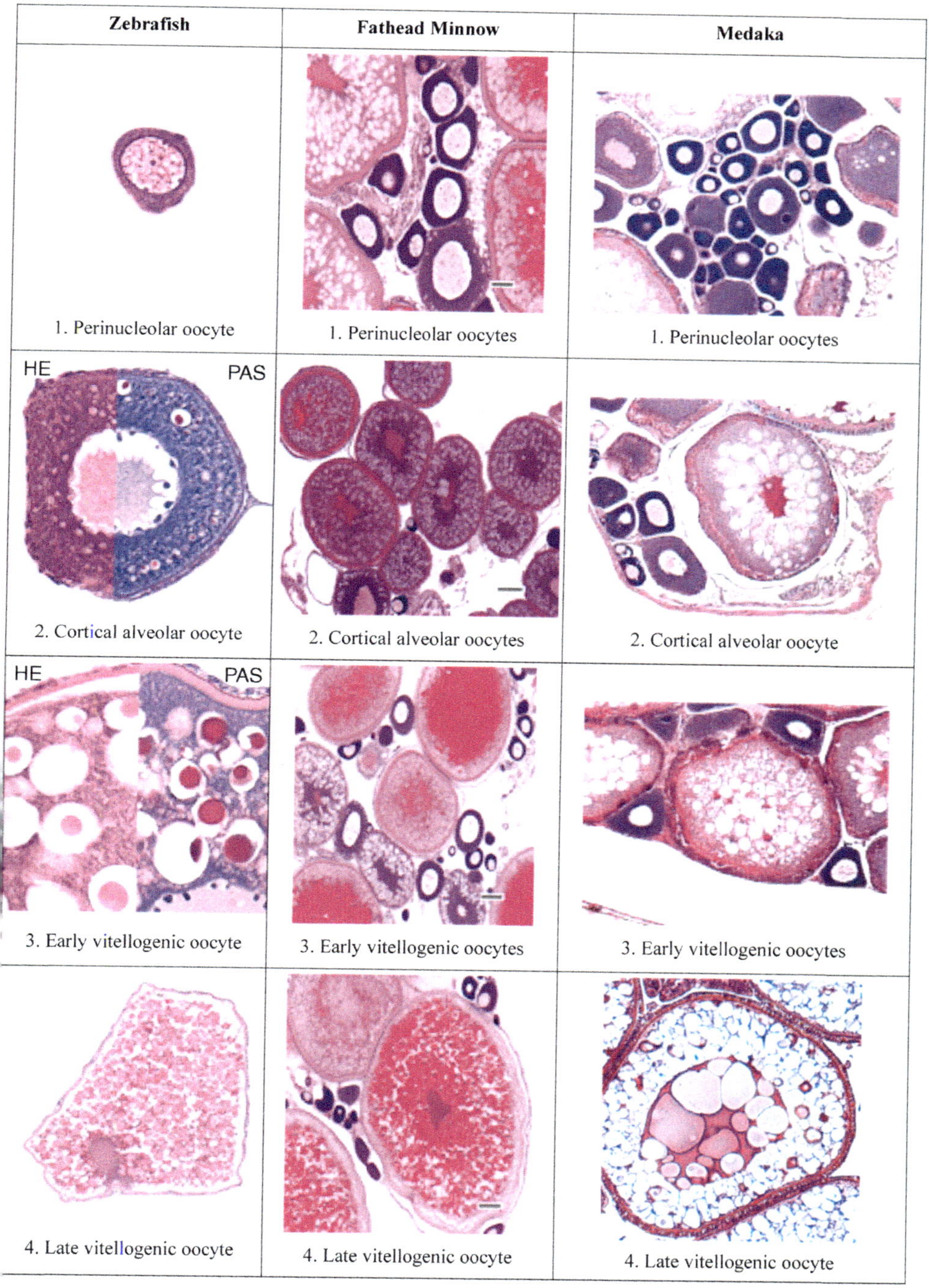
Zebrafish
Fathead Minnow
Medaka
1. Perinucleolar oocyte
1. Perinucleolar oocytes
1. Perinucleolar oocytes
HE
PAS
2. Cortical alveolar oocyte
2. Cortical alveolar oocytes
2. Cortical alveolar oocyte
HE
PAS
3. Early vitellogenic oocyte
3. Early vitellogenic oocytes
3. Early vitellogenic oocytes
4. Late vitellogenic oocyte
4. Late vitellogenic oocyte
4. Late vitellogenic oocyte

FIGURE 4.5

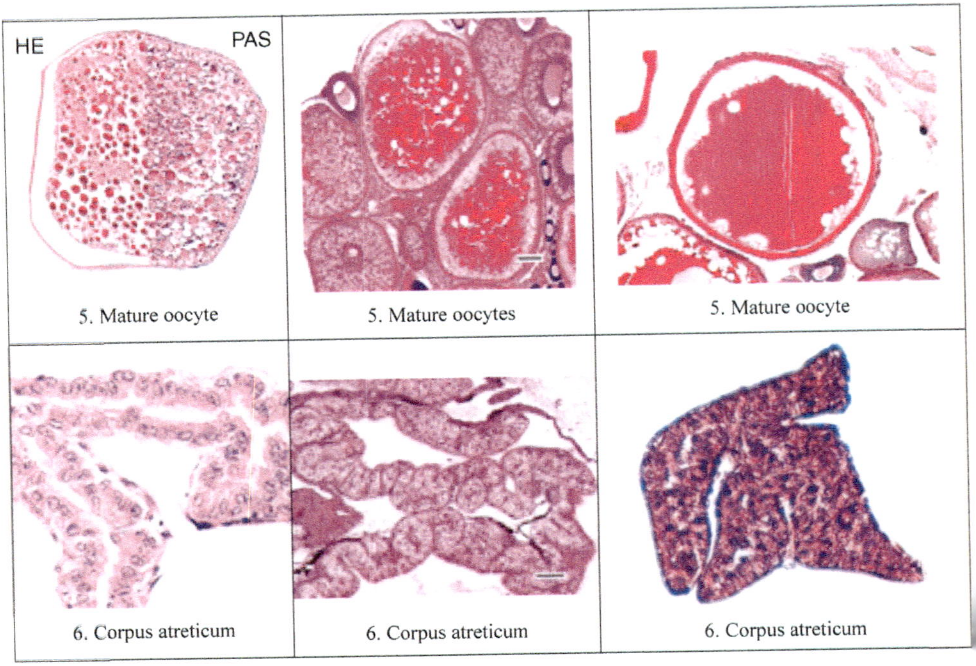

FIGURE 4.5 (*Continued*)

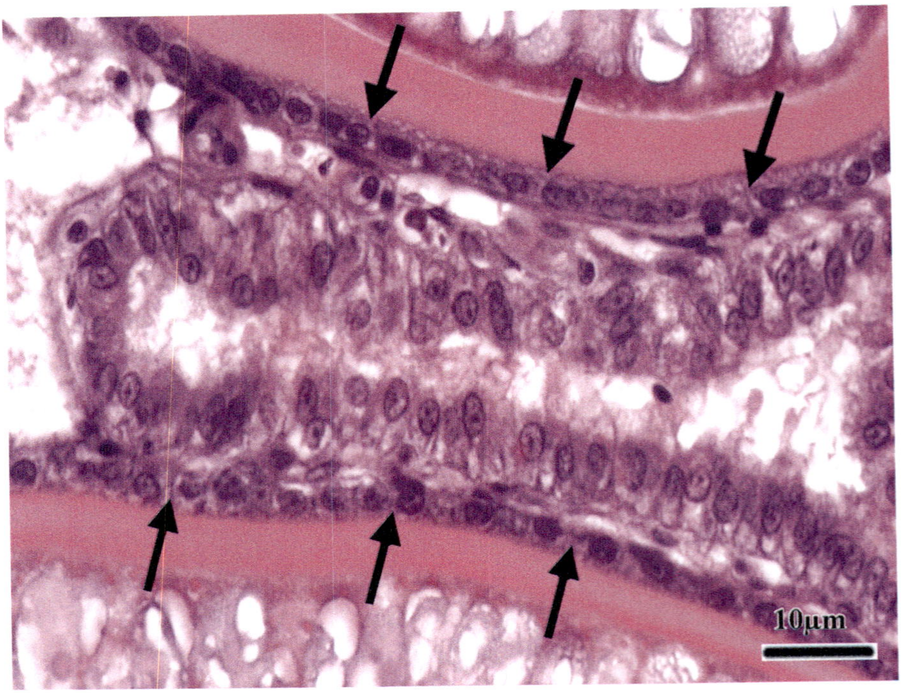

FIGURE 4.7

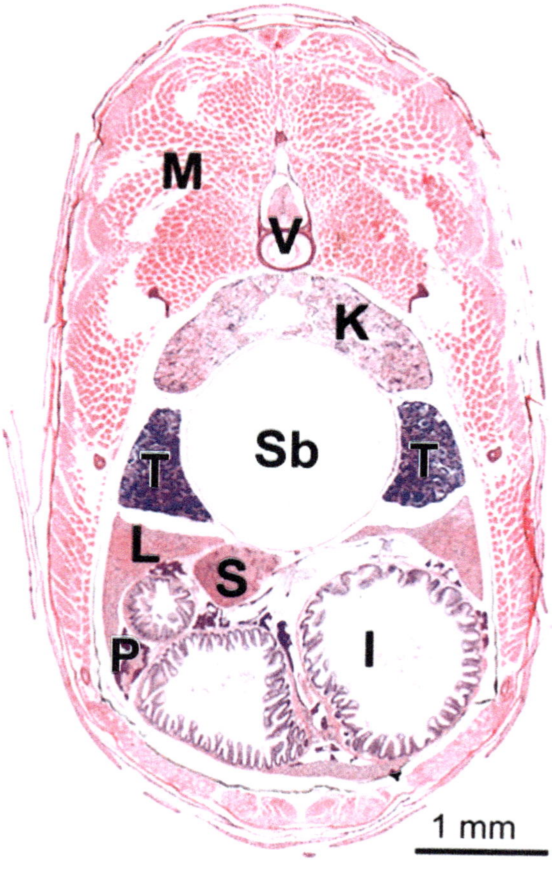

FIGURE 5.1

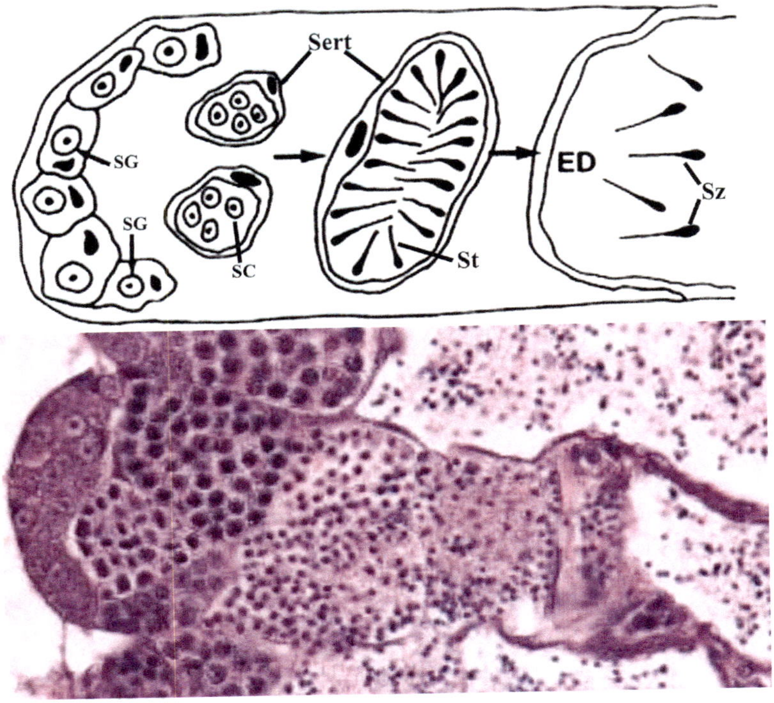

FIGURE 5.4

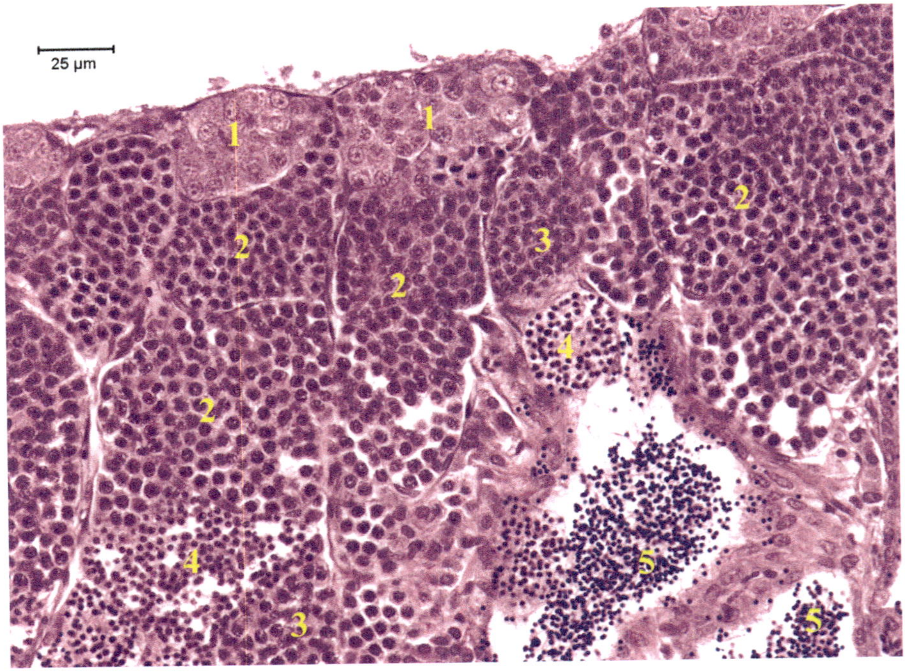

FIGURE 5.7

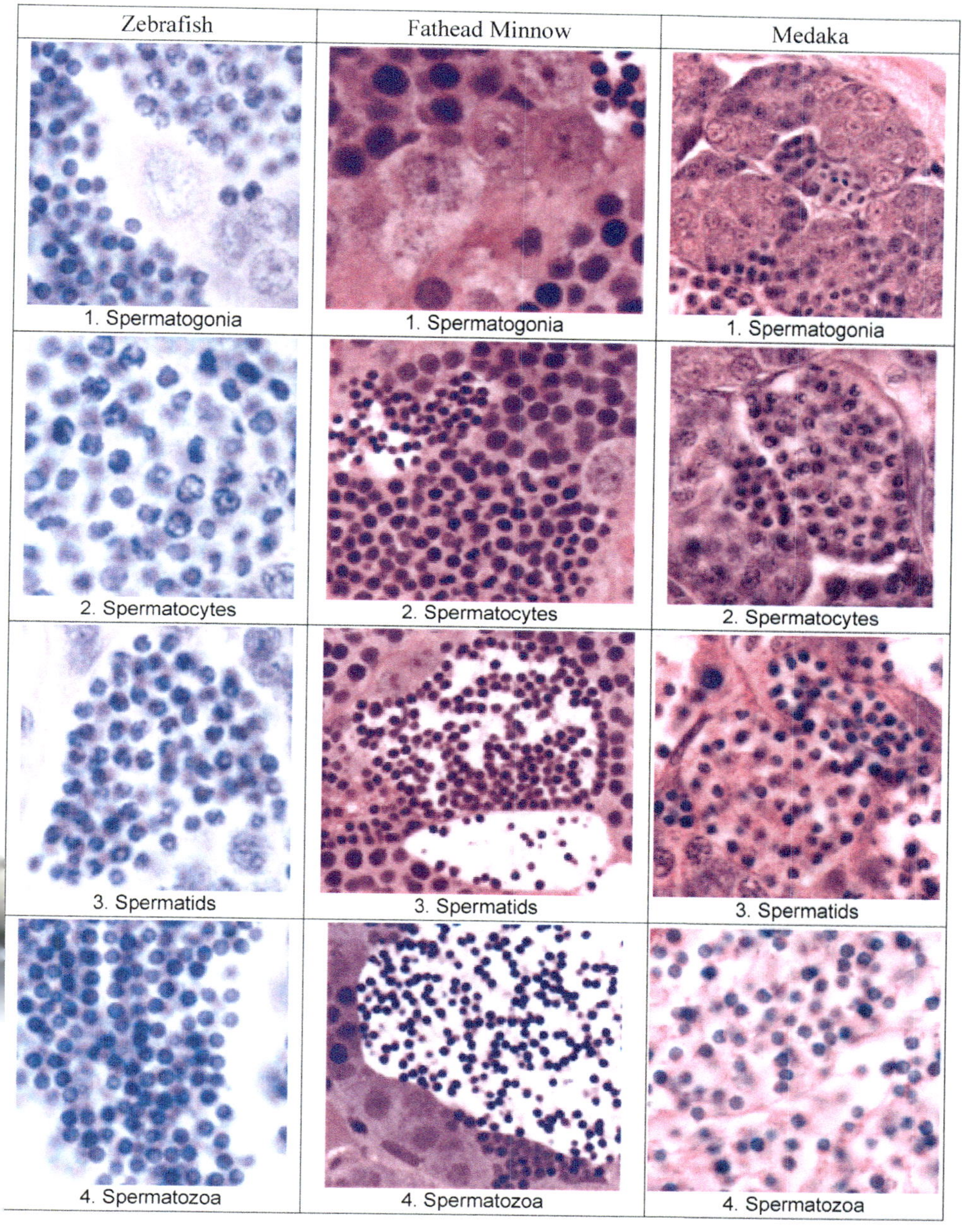
Zebrafish
Fathead Minnow
Medaka
1. Spermatogonia
1. Spermatogonia
1. Spermatogonia
2. Spermatocytes
2. Spermatocytes
2. Spermatocytes
3. Spermatids
3. Spermatids
3. Spermatids
4. Spermatozoa
4. Spermatozoa
4. Spermatozoa

FIGURE 5.9

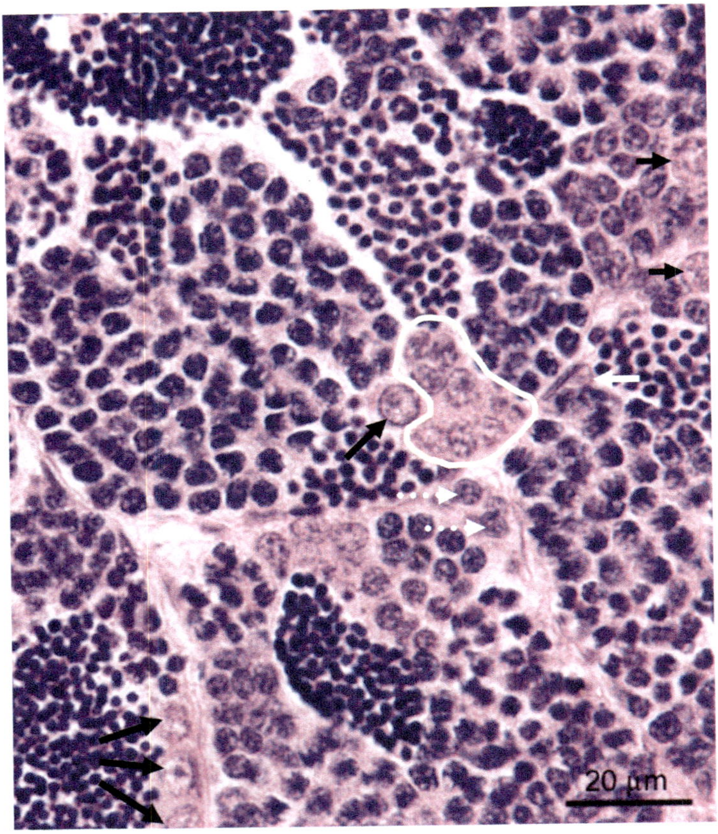

FIGURE 5.10

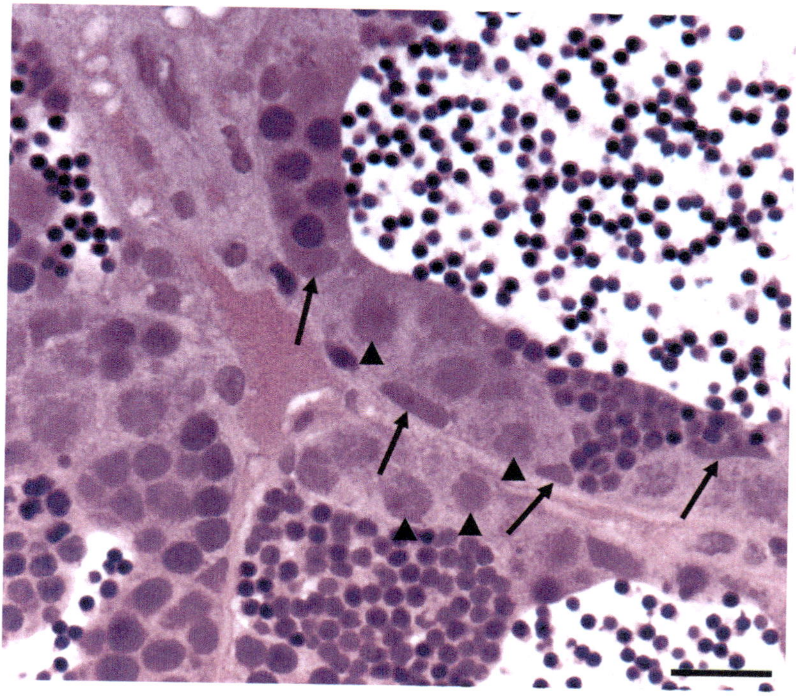

FIGURE 5.11

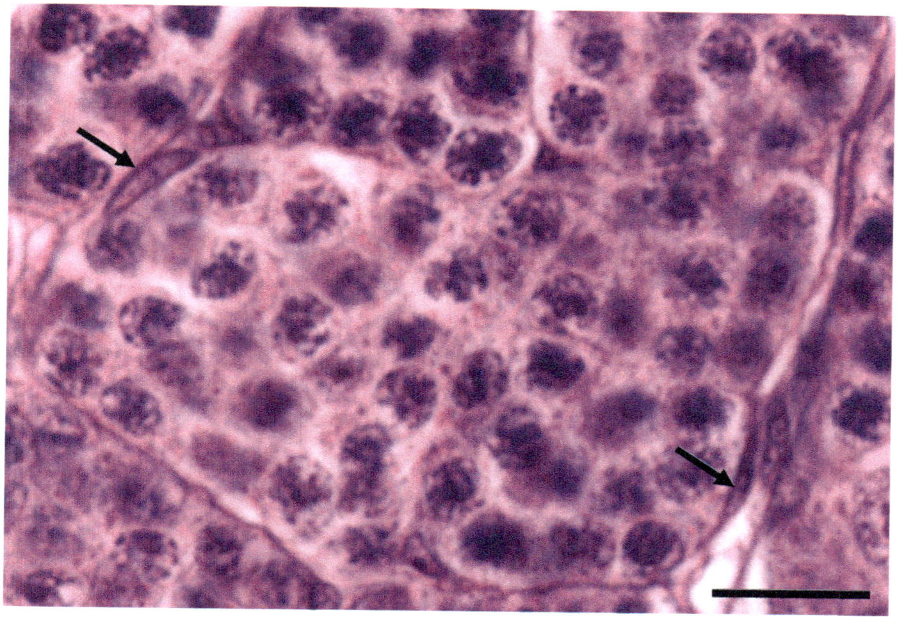

FIGURE 5.12

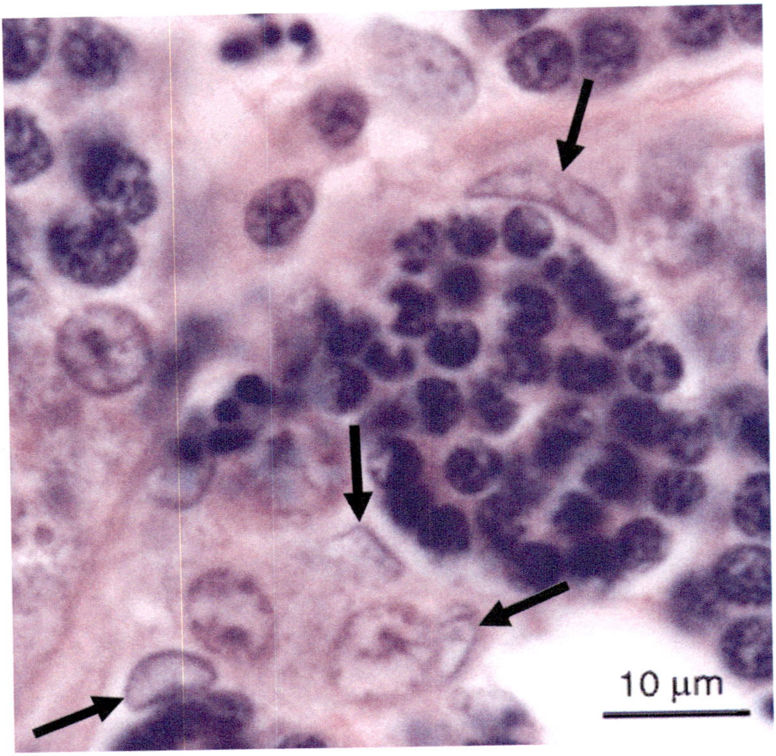

FIGURE 5.13

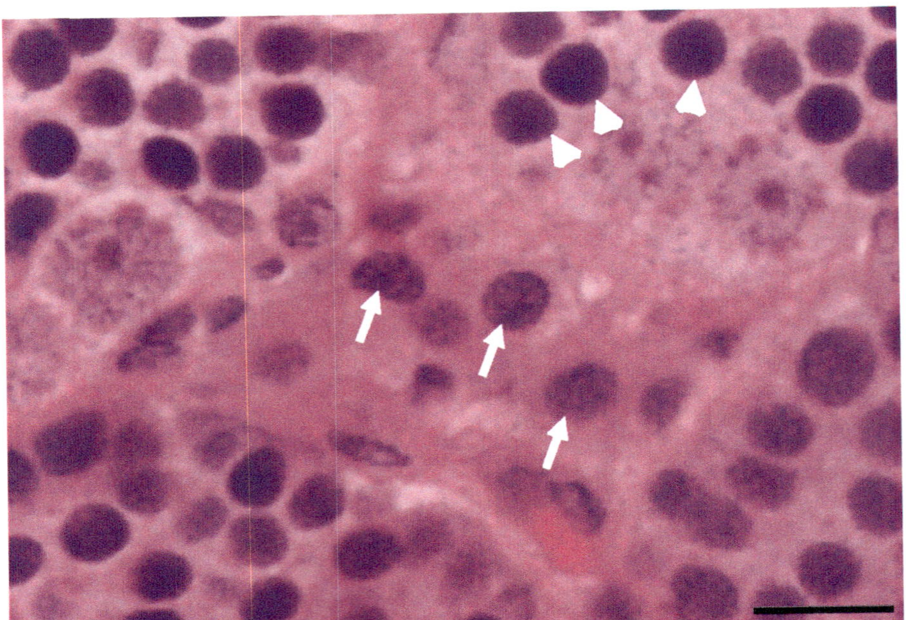

FIGURE 5.15

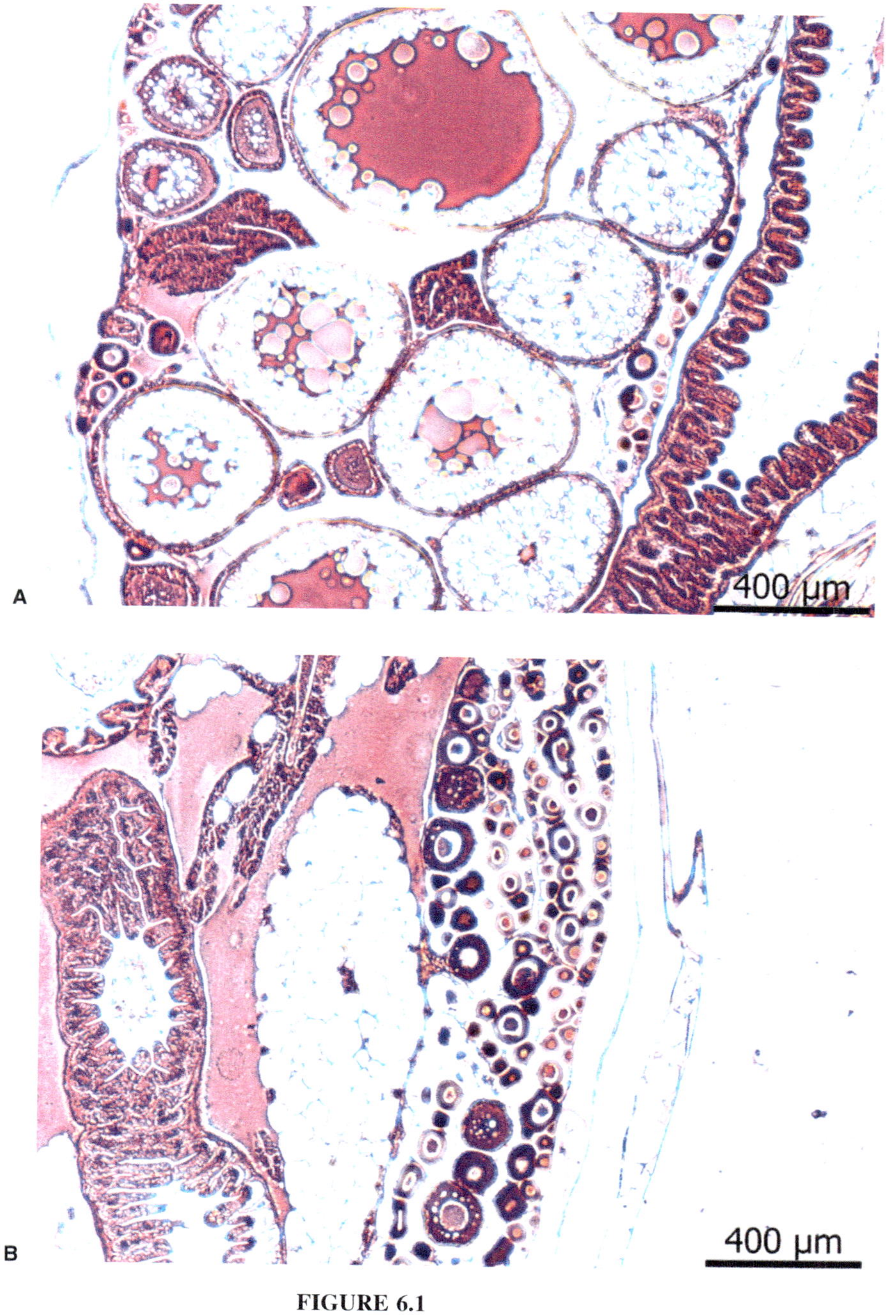

FIGURE 6.1

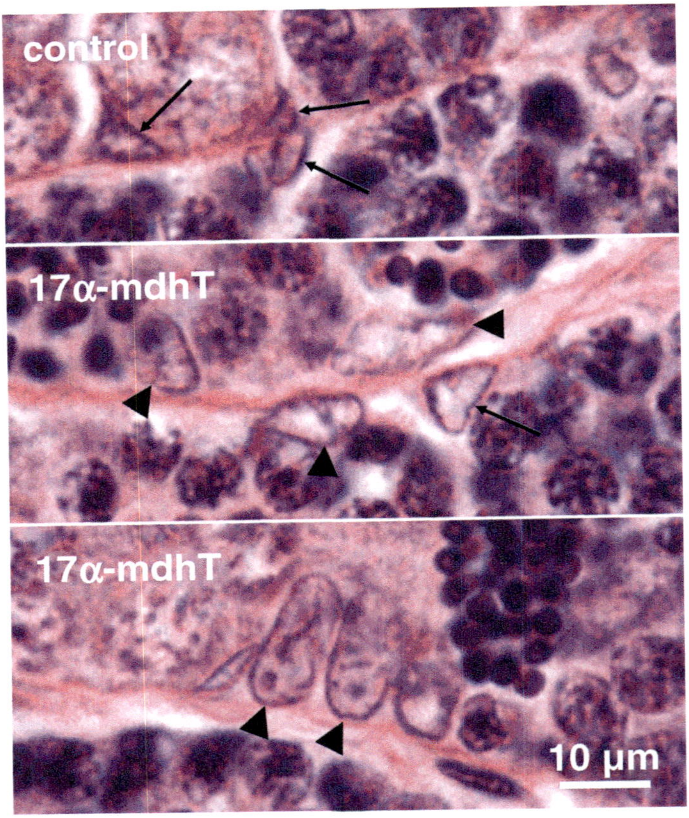

FIGURE 6.4

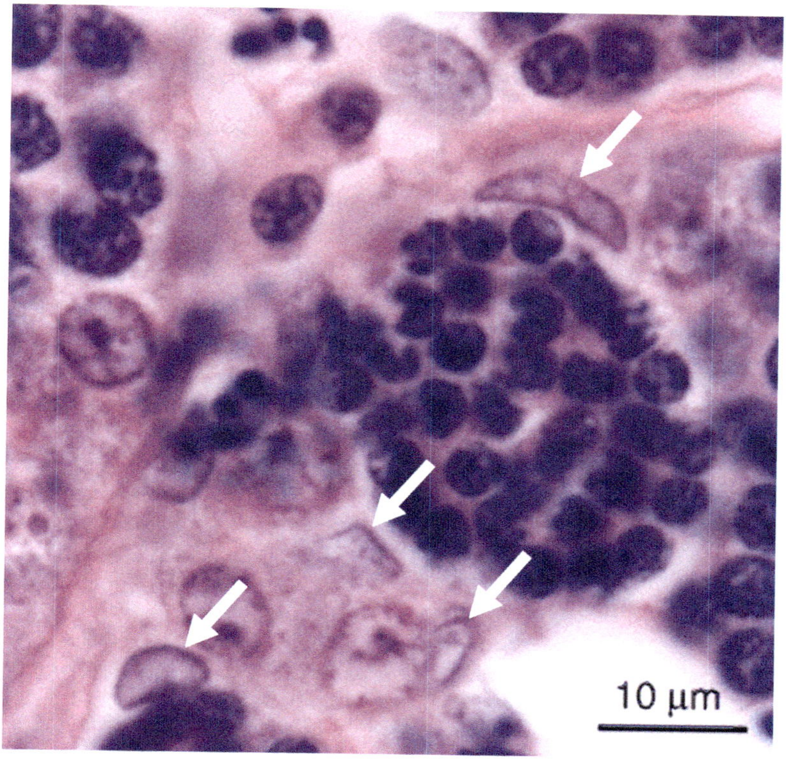

FIGURE 6.6

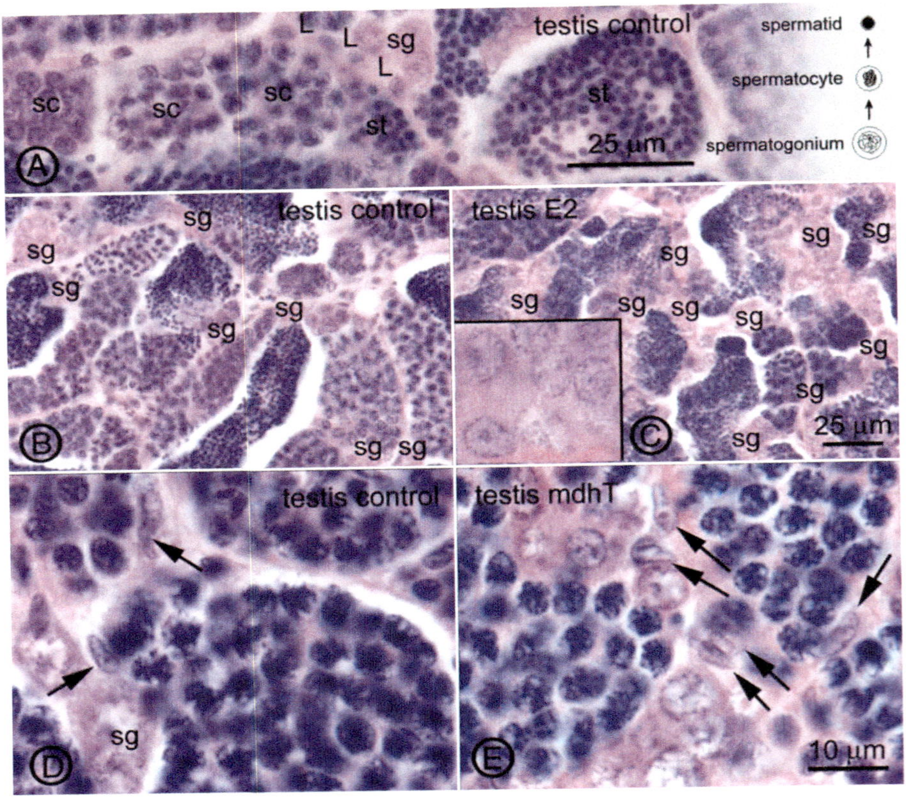

testis control
spermatid
spermatocyte
spermatogonium
sc
sc
sc
L
L
sg
L
st
st
25 µm
A
testis control
sg
sg
sg
sg
sg
sg
sg
B
testis E2
sg
sg
sg
sg
sg
sg
sg
sg
sg
sg
sg
25 µm
C
testis control
sg
D
testis mdhT
10 µm
E

FIGURE 6.7

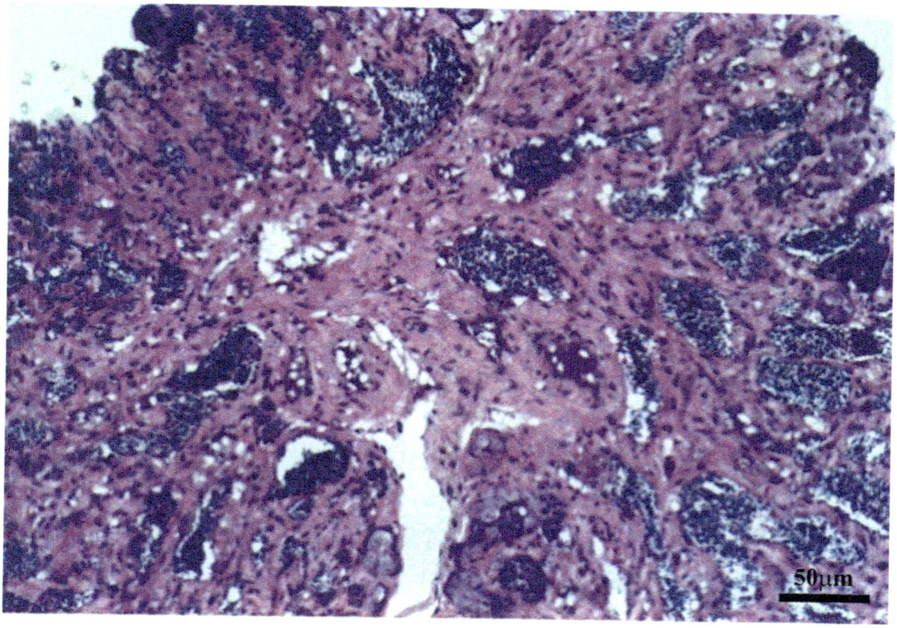

FIGURE 6.8

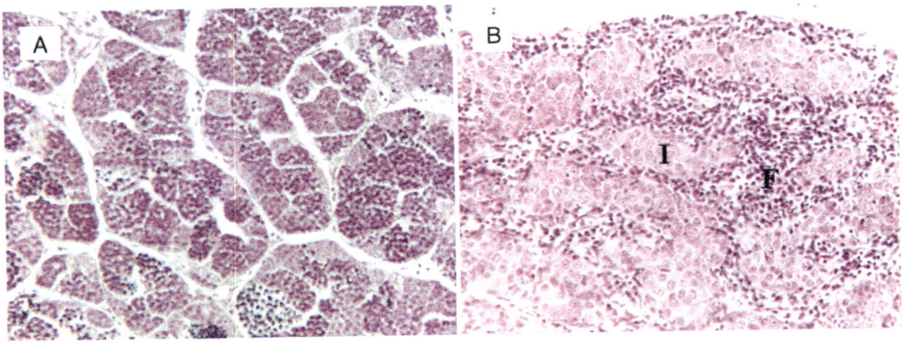

FIGURE 6.9

FIGURE 7.1

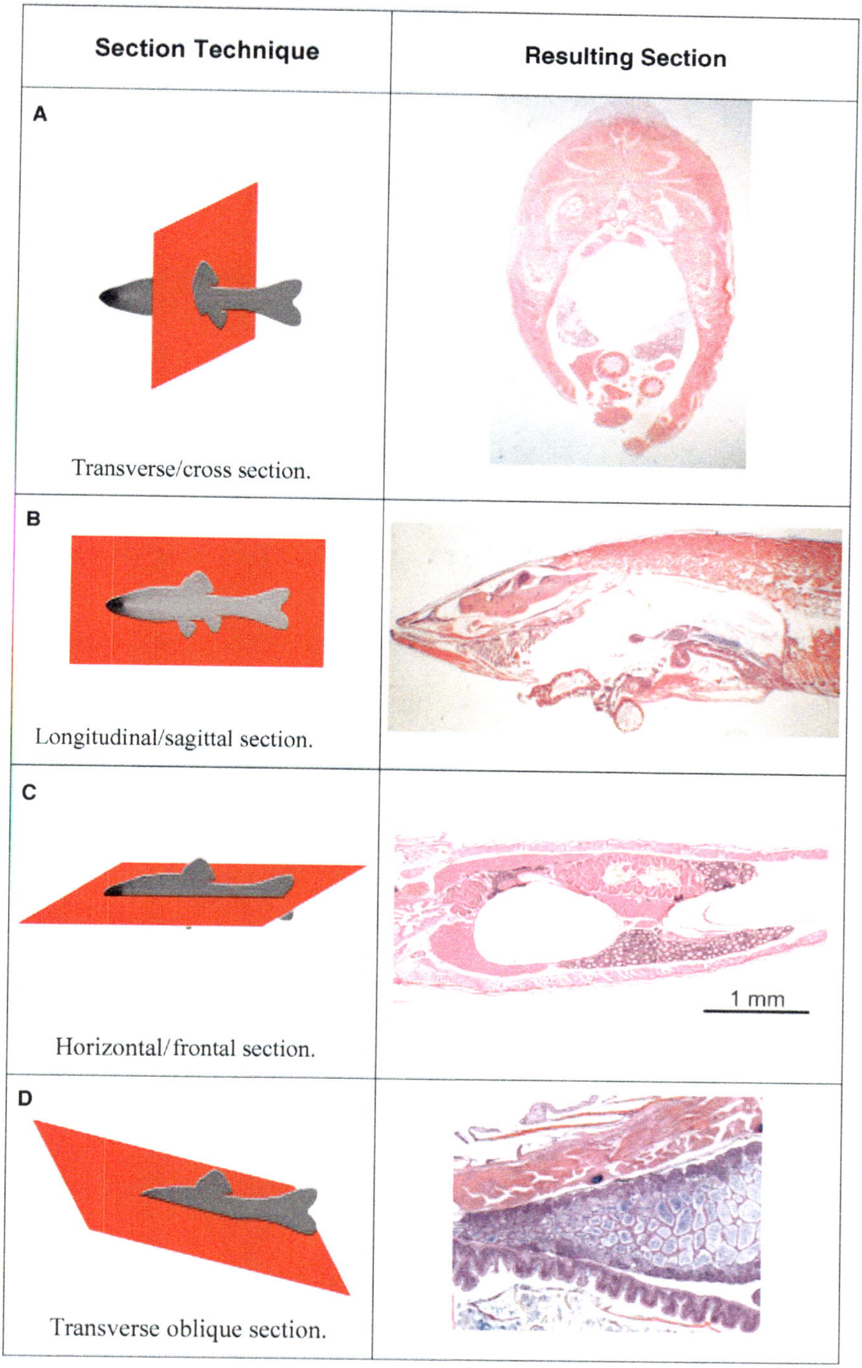
Section Technique
Resulting Section
A
Transverse/cross section.
B
Longitudinal/sagittal section.
C
Horizontal/frontal section.
1 mm
D
Transverse oblique section.

FIGURE 7.2

Section Technique	Resulting Section
A 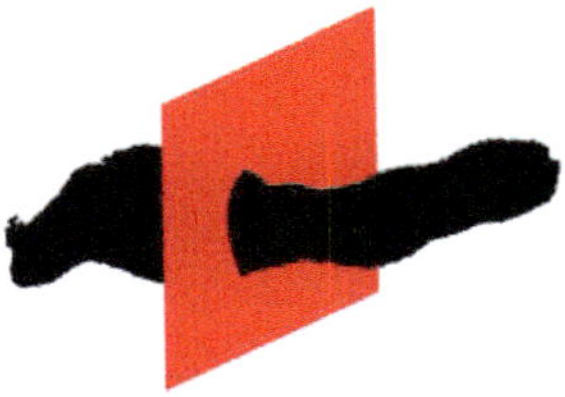 Transverse/cross section.	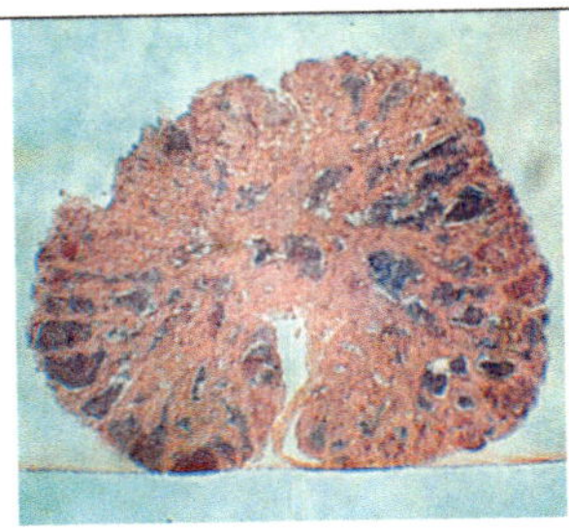
B 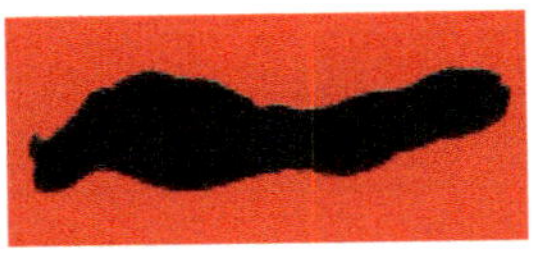 Longitudinal/horizontal section.	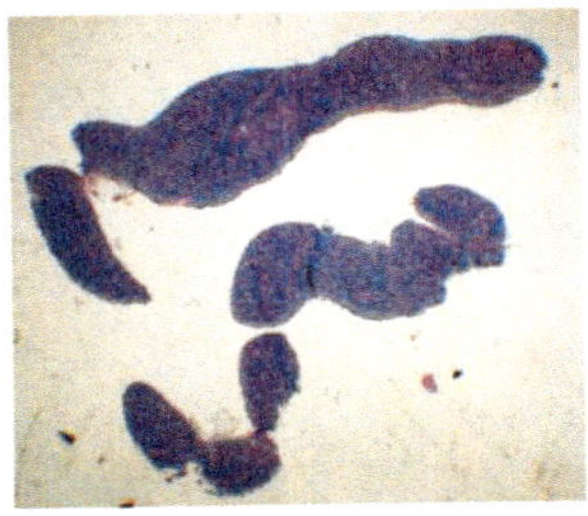

FIGURE 7.3

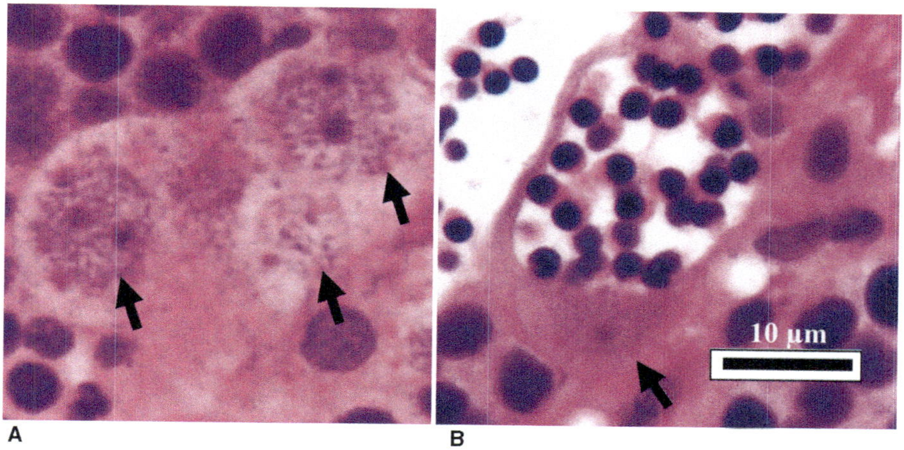

FIGURE 8.1

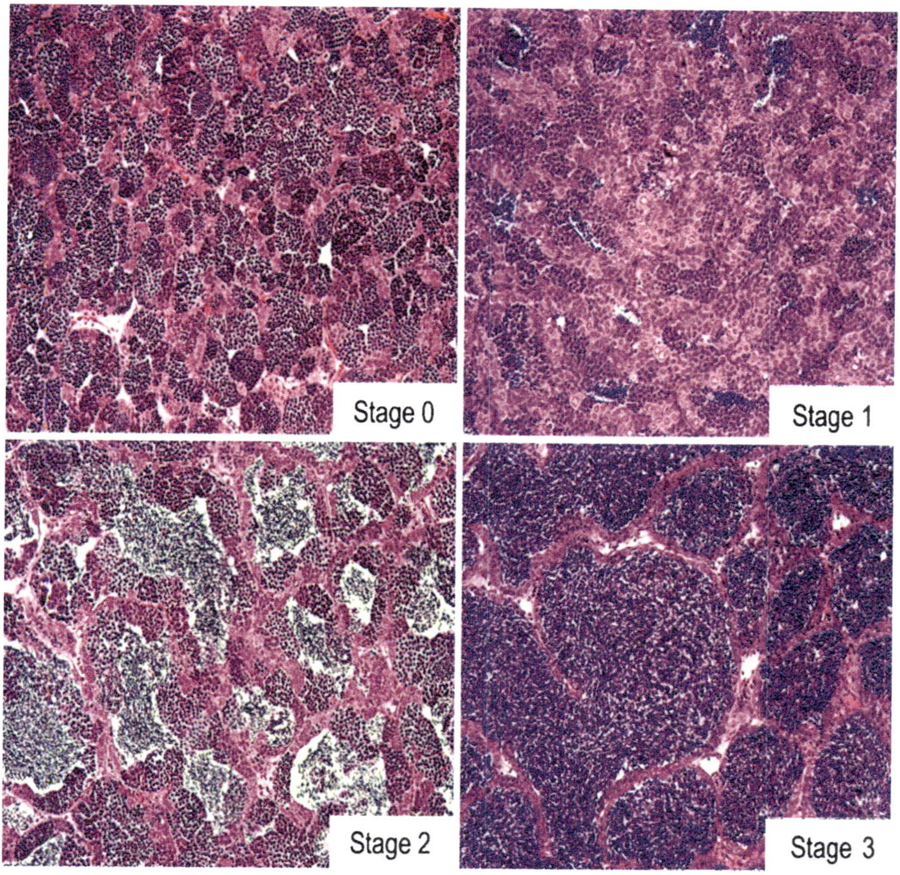

FIGURE 8.2

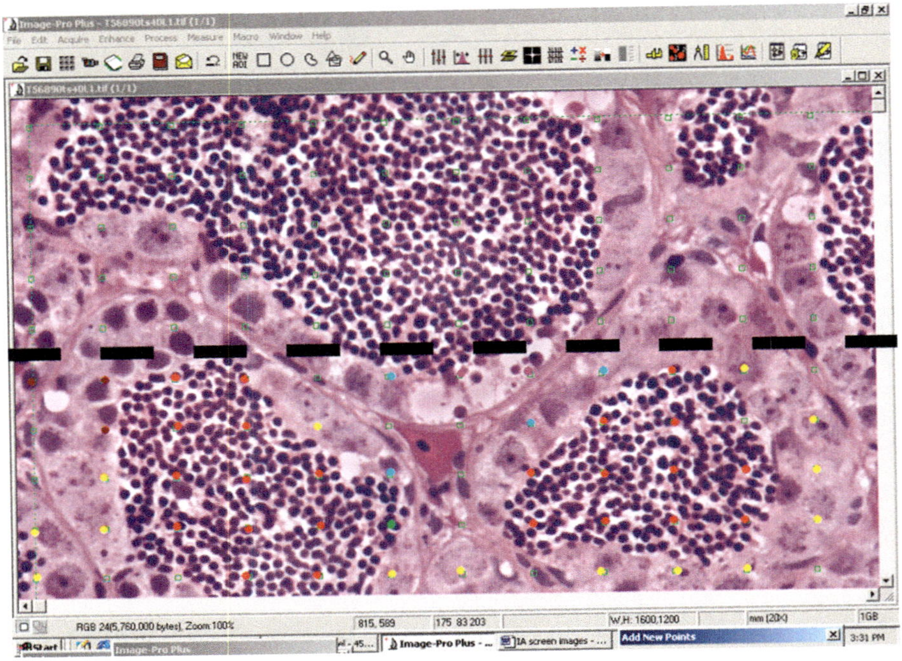

FIGURE 8.3

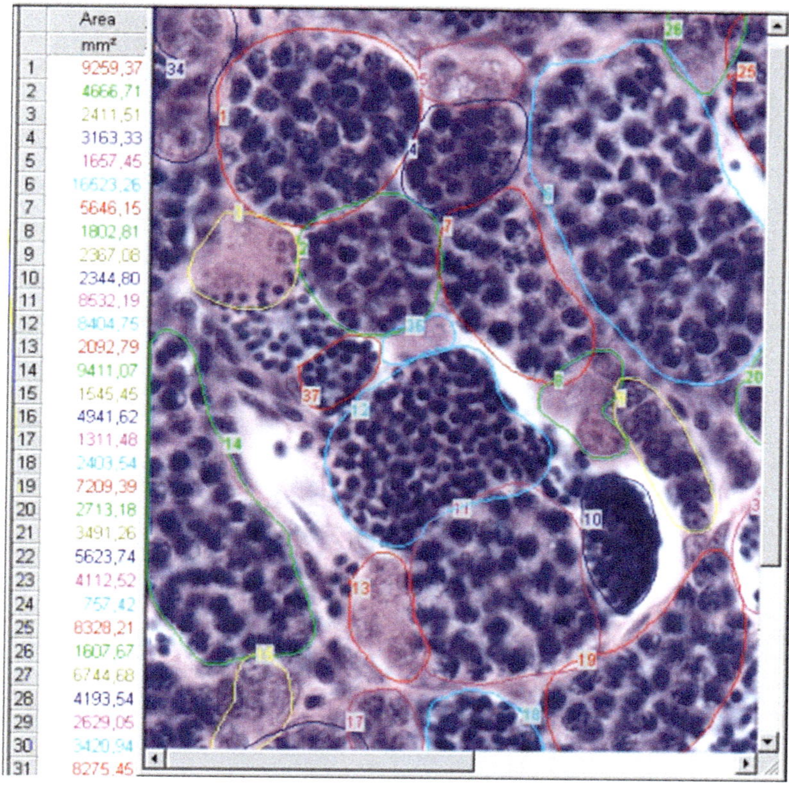
Area
mm²
1 9259,37
2 4666,71
3 2411,51
4 3163,33
5 1657,45
6 16623,26
7 5646,15
8 1802,81
9 2367,08
10 2344,80
11 8532,19
12 8404,75
13 2092,79
14 9411,07
15 1545,45
16 4941,62
17 1311,48
18 2403,54
19 7209,39
20 2713,18
21 3491,26
22 5623,74
23 4112,52
24 757,42
25 8328,21
26 1607,67
27 6744,68
28 4193,54
29 2629,05
30 3420,94
31 8275,45

FIGURE 8.4

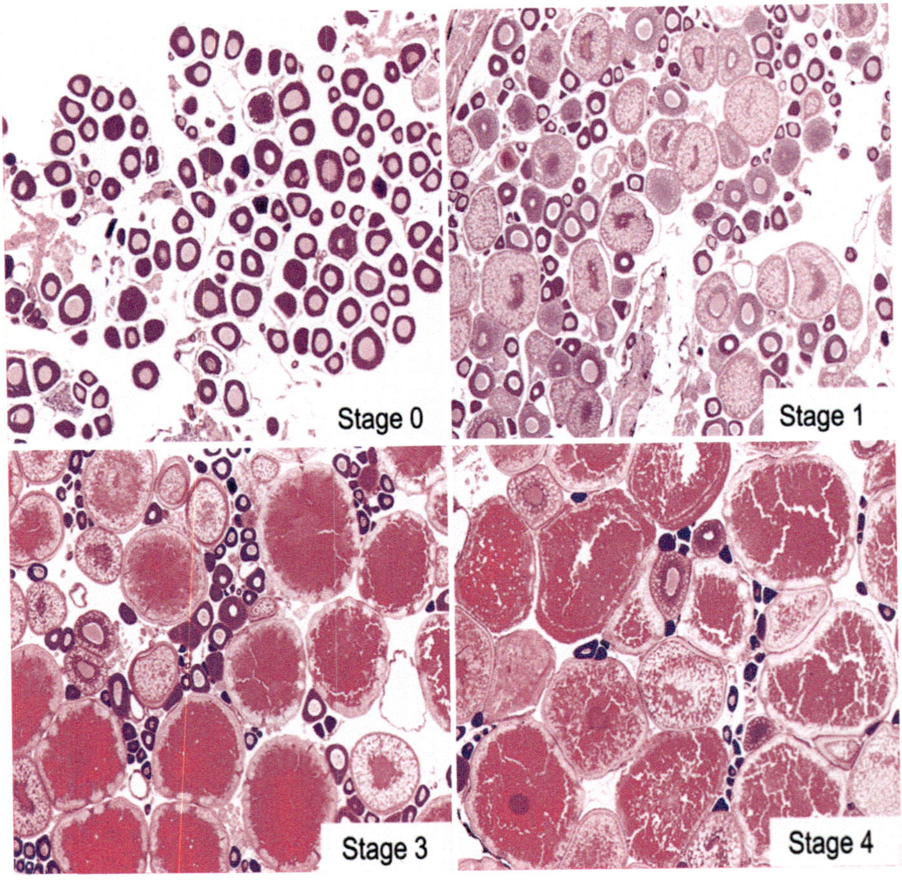

FIGURE 8.5

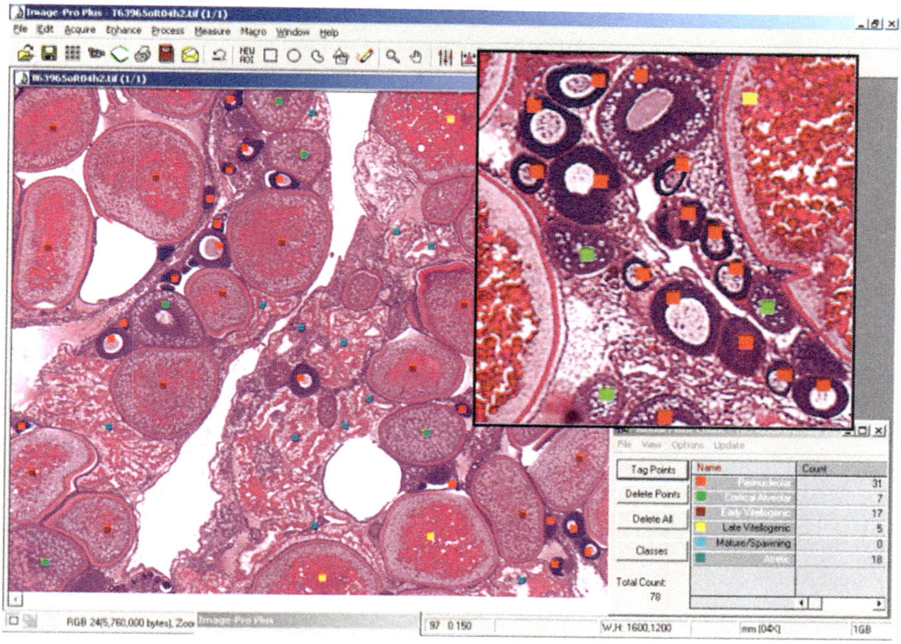
Image-Pro Plus
File
Edit
Acquire
Enhance
Process
Measure
Macro
Window
Help
File
View
Options
Update
Tag Points
Delete Points
Delete All
Classes
Name
Count
31
7
17
Late Vitellogenic
5
Mature/Spawning
0
18
Total Count
78
W,H: 1600,1200
1GB

FIGURE 8.6

APPENDIX: Microdissection of Organs

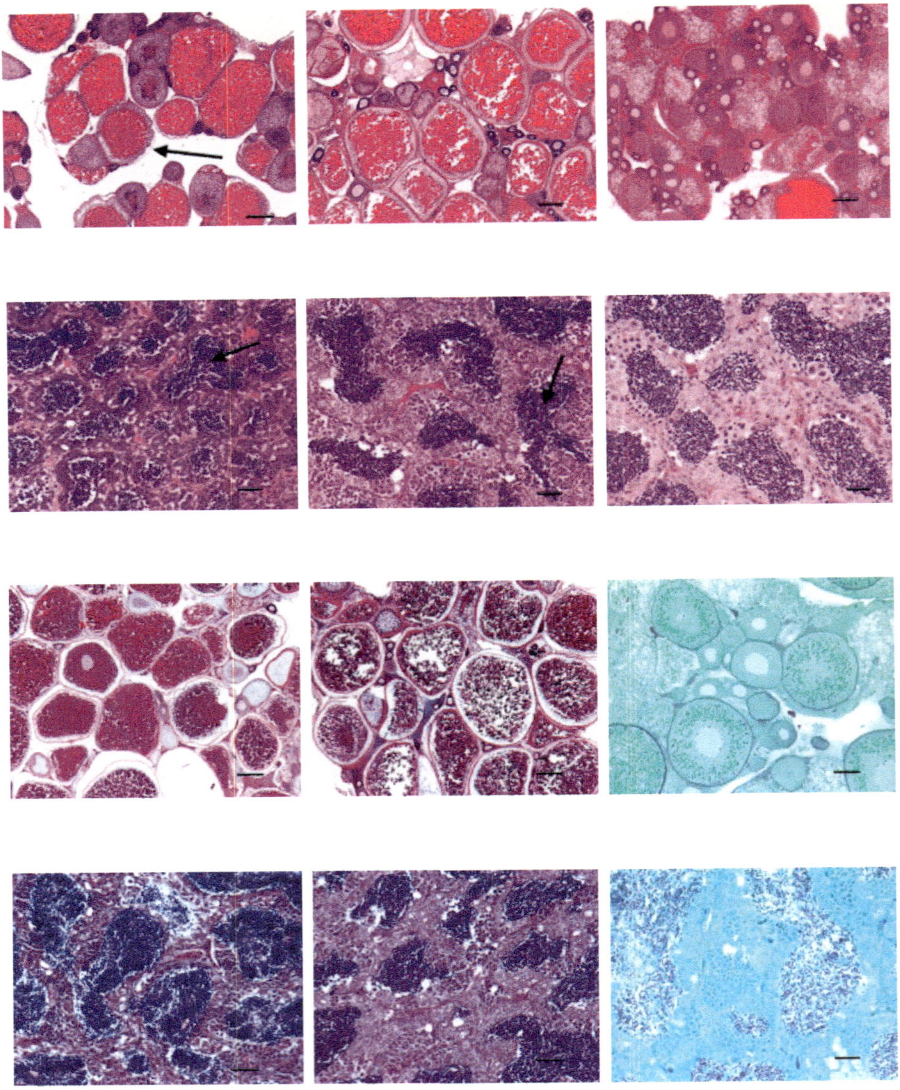

APPENDIX: Examples of Fish Preparation and Staining

However, a comparison of the more detailed investigated EDC effects observed in Leydig cells (see below) with those observed in Sertoli cells suggests that at least some of the EDC-induced hypertrophic and hyperplastic Sertoli cell responses observed may stem from abrogated EDC-mediated signalling between Sertoli cells [73] and gametogenic cells within spermatocysts and Leydig cells.

6.1.5 Leydig Cells Hypertrophy and Hyperplasia

Commensurate with the function as a steroidogenic cell (see Section 5.2.5.2) [88–95], 3β-HSD has been demonstrated histochemically in the Leydig (interstitial) cells of the testis of numerous teleosts [59]. Exposure of male fathead minnows to ketoconazole, known to inhibit the cyp P450-mediated step in ergosterol biosynthesis, caused a marked proliferation of Leydig cells [96]. In addition, mRNA transcripts for two key steroidogenic enzymes, cytochrome P450 side-chain cleavage (CYP11A) and cytochrome P450 c17α hydroxylase/17,20-lyase (CYP17), were elevated. The changes in transcript levels and proliferation of Leydig cells both represent potential adaptive or compensatory responses to impaired steroidogenic capacity. This interpretation would also be supported by the Leydig cell hyperplasia observed in Nile tilapia (*Oreochromis niloticus*) exposed to high dietary doses of Aroclor 1254 for 40 days [97], whereby Aroclor 1254 is assumed to induce a general adaptive response not restricted to Leydig cells by virtue of its capacity to increase the expression of an array of genes involved in steroidogenesis and metabolism (CYPs) in most organs, as shown for example, by the significantly lowered thyroxine plasma concentration. Similarly, medaka exposed from hatch until sexual maturity to 10 ng/L EE_2 [98] presented with a sixfold increased rate of Leydig and Sertoli cell apoptosis as well as of apoptotic spermatocytes and spermatids, suggesting that the high concentration of the estrogenic EE_2 resulted in a repressed androgen production in Leydig cells and therefore abrogated maintenance of the sufficiently high levels of 11-ketotestosterone required for Leydig and Sertoli cell homeostasis as well as spermatogenesis.

6.1.6 Fibrosis

Fibrosis and abnormal connective tissue in testes and ovaries following exposure to natural and synthetic hormones as well as endocrine disruptors have been reported in several fish species, including zebrafish, fathead minnows, medaka, carp (*Carassius auratus*), sheepshead minnow (*Cyprinodon variegates*), and sea bass (*Dicentrarchus labrax*) (see Tables 6.3 and 6.9). Generally, there are fewer reports of fibrosis in female than in male fish. The reason for this may be that at least in the earlier EDC research, female fish were investigated less frequently or as with many other previous pathological denominations or lesions, clear morphological descriptions and criteria of what was meant by the term *fibrosis*, abnormal connective tissue, connective tissue proliferation, or proliferation of interstitial stroma [99], are missing in most publications, making the comparison and understanding of the etiology and thus the mechanism(s) underlying these lesions difficult. Morphologically, fibrosis can be described as an increased proliferation of fibroblasts followed by an increased deposition of collagen, resulting in a thickened appearance of the connective tissue. Collagen can be visualized histochemically (e.g., by employing Snook's silver stain). Biochemically, fibrosis can be defined as an increased amount of collagen, which can be estimated from the hydroxyproline content of the organ of interest or differences in relative concentration of the various collagen types [100].

In some cases, fibrosis in fish is described in conjunction with inflammation and/or degenerative lesions (e.g., degeneration of germ cells; see Section 6.1.2). Within this context, fibrosis can be assumed to be a scarring (wound-healing) reaction aimed at maintaining the overall structure of the organ, whereby tissue injury (necrosis) is followed by an infiltration of inflammatory cells (macrophages and occasional neutrophils), that phagocytose and lyse the necrotic cellular debris (see Section 6.2.1.6 and 6.2.2.6), followed by the recruitment of fibroblasts by inflammatory cells, fibroblast proliferation, and collagen deposition. Thus, fibrosis would be secondary to a primary effect (e.g., necrosis) in the tissue following chemical exposure, whether the tissue affected is the gonad or another organ. Indeed, in studies with high doses of 17β-estradiol (E_2) (Figure 6.8), 17α-ethinylestradiol (EE_2), cyproterone acetate, vinclozolin, octylphenol, or nonylphenol, fibrosis of the gonads and other organs was observed concurrently with inflammation or cell necrosis and degeneration [4,26,27,29,99,101]. This would imply that overt cell toxicity and subsequent inflammation could provide sufficient tissue damage to promote regenerative repair through fibrosis.

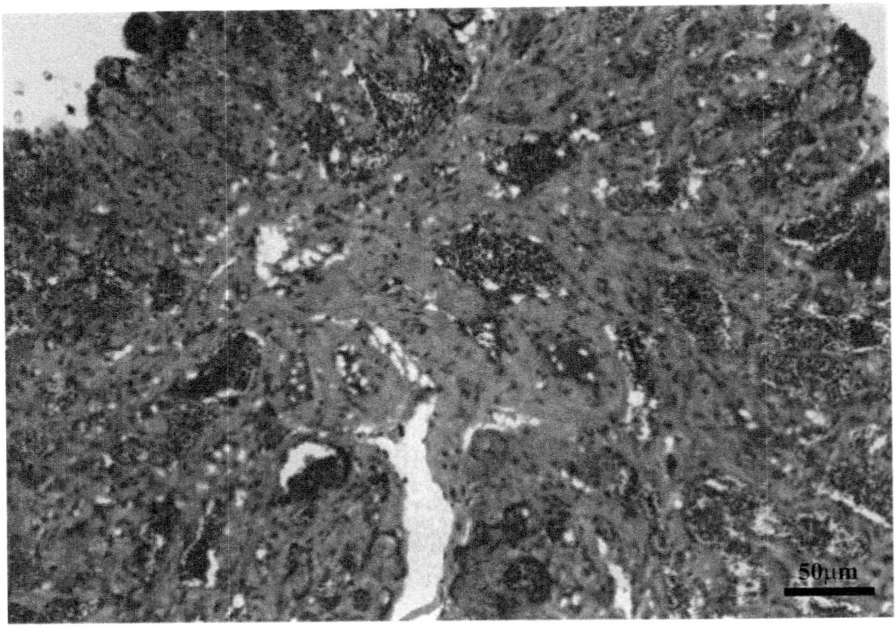

FIGURE 6.8 Abnormal testicular conditions of male medaka following 17β-estradiol (E_2) treatment. Development of connective tissue in the testis of a medaka exposed to 463 ng/L E_2 for 25 days. The arrow indicates connective tissues (C). H&E staining. (From [4], with permission.) (See insert for color representation.)

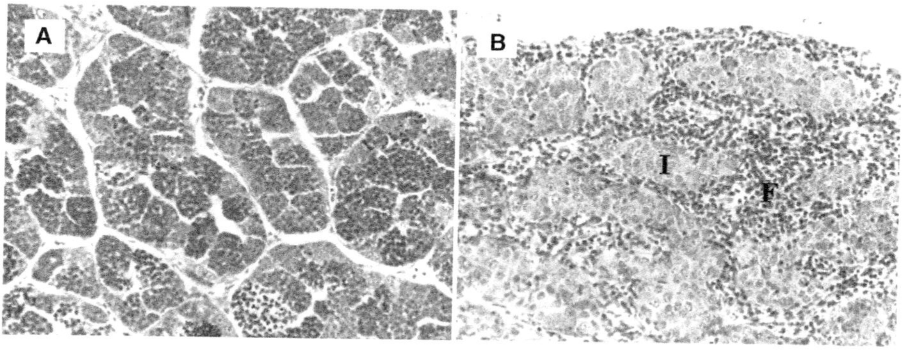

FIGURE 6.9 Testis of fathead minnow (A) from a reference site; (B) exposed to 6.0 ± 2.8 ng/L ethinylestradiol for 113 days in an experimental lake. Exposed animals show F = fibrosis, and I = inhibited development of lobules. (From [33], with permission.) (See insert for color representation.)

Fibrosis is however, also suggested to occur by mere stimulation of fibroblast proliferation without other obviously detectable concurrent pathological alterations (e.g., necrosis and inflammation) within the organ. Prolonged exposure of adult fathead minnows to 8 ng/L EE_2 (Figure 6.9) [33], 21-day exposure of adult medaka to high concentrations of EE_2 (488 ng/L) [1], 36-day exposure of adult male medaka to high concentrations of E_2 (100 μg/L) [57], or feeding of sea bass with food pellets containing high concentrations (5 to 10 mg/kg food) of E_2 or EE_2 from 46 to 426 dpf [50] resulted in nearly complete reduction (compression) of testicular germinal tissue due to interstitial fibrosis. A similar, yet less pronounced effect was observed in medaka exposed to high concentrations (200 μg/L) of bisphenol A (BPA) from 1 to 110 dph [102], or in medaka exposed to very high concentrations of 4-nonylphenol (183 μg/L) or BPA (1820 μg/L) from the fertilized eggstage through 102 dph [103,104]. This was most prevalent in male fish presenting with testis–ova [50,103,104]. Contrary to the findings by Yokota et al. [103,104], shorter exposure (36 days) of medaka to 4-nonylphenol or BPA, even when employing higher concentrations (up to 3120 μg/L BPA), resulted in no detectable fibrosis despite the fact that testis–ova were observed [3,34]. This could suggest that the time (only juveniles) and prolonged duration of exposure is crucial for the formation of fibrosis in conjunction with the exposure of small fish to endocrine-disrupting compounds (e.g., nonylphenol, octylphenol, or BPA). In contrast, natural or synthetic estrogens, E_2 or EE_2, appear to be able to induce fibrosis irrespective of the developmental stage of the fish. Only the concentrations of the estrogens and the duration of exposure appear to play a role in whether fibrosis can be induced: Lower concentrations and prolonged exposure inducing fibrosis in undeveloped or juvenile fish, high concentrations and short durations of exposure inducing fibrosis in adult fish. The latter observation could suggest that the induction of fibrosis (replacement of testicular and ovarian germinal with connective tissue) could be mediated by a mechanism other than mere regeneration and scarring.

In a medaka study using repeated implantations of estrone pellets, Egami noted a remarkable increase in the amount of connective tissue after five weeks of exposure, and in some of the testes the germinal tissue had been "practically converted into masses of connective tissue" [105]. The question thus arises whether fibroblast proliferation (fibrosis) could be effected through the direct action of natural and synthetic estrogens or endocrine disruptors on the fibroblasts without a preceding or concurrent inflammatory response. Although no information is available with regard to estrogen regulation of fibroblasts in fish, some findings from human mesenchymal cells may offer some explanation. Indeed, fibroblast growth factor-9 (FGF-9) is a steroid-regulated mitogen and survival factor for human nerve and mesenchymal cells (e.g., endometriotic stromal cells). Concordant treatment of these endometriotic stromal cells with 4-hydroxyandrostenedione (an aromatase inhibitor) or ICI 182,870 inhibited their proliferation, whereas proliferation was reversed by co-administration of E_2 or FGF-9 [106]. The near-exclusivity of estrogens in inducing enhanced endometrial proliferation, specifically epithelial and stromal cells, was also observed in the ovariectomized macaque model of postmenopausal women [107]. Furthermore, Cline et al. [107] were able to demonstrate that putative selective estrogen receptor modulators and soybean phytoestrogens were able to antagonize the E_2-induced proliferation. Conversely, E_2 was demonstrated to have an ameliorative effect on hepatic fibrosis, whether chronic as a result of a virus infection (hepatitis C) [108], or acute following a toxic insult with CCl_4 [109]. Thus, estrogen-induced fibroblast proliferation appears to be a specific characteristic restricted to the reproductive tract and not a general phenomenon.

In addition, the observation that putative selective estrogen receptor modulators can antagonize the effect of E_2 may explain why compounds other than E_2 or EE_2 were not shown to induce fibrosis in the absence of necrosis and inflammation in the gonads of exposed fish. To be substantiated, however, the latter hypothesis needs a great deal more scientific support, including a thorough mechanistic understanding of the underlying processes. It is interesting to note that despite the relative paucity of publications reporting fibrosis in fish exposed to endocrine disrupters, the predominant portion represent reports in medaka. Whether this is a reflection of a species-specific susceptibilities or a bias toward more detailed mechanistic investigations in medaka cannot be ascertained at present. It is noteworthy though, that in a study exposing both sexes of medaka to 100 μg/L octylphenol up to three months, exclusively male medaka presented with fibrosis in the testes while very few females presented with some stromal proliferation but not explicitly fibrosis [57]. On the other hand, Schafers et al. [101] also reported fibrosis in ovaries and testes of zebrafish exposed to high concentrations of EE_2, thus raising the question of whether the male specificity of gonadal fibrosis was observed by chance or is a medaka- and/or octylphenol-specific issue.

6.1.7 Gonadal Duct Formation

Gonadal duct formation was reported to be partially or completely inhibited in fish (e.g., carp [30,110], fathead minnows [111], and the genus *Oreochromis,* including tilapia (*Tilapia mossambica*) [112]), exposed to endocrine-active compounds throughout their developmental and sexual differentiation stages. As discussed in Section 3.3, morphological (somatic cell) differentiation precedes cytological (primordial germ cell) differentiation in some species, especially in cyprinids, gobeids, and cichlids. It is therefore not surprising that species belonging to the cyprinid family [e.g., roach (*Rutilus rutilus*) and gudgeon (*Gobio gobio*)] [24,25,113,114] taken from rivers with a chronic impact of sewage treatment effluent containing high concentrations of natural and synthetic estrogens have been reported to present with undeveloped or nonfunctional sperm ducts.

Sexual differentiation appears to occur much earlier in female than in male fish (see Section 3.3). Consequently, if sufficiently high concentrations of endocrine-disruptive compounds (estrogenic compounds) [10,24,52,102,115,116] are present during sexual differentiation of developing fish, normal development of sperm ducts can be inhibited. Masculinization, including abnormal development ovarian ducts of female fish is more rarely reported [117–120], probably as a result of the lower abundance of androgenic compounds in sewage treatment effluents [117]. Non-existent or non functional sperm ducts were reported in wild roach in both the absence and presence of simultaneous testis–ova or other testicular abnormality [24], indicating that at least for this species, abnormal duct formation may serve as an endpoint in its own right, which needs to be assessed in order to fully grasp the potential environmental effects of endocrine disrupters. Indeed, Jobling et al. [24] were able to demonstrate that some of the male roach presenting with nonfunctional sperm ducts had fully mature and functional testes able to produce sperm. Although these findings may suggest that abnormal gonadal duct formation could indeed potentially affect the sustainability of a fish population, the actual incidence of fish with this specific abnormality ranged only between 8.3 and 38% of all males inspected. It is important to note that only a small fraction of the abnormally developed ducts were actually nonfunctional [24]. Generally speaking, very few histopathologically oriented investigations of endocrine disruption also inspect gonadal duct formation. It is therefore quite possible that the occurrence of duct abnormality (presumably restricted to testes) is found in many more species than is reflected by the literature at present.

6.2 EFFECTS ASSOCIATED WITH EXPOSURE TO SPECIFIC COMPOUNDS OR COMPOUND CLASSES

To sort the various effects reported for EDCs in fish, all of the currently available literature on EDCs and associated effects in fish (approximately 1200 publications) was scanned for specific histopathological findings in male (Tables 6.1 to 6.6) and female fish (Tables 6.7 to 6.12) and listed according to their respective occurrence in the three main laboratory fish species (fathead minnows, zebrafish, and medaka) and other species, as well as with regard to the compound or compound type. Although the tables compiled are certainly far from including absolutely everything published in the international literature to date, it should represent a major proportion of information available. The tables should thus foster some insight as to the occurrence of certain effects in conjunction with a given species, exposure type, and/or compound or compound type.

TABLE 6.1 Overt Testicular Changes in Male Fish

Concentration	Stage	Duration	Route of Exposure	Histopathological Observations	Ref.
		17α-ethinylestradiol: Fathead Minnow (*Pimephales promelas*)			
0.2, 1, 4, 16, and 64 ng/L	<24 hr	305 d	Aqueous (flow-through)	4 ng/L: males failed to develop normal secondary sexual characteristics; 56 dph, 4 ng/L: ♀ : ♂ ratio = 84 : 5, testicular atrophy and apoptosis; 176 dph, 4 ng/L: no testicular tissue in any of the fish	[68]
0.01 μg/L	Early blastula embryos	Up to 114 dph	Aqueous	Dysplasia of gonadal tissue, 100% doubly attached gonads, feminization	[124]
10 ng/L	Various stages during sexual differentiation	dph: 0–5 5–10 10–15 15–10 0–20	Aqueous	Ovarian-like cavity at 100 dph: <10% 40% 65% 45% 20%	[111]
10 ng/L	<24 hpf	Up to 107 dph	Aqueous (flow-through)	No fish with male gonadotype, sexually undifferentiated fish with two mesentarial attachments and ovarian-like cavity	[125]

17α-ethinylestradiol: Medaka (*Oryzias latipes*)					
0.0001, 0.001, 0.01, 0.1, and 1 μg/L	1 dph	85–110 d	Aqueous	Shift in sex ratio toward females; highest concentration toxic, surviving fish all female	[102]
1, 10, and 100 ng/L	Juvenile	2 m, 6 w recovery	Aqueous	100 ng/L: complete sex reversal in genetic males	[126]
0.005, 0.05, 0.5, 2.5, or 5.0 ng/egg (5 ng/egg lethal)	Egg stage	To adulthood	Nano-injection	0.5 and 2.5 ng/egg: complete sex reversal in genetic males (d-rR strain); typical female ducts with separate opening for the genital and urinary tracts; non-sex reversed males had normally fused urinary and gonadal duct forming a single urogenital opening	[127]
17α-ethinylestradiol: Zebrafish (*Danio rerio*)					
5 ng/L	Egg stage	210 d (lifelong)	Aqueous (flow-through)	Suppression of male sexual differentiation (43% had undifferentiated testes, resembling primary stage ovary-type)	[128]
1, 10, and 100 ng/L	2 dph	56 d	Aqueous	100% phenotypic females (100 ng/L), inhibition of gonadogenesis	[129]
1, 2, 5, 10, and 25 ng/L	20 dph	40 d	Aqueous	>2 ng/L: all females, complete sex reversal	[130]
1 and 10 ng/L	2 dph	Up to 60 dph	Aqueous	10 ng/L: no testes	[26]

(*continued*)

TABLE 6.1 (*Continued*)

Concentration	Stage	Duration	Route of Exposure	Histopathological Observations	Ref.
9.3 ng/L	0 dph	177 dph	Aqueous (flow-through)	All individuals displayed gonads with ovarian morphology; after 3 m of depuration: 5 out of 20 fish with testes	[101]
15.4 ng/L	Different stages from fertilization to 60 dph	Up to 60 dph	Aqueous	Exposure from 20 to 60 dph or from hatch to 60 dph: 100% feminization, females with small ovaries containing few primary oocytes	[131]
3 ng/L	Fertilized egg	118 d	Aqueous (flow-through)	After 118 d: all fish possess ovaries After 176 d: 6 fish with testes, 20 with ovaries	[132]
25 ng/L	Adult (2.5–4 cm)	21 d	Aqueous (flow-through)	No signs of feminization	[54]
6.2 and 10 ng/L	20 dph	40 d	Aqueous (semistatic)	>90% and 100% feminization	[133]
		17α-ethinylestradiol: Other Species			
European sea bass (*Dicentrarchus labrax*)					
5 mg/kg food	From 60 dpf	To 260 dpf	Diet	Approximately 80% females and 20% with suppressed gonadal development	[50]
Flounder (*Platichthys flesus*)					
10 ng/L	Adult	3 w	Aqueous (flow-through)	No serious abnormalities in gonads	[134]

Guppy (*Poecilia reticulata*)					
10, 50, and 200 ng/L	Birth	3.5 m (to adulthood)	Aqueous (flow-through)	50 ng/L: increased interstitial tissue; 200 ng/L: increased interstitial tissue, drastically shrunken testes, absence of gonads, feminization	[87]
Three-spined stickleback (*Gasterosteus aculeatus*)					
0.05 μg/L	0 dph	To 39–58 dph	Aqueous (semistatic)	Gonadal feminization of males in 1-and 10-μg/L treatment	[136]
1.75 ± 0.4 ng/L and 27.7 ± 1.4 ng/L	Hatched	4 w	Aqueous (flow-through)	After reaching adulthood: Low concentration: no effects observed on gonads High concentration: sex reversal toward females	[137]
Rare minnow (*Gobiocypris rarus*)					
0.2, 1, 4, 16, and 64 ng/L (16 and 64 ng/L: toxic)	2 hpf	180 d	Aqueous (flow-through)	0.2, 1, and 4 ng/L: sex ratio skewed toward females (30 : 48, 14 : 57, and 0 : 30)	[138]
4 ng/L	Adult	21 d	Aqueous	Male feminization	[135]
17β-estradiol (E_2), Estrone, Stilbestrol: Medaka (*Oryzias latipes*)					
1 mg/L E_2	0–10 dpf	To adulthood (>30 mm)	Aqueous	Sex reversal from genetic males to phenotypic females	[139]
5 μg/L E_2	Embryo stage 10, 1 dph, 7 dph, and 21 dph	24 h	Aqueous	Biased sex ratio toward females	[140]

(*continued*)

TABLE 6.1 (*Continued*)

Concentration	Stage	Duration	Route of Exposure	Histopathological Observations	Ref.
0.001, 0.01, 0.1, and 1 μg/L E_2	1 dph	85–110 d	Aqueous	Shift in sex ratio toward females	[102]
0.01, 0.1, 1, and 10 μg/L estrone	1 dph	85–110 d	Aqueous	Shift in sex ratio toward females (except 0.1 μg/L treatment: significantly more males)	[102]
0.01, 0,12, and 1.66 μg/L E_2	5–8 days	1 m	Aqueous	All females, no testis–ova	[141]
33.5 and 140.6 ng/L E_2	Fertilized eggs	20 d	Aqueous (flow-through)	50% sex reversal of genetic males after 20 d (high dose only)	[142]
125 μg/g estrone; 62.5 μg/g stilbestrol	Newly hatched fry	7 m	Diet	Sex reversal toward phenotypical and functional female fish	[143]
17β-estradiol (E_2), Estrone, Estriol: Zebrafish (*Danio rerio*)					
5, 25, and 100 ng/L E_2	0 dpf 21 dpf >200 dpf	To 21 dpf To 42 dpf 3 w	Aqueous	Observations at 42 dpf: increased proportion of presumptive males with feminized gonadal ducts (two points of attachments forming an ovarian cavity)	[144]
4, 8, 14, 24, 54, 96, and 250 ng/L E_2	20 dph	40 d	Aqueous (semistatic)	14–54 ng/L: undifferentiated gonads 54 ng/L: >90% females 96 and 250 ng/L: 100% females	[133]
14, 20, 35, 50, and 98 ng/L estrone	20 dph	40 d	Aqueous (semistatic)	>50 ng/L: significantly increased proportion of females 98 ng/L: some undifferentiated gonads	[133]

0.1, 0.3, 0.6, 3, 6.7, and 21.7 μg/L estriol	20 dph	40 d	Aqueous (semistatic)	0.6 μg/L: few undifferentiated gonads 3 and 6.7 μg/L: no findings 21.7 μg/L: 100% feminization	[133]
		17β-estradiol: Other Species			
Baltic salmon (*Salmo salar*)					
0.94 or 9.4 μg/g	Fry	8, 13, 17, and 41 w	Diet	100% feminization, lack of ovarian lamellae	[145]
Carp (*Cyprinus carpio*)					
10 μg/L	50 dph	90 d	Aqueous (flow-through)	All phenotypic females	[146]
10 and 100 μg/L	50 dph	90 d	Aqueous (flow-through)	Testes with oviducts	[30]
European sea bass (*Dicentrarchus labrax*)					
5 and 10 mg/kg food	48 dpf	To 426 dpf	Diet	Suppressed gonadal development; intersex gonad with ovarian cavity, some pre-vitellogenic oocytes and testicular lobules; some sterile gonads without germinal tissue	[50]
Summer flounder (*Paralichthys dentatus*)					
0.1, 1.0, and 10.0 mg/kg	Juvenile (2 y)	4 w	Two injections, second after 2 w	Inhibition of testicular growth, atrophy and shrinkage of testis, 1 sex-reversed female (out of 7) with stage VI oocytes	[147]

(continued)

TABLE 6.1 (*Continued*)

Concentration	Stage	Duration	Route of Exposure	Histopathological Observations	Ref.
Three-spined stickleback (*Gasterosteus aculeatus*)					
0.01, 1.0 and 10.0 μg/L	0 dph	39–58 d	Aqueous (semistatic)	Gonadal feminization of males in 1- and 10 μg/L groups	[136]
1 and 10 μg/L	0 dph	14 d			
10 μg/L	14 dph	39–58 d			
10 μg/L	1 dpf	To hatch			
Tilapia (*Tilapia mossambica*)					
50 μg/g	6 dph		Diet	Examination at 30, 40, 60, 70 dph:	[112]
		From 6 to 25 dph		Complete feminization of gonad of genetic males, similar to ovaries of females	
		From 6 to 15 dph		No signs of modifications, no intersex gonad	
		From 11 to 20 dph		Incomplete feminization of genetic males, intersex gonads (testis in posterior region, ovary in anterior region, including ovarian cavity)	
Androgens: Fathead Minnow (*Pimephales promelas*)					
Methyldehydrotestosterone					
0, 0.10, 0.32 and 1.0 μg/L	Early blastula embryos	Up to 114 dph	Aqueous	Mixed sex gonads (10–25%), smaller testes, dysplasia of gonadal tissue	[124]

17α-methyltestostosterone					
0.1, 1, 5, and 50 µg/L	Adult (10–15 m)	3 w	Aqueous (flow-through)	No effects observed	[148]
		Androgens: Zebrafish (*Danio rerio*)			
17α-methyltestosterone					
10 µg/L	Fertilized egg	35–71 d	Aqueous	100% feminization	[132]
26, 50, 100, 260, 500, and 1000 ng/L	20 dph	40 d	Aqueous	1000 ng/L: some fish with one gonad occupied by oocytes, and one gonad consisted of testicular tissue	[130]
		Androgens: Other Species			
Arctic char (*Salvelinus alpinus*); 17α-mthyldihydrotestosterone					
0.5, 1, 3, 5, and 10 mg/L	Hatched	140°C dph	Aqueous (flow-through)	Increasing proportion of "not females"	[149]
0.5 mg/kg	140°C dph	600°C dph	Diet	Increasing proportion of "not females"; no "affected gonads"	
Three-spined stickleback (*Gasterosteus aculeatus*); 17α-methyltestosterone					
1 µg/L	0 dph	39–51 d	Aqueous (semistatic)	Testis–ova in genetic males and females	[136]
	0 dph	14 d		Development of large branched cavities with unidentified origin	
	14 dph	39–51 d		No effects observed	
Tilapia (*Tilapia mossambica*);17α-methyltestosterone					
50 µg/g	7 dph	19 d	Diet	No effects observed on males	[112]
1000 µg/g	7 dph	19 d		Testes with ovarian cavity	
1000 µg/g	7 dph	44 d		Testes with ovarian cavity	

(*continued*)

TABLE 6.1 (*Continued*)

Concentration	Stage	Duration	Route of Exposure	Histopathological Observations	Ref.
		Anti-androgens: Fathead Minnow (*Pimephales promelas*)			
Flutamide 62.7 ± 6 and 651 ± 45 μg/L	Adult (6 m)	21 d	Aqueous (flow-through)	Less mature testes in 1 out of 3 males (high treatment)	[150]
		P450-Aromatase Inhibitors: Fathead Minnow (*Pimephales promelas*)			
Fadrozole 2, 10, and 50 μg/L	Reproductive stage	21 d	Aqueous	Enlargement of seminiferous tubules filled with sperm	[64]
		Bisphenol A: Fathead Minnow (*Pimephales promelas*)			
0, 1, 16, 160, 640, and 1280 μg/L	Reproductive stage	43, 71, and 164 d	Aqueous (flow-through)	No effects observed	[6]
0, 1, 16, 64, 160, and 640 μg/L	Reproductive stage	164 d	Aqueous (flow-through)	Slightly enlarged to enlarged testes with intravascular proteinaceous fluid at 160 and 640 μg/L	[23]
		Bisphenol A: Medaka (*Oryzias latipes*)			
0, 10, 50, 100, and 200 μg/L	1 dph	85–110 d	Aqueous	Higher concentrations: loss of testicular structure	[102]
2.28, 13, 91.2, 355, and 1820 μg/L	Fertilized egg stage	60 d	Aqueous	Shift toward phenotypical females; highest concentration group: 100% phenotypical females	[103]

Bisphenol A: Carp (*Cyprinus carpio*)					
0, 1, 10, 100, and 1000 µg/L	1 y	14 d	Aqueous (flow-through)	Loss of testicular structure, degenerated spermatozoa, interlobular alterations	[151]
Pentylphenol, Nonylphenol: Fathead Minnow, Medaka, Zebrafish					
Fathead minnow (*Pimephales promelas*); 4-*tert*-pentylphenol					
56, 180, and 560 µg/L	<24 hpf	Upto 107 dph	Aqueous (flow-through)	180 µg/L: formation of an ovarian like cavity; 560 µg/L: complete absence of testicular development	[125]
Medaka (*Oryzias latipes*); *p*-nonylphenol					
10, 50, and 100 µg/L	1 dph	3 m	Aqueous	♂: ♀ ratio (control): 2: 1 ♂: ♀ ratio (100 µg/L): 1: 2	[57]
Zebrafish (*Danio rerio*); 4-nonylphenol					
10, 30, and 100 µg/L	2 dph	58 dph	Aqueous	90% phenotypic females (100 µg/L)	[129]
Pentylphenol, Octylphenol: Other Species					
Carp (*Cyprinus carpio*); 4-*tert*-pentylphenol					
100, 320, and 1000 µg/L	50 dph	90 d	Aqueous (flow-through)	All testes with oviduct	[146]
100, 320, and 1000 µg/L	50 dph	90 d	Aqueous (flow-through)	Testes with oviduct	[30]
140 µg/L	−3, 0, and 3 dph; 24 dph	Upto 110 dph	Aqueous	All testes with oviduct	[110]

(*continued*)

TABLE 6.1 (*Continued*)

Concentration	Stage	Duration	Route of Exposure	Histopathological Observations	Ref.
Eelpout (*Zoarces viviparus*); 4-*tert*-octylphenol					
10, 50, and 100 μg/L	Adult	3 w	Aqueous (flow-through)	50 and 100 μg/L: degenerated testes	[152]
Industrial Chemicals: Other Species					
Nile tilapia (*Oreochromis niloticus*); Aroclor 1254					
0.05 g/kg	10 dph	40 d	Diet	After 18 months: decrease of the germinal epithelium layer	[97]
Pesticides: Medaka, Zebrafish					
Medaka (*Oryzias latipes*); o,p'-DDT					
1, 5, 10, and 50 μg/L	1 dph	100 d	Aqueous	Trend toward higher proportion of females	[153]
Zebrafish (*Danio rerio*); Prochloraz					
0, 16, 64, and 202 μg/L	24 hpf	60 d	Aqueous (flow-through)	202 μg/L: masculinization	[154]
Pesticides: Other Species					
Bluegill sunfish (*Lepomis macrochirus*); endosulfan					
1 μg/L	Adult (2 y)	1, 2, 3, and 4 d	Aqueous (semistatic)	1 d: testes structure slightly disrupted 2 d: continued breakage of connective tissue	

		1 and 2 w		3 d: more damaged connective tissue, 4 d: poorly visible connective tissue 1 w: more disorganized structure, missing connective tissue 2 w: connective tissue missing in certain areas	[123]
Goldfish (*Carassius auratus*); atrazin					
100 and 1000 μg/L	Adult	21 d	Aqueous (flow-through)	High dose: progressive increase in gaps in the interstitium between lobules	[155]
Nile tilapia (*Oreochromis niloticus*); paraquat					
0.5 mg/L at 17°C or 27°C	Adult	45 d	Aqueous (flow-through)	General testicular eosinophilia	[156]
Redbelly tilapia (*Tilapia zillii*); metronidazole (antiparasitic, antibacterial)					
5, 10, and 20 mg/L	~75 g	2, 7, and 14 d	Aqueous (flow-through)	10 and 20 mg/L: testicular degeneration; simultaneous administration of 200 g/kg vitamin C abolished all effects	[157]
Summer flounder (*Paralichthys dentatus*); *o,p'*-DDT, *p,p'*-DDE					
15, 30, and 60 mg/kg	Juvenile (2 y)	8 w	Two injections, second after 2 w	DDT: testicular atrophy; DDE: no effects observed	[147]
		Field: Fathead Minnow (*Pimephales promelas*)			
Michigan wastewater treatment plant effluent					
n.d.	Adult	3 w	Field	No effects observed	[158]

(continued)

TABLE 6.1 (*Continued*)

Concentration	Stage	Duration	Route of Exposure	Histopathological Observations	Ref.
		Field: Other Species			
Carp (*Cyprinus carpio*); Anoia River (estrogenic compounds)					
n.d.	Adult	Lifelong	Field	Testicular atrophy	[159]
Crucian carp (*Carassius carassius*); treated sewage effluent					
25%, 50%, and 100%	Adult (1–2 y)	28 d	Field	Degenerated tissue	[48]
Redear sunfish (*Lepomis microlophus*); Martin Lake (selenium contaminated)					
n.d.	Adult	Lifelong	Field	Mostly immature testes	[160]
Roach (*Rutilus rutilus*); treated sewage effluent					
100% contains 37 ng/L estrone, 5.9 ng/L E_2, and <1 ng/L EE_2	50 dph (during sexual differentiation)	50 or 100 d	Aqueous	After 100 d: increased incidence of "male" fish with feminized gonadal ducts (two points of mesentarial attachments: 100% incidence at 100% effluent, 70% at 47.4% effluent) After 350 d: 100% incidence at 100% effluent	[114]

White suckers (*Catostomus commersoni*); Colorado River basin					
n.d.	Adult	Lifelong	Field	Gonad deformities, intersex gonads	[161]
Whitefish (*Coregonus sp.*); Lake Thun					
n.d.	1–6 y; highest prevalence in 3- to 5-y old fish	n.d.	Field	35% affected gonads (281 out of 808): adhesions/fusions to peritoneal wall and lateral trunk musculature; asymmetry, atrophy, and constrictions of gonads Hermaphrodism: Consecutive sequence of testicular and ovarian lobes on the same gonad strand One gonad strand ovarian tissue and the other testicular tissue	[162]

Compounds and compound classes observed to induce overt testicular changes (e.g., sex reversal, real intersex (large proportion of gonadal tissue of the other gender), undifferentiated gonad, immature gonad, lacking gonad, an ovarian cavity, oviducts, two mesentarial attachments, testicular atrophy, etc.) in conjunction with a defined concentration, developmental stage, duration, and form/route of exposure. Tables are ordered according to the three main laboratory fish species and other species and distinguish laboratory experiments from field observations. Included are field trials with "contaminated" water of unknown chemical composition. The most prominent findings are listed along with the respective reference. d, days; dpf, days post-fertilization; dph, days post-hatch; °C dph, accumulated daily degrees °C post-hatch; h, hatch; hpf, hours post-fertilization; hr, hours; n.d., not defined or specifiable; m, months; w, weeks; y, years.

TABLE 6.2 Testis–Ova in Male Fish

Concentration	Stage	Duration	Route of Exposure	Histopathological Observations	Ref.
		17α-ethinylestradiol: Fathead Minnow (*Pimephales promelas*)			
5–6 ng/L	All stages	7 y	Experimental lake—field trial	Testis–ova	[7]
0.2, 1, 4, 16, and 64 ng/L	<24 hr	305 d	Aqueous (flow-through)	Testis–ova	[68]
		17α-ethinylestradiol: Medaka (*Oryzias latipes*)			
0.0001, 0.001, 0.01, 0.1, and 1 μg/L	1 dph	85–110 d	Aqueous	Testis–ova	[102]
32.6, 63.9, 116, 261, and 488 ng/L	Adult	21 d	Aqueous (flow-through)	>63.9 ng/L: Testis–ova	[1]
		17α-ethinylestradiol: Zebrafish (*Danio rerio*)			
5 ng/L	Egg stage	210 d (lifelong)	Aqueous (flow-through)	Low incidence of testis–ova	[128]
15.4 ng/L	Fertilization to h, h to 10 dph	Up to 60 dph	Aqueous	All groups with testis–ova	[131]
	10–20 dph, 20–30 dph			No effect	
	h to 25 dph, 20–40 dph			Testis–ova with atretic oocytes	
3 ng/L	Fertilized egg	118 d	Aqueous (flow-through)	After 176 d: 1 fish out of 27 with testis–ova	[132]

		17α-ethinylestradiol: Other Species			
Rare minnow (*Gobiocypris rarus*)					
0, 1, 5, and 25 ng/L	Adult (7 m)	28 d	Aqueous (flow-through)	Testis–ova	[165]
4 ng/L	Adult (9 m)	21 d	Aqueous (flow-through)	Testis–ova (6 out of 8)	[138]
0.2, 1, 4, 16, and 64 ng/L (16 and 64 ng/L: toxic)	2 hpf	180 d	Aqueous (flow-through)	0.2 and 1 ng/L: Testis–ova (22% and 64%)	[138]
Sheepshead minnow (*Cyprinodon variegates*)					
0.2, 2, 20, 200, 400, 800, 1600, and 3200 ng/L	Subadult	Up to 59 d	Aqueous (flow-through)	Testis–ova	[69]
Three-spined stickleback (*Gasterosteus aculeatus*)					
0.05 μg/L	0 dph	To 39–58 dph	Aqueous (semistatic)	High incidence of testis–ova	[136]
1.75 ± 0.4 ng/L	Hatched	4 w	Aqueous (flow-through)	No effects in gonads observed after reaching adulthood	[137]
27.7 ± 1.4 ng/L				Testis–ova	
		17β-estradiol: Fathead Minnow (*Pimephales promelas*)			
1 μM	5 dph	20 d	Aqueous	Testis–ova after 5 months	[61]
		17β-estradiol, Estrone, Estriol: Medaka (*Oryzias latipes*)			
29.3, 55.7, 116, 227, and 463 ng/L E_2	Adult (4 m)	21 d	Aqueous	Testis–ova in all treatment groups, not concentration dependent	[4]
5 μg/L E_2	Embryo stage 10, 1 dph, 7 dph, and 21 dph	24 h	Aqueous	After 5 months: Testis–ova	[140]

(*continued*)

TABLE 6.2 (*Continued*)

Concentration	Stage	Duration	Route of Exposure	Histopathological Observations	Ref.
0.001, 0.01, 0.1, and 1 μg/l E_2	1 dph	85–110 d	Aqueous	Testis–ova (effect not present in lowest concentration)	[102]
0.939, 2.86, 8.66, 27.9, and 92.4 ng/L E_2	12 hpf	101 d	Aqueous	≥8.66 ng/L: Testis–ova	[166]
0.01, 0.1, 1, and 10 μg/L estriol	1 dph	85–110 d	Aqueous	10 μg/L: Testis–ova	[102]
0.01, 0.1, 1, and 10 μg/L estrone	1 dph	85–110 d	Aqueous	Testis–ova	[102]
17β-estradiol, Estrone: Medaka (*Oryzias latipes*), Red Variety Strain					
Estrone pellet with unknown dose	Adult	5 w	Repeated subcutaneous pellets (1/w)	2 d after last implant: mild testis–ova 35 d after last implant: well-developed testis-ova	[105]
10-μg estrone crystals	Adult	—	Single intraperitoneal injection	34 d after injection: Testis–ova in the peripheral part of the testis	[105]
Unknown (estrone pellet in food)	Adult	35 d	Diet, single dose	Testis–ova	[105]
10–800 μg/L estrone	Adult	35 d	Aqueous	200 and 400 μg/L: Testis–ova	[105]
4–320 μg/L E_2	Adult	31–37 d	Aqueous	40–320 μg/L: Testis–ova	
1.5 mg/L estrone crystals	Adult	30 d	Aqueous, static, single dose	5 out of 15 males had developed testis–ova	[105]
Estrone pellet with unknown dose	Adult	8 w	Single subcutaneous pellet	Testis–ova, increasing severity over time	[105]
17β-estradiol, Estriol: Zebrafish (*Danio rerio*)					
4, 8, 14, 24, 54, 96, and 250 ng/L E_2	20 dph	40 d	Aqueous	8–24 ng/L: Testis–ova	[133]

0.1, 0.3, 0.6, 3, 6.7, and 21.7 μg/L estriol	20 dph	40 d	Aqueous (semistatic)	0.1 and 0.3 μg/L: few fish with testis–ova	[133]
		17β-estradiol: Other Species			
Carp (*Cyprinus carpio*)					
0.1 and 1 μg/L	Sexually mature (<210 dph)	3 m	Aqueous (flow-through)	Testis–ova	[29]
Eelpout (*Zoarces viviparus*)					
0.5 μg/L	Adult	3 w	Aqueous (flow-through)	Testis–ova	[152]
Three-spined stickleback (*Gasterosteus aculeatus*)					
0.01, 1.0, and 10.0 μg/L	0 dph	To 39–58 dph	Aqueous (semistatic)	High incidence of testis–ova	[136]
1 and 10 μg/L	0 dph	To 14 dph		Few testis–ova (significant)	
10 μg/L	14 dph	To 39–58 dph		Few testis–ova (significant)	
10 μg/L	1 dpf	To hatch (6–7 d)		Low incidence of testis–ova (1 out of 43)	
		Androgens: Fathead Minnow, Medaka, Zebrafish			
Fathead minnow (*Pimephales promelas*); methyltestosterone					
22.5, 46.8, 88.1, 188, and 380 ng/L	Adult	3 w	Aqueous (flow-through)	380 ng/L: Testis–ova (2 out of 6)	[167]
Medaka (*Oryzias latipes*); testosterone					
100 μg/L	Embryo stage 10, 1 dph, 7 dph, and 21 dph	6 d	Aqueous	After 5 months: Testis–ova (not in medaka treated at 21 dph)	[140]

(continued)

TABLE 6.2 (*Continued*)

Concentration	Stage	Duration	Route of Exposure	Histopathological Observations	Ref.
Zebrafish (*Danio rerio*); 17α-methyltestosterone					
26–1000 ng/L	20 dph	40 d	Aqueous	27% testis–ova	[130]
Androgens: Other Species					
Arctic char (*Salvelinus alpinus*); 17α-methyldihydrotestosterone					
0.5, 1, 3, 5, and 10 mg/L	Hatched	140°C dph	Aqueous (flow-through)	Increasing proportion of "affected gonads" (testis–ova-like appearance);	[149]
0.5 mg/kg	140°C dph	600°C dph	Diet	No effects observed	
European sea bass (*Dicentrarchus labrax*); 17α-methyltestosterone					
10 mg/kg	126 dpf	100 d	Diet	At 3 years of age: no effects observed	[168]
	226 dpf	100 d			
	326 dpf	100 d			
	126–426 dpf	300 d		7% with testis–ova	
Three–spined stickleback (*Gasterosteu aculeatus*); 17α-methyltestosterone					
1 μg/L	0 dph	To 39–51 dph	Aqueous (semistatic)	Testis–ova	[136]
	0 dph	To 14 dph			
	14 dph	To 39–51 dph		No effects observed	
Anti-androgens: Medaka (*Oryzias latipes*)					
Flutamide					
0.032, 0.1, 0.32, 1.0, and 3.2 mg/L	1 w	96 h	Aqueous (semistatic)	After 3 m: Testis–ova in males (Qurt strain)	[44]

Cyproterone acetate					
1 and 10 μg/L	1 dph	100 d	Aqueous	Testis–ova in males (golden strain)	[27]
Flutamide					
0.101, 0.202, 0.397, 0.787, and 1.56 mg/L	3 m (~300 mg; ~33 mm)	3 w	Aqueous (flow-through)	0.202, 0.397, and 0.787 mg/L: Testis–ova (1–2 out of 6)	[169]
P450-Aromatase Inhibitors: Other Species					
Nile tilapia (*Oreochromis niloticus*); fadrozole					
0, 50, 75, and 100 mg/kg	9 dph	15 and 30 d	Diet	At 80 dph: 1 fish out of 60 with testis–ova (50 mg/kg for 30 d)	[170]
Bisphenol A: Fathead Minnow (*Pimephales promelas*)					
0, 1, 16, 160, 640, and 1280 μg/L	Reproductive stage	43, 71, and 164 d	Aqueous (flow-through)	No effects observed	[6]
0, 1, 16, 64, 160 and 640 μg/L	Reproductive stage	164 d	Aqueous (flow-through)	No effects observed	[23]
Bisphenol A: Medaka (*Oryzias latipes*)					
837, 1720, and 3120 μg/L	Adult (4 m)	21 d	Aqueous	Testis–ova in all treatment groups, not concentration dependant	[3]
0, 10, 50, 100, and 200 μg/L	1 dph	85–110 d	Aqueous	Few testis–ova in the 10-μg/L group	[102]
2.28, 13, 91.2, 355, and 1820 μg/L	Fertilized egg stage	60 d	Aqueous	Highest concentration group: Testis–ova	[103]

(*continued*)

TABLE 6.2 (*Continued*)

Concentration	Stage	Duration	Route of Exposure	Histopathological Observations	Ref.
		Bisphenol A: Other Species			
Carp (*Cyprinus carpio*) 0, 1, 10, 100, and 1000 μg/L	1 y	14 d	Aqueous (flow-through)	Previtellogenic oocytes in testis of males exposed to 1000 μg/L	[151]
		4-*tert*-Pentylphenol: Fathead Minnow (*Pimephales promelas*)			
56, 180, and 560 μg/L	<24 hpf	Up to 107 dph	Aqueous (flow-through)	180 μg/L: Testis–ova 560 μg/L: complete absence of testis	[125]
		Nonylphenol, Octylphenol: Medaka (*Oryzias latipes*)			
4-Nonylphenol					
4.2, 8.2, 17.7, and 51.5 μg/L	1 dph	60 d	Aqueous	≥ 17.7 μg/L: Testis–ova	[104]
24.8, 50.9, 101, and 184 μg/L	Adult (4 m)	21 d	Aqueous	Testis–ova in all treatment groups	[34]
Octylphenol					
100 μg/L	1, 3, 5, 7, 21, and 35 dph	Up to 100 dph	Aqueous	7–29% testis–ova (highest incidence when exposed from 3 dph)	[99]
100 μg/L	1 dph	1, 2, and 3 m	Aqueous	3 out of 50 developed testis–ova (3 m treatment only)	[99]
200 and 300 μg/L	Adult (~11 m)	18 and 36 d	Aqueous	1 out of 6 developed testis–ova (300 μg/L treatment for 36 d only)	[99]

p-Nonylphenol					
10, 50, and 100 μg/L	1 dph	3 m	Aqueous	50 μg/L: 50% testis–ova 100 μg/l: 86% testis–ova	[57]
4-Nonylphenol: Zebrafish (*Danio rerio*)					
10, 30, and 100 μg/L	2 dph	58 dph	Aqueous	Testis–ova	[129]
Pentylphenol, Nonylphenol, Octylphenol: Other Species					
Carp (*Cyprinus carpio*); 4-*tert*-pentylphenol					
100, 320, and 1000 μg/L	50 dph	90 d	Aqueous (flow-through)	Testis–ova	[146]
Eelpout (*Zoarces viviparus*); 4-*tert*-octylphenol					
10, 50, or 100 μg/L	Adult	3 w	Aqueous (flow-through)	10 μg/L and 50 μg/L: Testis–ova	[152]
50 μg/L plus 20 μg/g ZM 189,154 (anti-estrogen) per week				50 μg/L plus ZM 189,154: Testis–ova	
Rare minnow (*Gobiocypris rarus*); nonylphenol					
5, 10, and 20 μg/L	Adult (9 m)	21 d	Aqueous (flow-through)	10 μg/L: Testis–ova (1 out of 8); 20 μg/L: Testis–ova (3 out of 8)	[135]
Industrial Chemicals: Other Species					
Nile tilapia (*Oreochromis niloticus*); Aroclor 1254					
0.05 g/kg	10 dph	40 d	Diet	After 18 months: Testis–ova	[97]

(continued)

TABLE 6.2 (*Continued*)

Concentration	Stage	Duration	Route of Exposure	Histopathological Observations	Ref.
		Pesticides: Medaka (*Oryzias latipes*)			
o,p-DDT 1, 5, 10, and 50 μg/L	1 dph	100 d	Aqueous	Testis–ova (same stage as oocytes in females)	[153]
Pentachlorophenol 10, 20, 50, 100, and 200 μg/L	Adult	28 d	Aqueous (flow-through)	Testis–ova in all treatment groups (except 10 μg/L)	[171]
Vinclozolin (Ronilan) 1 and 5 mg/L	1 dph	100 d	Aqueous	Testis–ova in males (golden strain)	[27]
		Pesticides: Zebrafish (*Danio rerio*)			
Prochloraz 0, 16, 64, and 202 μg/L	24 hpf	60 d	Aqueous (flow-through)	Testis–ova	[154]
		Field: Fathead Minnow (*Pimephales promelas*)			
Michigan wastewater treatment plant effluent n.d.	Adult	3 w	Field	No finding	[158]
		Field: Other Species			
Barbel (*Barbus plebejus*); River Po, Italy n.d.	Adult	Lifelong	Field	8 out of 16 males with testis–ova	[172]

Bream (*Abramis brama*); Dommel (small contaminated river in The Netherlands)					
n.d.	Adult	Lifelong	Field	Testis–ova in 37% of male bream; no testis–ova observed in flounder (*Platichthys flesus*)	[173]
Carp (*Cyprinus carpio*)/goldfish (*Carassius auratus*); Hamilton Harbour, Lake Ontario, Canada					
n.d.	Adult	Lifelong	Field	No finding	[174]
Carp (*Cyprinus carpio*); Anoia River (estrogenic compounds)					
n.d.	Adult	Lifelong	Field	Testis–ova	[159]
Catfish (*Pangasianodon hypophthamus*); Mekong River water (containing heavy metals and POPs)					
n.d.	0 dph	Lifelong	Field	No finding	[163]
Common carp (*Cyprinus carpio*)/black bass (*Micropterus* spp.)/channel catfish (*Ictalurus punctatus*); Colorado River basin (dioxins, PCBs, heavy metals, pesticides)					
n.d.	Adult	Lifelong	Field	Low to moderate testis–ova in fish from 7 out of 14 sampling sites; at one sampling site, male smallmouth bass had an incidence of 70% testis–ova	[164]
Largemouth bass (*Micropterus salmoides*); Rio Grande basin (dioxins, PCBs, heavy metals, pesticides)					
n.d.	Adult	Lifelong	Field	Testis–ova; no effects observed in common carp (*Cyprinus carpio*), channel catfish (*Ictalurus punctatus*), or smallmouth bass (*Micropterus dolomieu*)	[49]

(*continued*)

TABLE 6.2 (*Continued*)

Concentration	Stage	Duration	Route of Exposure	Histopathological Observations	Ref.
Crucian carp (*Carassius carassius*); treated sewage effluent					
25%, 50%, and 100%	Adult (1–2 y)	28 d	Field	Testis–ova in 20% of males exposed to 100%	[48]
Eelpout (*Zoarces viviparus*); coastal waters					
n.d.	Adult	n.d.	Field	Testis–ova (large strands of oocytes within testes)	[175]
Flounder (*Platichthys flesus*); UK estuarine and coastal waters					
n.d.	Adult	Lifelong	Field	Testis–ova (5 out of 30) at one sampling site out of 5	[134]
Flounder ♂ (*Pleuronectes yokohamae*); Tokyo Bay					
n.d.	Adult	Lifelong	Field	Testis–ova (3 out of 20)	[176]
Gudgeon (*Gobio gobio*)/stoneloach (*Barbatula barbatula*); treated sewage effluent					
n.d.	Adult	Lifelong	Field	Low incidence of testis–ova	[177]
Gudgeon (*Gobio gobio*); treated sewage effluent					
n.d.	Adult (1–5 y)	Lifelong	Aqueous	Testis–ova	[113]
Mirror carp (*Cyprinus carpio*); treated sewage effluent					
25%, 50%, and 100%	Adult (2–3 y)	28 d	Field	Testis–ova	[178]

Redear sunfish (*Lepomis microlophus*); Martin Lake (selenium contaminated)					
n.d.	Adult	Lifelong	Field	No finding	[160]
Roach (*Rutilus rutilus*); treated sewage effluent					
37 ng/L Estrone, 5.9 ng/L E_2, and <1 ng/L EE_2	50 dph	50 or 100 d	Aqueous	No finding	[114]
Sharptooth catfish (*Clarias gariepinus*); "estrogen-pollutioned water" (containing *p*-nonylphenol)					
4 μg/L	Adult (65–80 cm)	n.d.	Field	Testis–ova in 20% of examined fish	[179]
Shovelnose sturgeon (*Scaphirhynchus platyrynchus*); Mississippi River					
n.d.	Adult	Lifelong	Field	Testis–ova (ovigerous lamellae containing large oocytes) in 2 out of 7 specimens	[180]
Three-spined stickleback (*Gasterosteus aculeatus*)/perch (*Perca fluviatilis*); small rivers					
n.d.	Adult	n.d.	Field	Testis–ova (most advanced stage: late pre-vitellogenic oocyte)	[175]
White suckers (*Catostomus commersoni*); Colorado River basin					
n.d.	Adult	Lifelong	Field	Testis–ova	[161]

(*continued*)

TABLE 6.2 (*Continued*)

Concentration	Stage	Duration	Route of Exposure	Histopathological Observations	Ref.
White perch (*Morone americana*); lower Great Lakes region					
n.d.	Young of the year to ~2 y	Lifelong	Field	22–83% testis–ova; no findings in goldfish (*C. auratus*), common carp (*C. carpio*), gizzard shad (*D. cepedianum*), brown bullhead (*I. ameiurus*), pumpkinseed (*L. gibbosus*), and bluegill (*L. macrochirus*)	[181]
White sucker (*Catostomus commersoni*); bleached kraft pulp mill effluent					
n.d.	Adult	Lifelong	Field	No finding	[53]
n.d.	Adult	Lifelong	Field	After minimizing discharge of chemical: no findings	[52]
Whitefish (*Coregonus* sp.); Lake Thun					
n.d.	1–6 y	n.d.	Field	Testis–ova	[162]

Compounds and compound classes observed to induce testis–ova in conjunction with a defined concentration, developmental stage, duration, and form/route of exposure. Tables are ordered according to the three main laboratory fish species and other species and distinguish laboratory experiments from field observations. Included are field trials with "contaminated" water of unknown chemical composition. The most prominent findings are listed along with the respective reference. d, days; dpf, days post-fertilization; dph, days post-hatch; °C dph, accumulated daily degrees °C post-hatch; h, hatch; hpf, hours post-fertilization; hr, hours; n.d., not defined or specifiable; m, months; w, weeks; y, years.

TABLE 6.3 Testicular Fibrosis in Male Fish

Concentration	Stage	Duration	Route of Exposure	Histopathological Observations	Ref.
		17α-ethinylestradiol: Fathead Minnow, Medaka, Zebrafish			
Fathead minnow (*Pimephales promelas*)					
6.0 ± 2.8 ng/L	Adult	113 d	Aqueous	Widespread fibrosis	[7,33]
Medaka (*Oryzias latipes*)					
32.6, 63.9, 116, 261, and 488 ng/L	Adult	21 d	Aqueous (flow-through)	488 ng/L: almost whole testis composed of connective tissue (fibrosis)	[1]
Zebrafish (*Danio rerio*)					
1 and 10 ng/L	2 dph	Up to 60 dph	Aqueous	1 ng/L: Fibrosis and abnormal connective tissue in testes	[26]
9.3 ng/L	0 dph	177 dph	Aqueous (flow-through)	All individuals displayed gonads with fibrosis	[101]
		17α-ethinylestradiol: Other Species			
Sheepshead minnow (*Cyprinodon variegates*); 17α-ethinylestradiol					
0.2, 2, 20, 200, 400, 800, 1600, and 3200 ng/L	Subadult	Up to 59 d	Aqueous (flow-through)	Fibrosis	[69]
		17β-estradiol, Estrone: Medaka (*Oryzias latipes*)			
29.3, 55.7, 116, 227, and 463 ng/L E_2	Adult (4 m)	21 d	Aqueous	Replacement of testicular tissue by connective tissue	[4]

(continued)

TABLE 6.3 (*Continued*)

Concentration	Stage	Duration	Route of Exposure	Histopathological Observations	Ref.
0.939, 2.86, 8.66, 27.9 and 92.4 ng/L E_2	12 hpf	101 d	Aqueous	≥8.66 ng/L: Fibrosis	[166]
Pellets with unknown estrone dose	Adult	5 w	Repeated subcutaneous pellets (1/wk)	2 d after last implant: increase in connective tissue, "some testes had been practically converted into masses of connective tissue"	[105]
100 μg/L E_2	Adult (~11 m)	18 and 36 d	Aqueous	Severe fibrosis	[99]
17β-estradiol, Estrone, Estriol: Other Species					
Carp (*Cyprinus carpio*) 0.1 and 1 μg/L	Sexually mature (<210 dph)	3 m	Aqueous (flow-through)	Fibrosis	[29]
Bisphenol A: Fathead Minnow (*Pimephales promelas*)					
0, 1, 16, 160, 640, and 1280 μg/L	Reproductive stage	43, 71, and 164 d	Aqueous (flow-through)	No effects observed	[6]
0, 1, 16, 64, 160, and 640 μg/L	Reproductive stage	164 d	Aqueous (flow-through)	No effects observed	[23]
Bisphenol A: Medaka (*Oryzias latipes*)					
0, 10, 50, 100, and 200 μg/L	1 dph	85–110 d	Aqueous	>10 μg/L group: Fibrosis	[102]
2.28, 13, 91.2, 355, and 1820 μg/L	Fertilized egg stage	60 d	Aqueous	Abnormal connective tissue in testes	[103]

Nonylphenol, Octylphenol: Medaka (*Oryzias latipes*)					
4-Nonylphenol					
4.2, 8.2, 17.7, and 51.5 μg/L	1 dph	60 d	Aqueous	≥17.7 μg/L: abnormal connective tissue in testes	[104]
Octylphenol					
100 μg/L	1, 3, 5, 7 21, and 35 dph	Up to 100 dph	Aqueous	Fibrosis	[99]
100 μg/L	1 dph	1, 2, and 3 m	Aqueous	Fibrosis	[99]
200 and 300 μg/L	Adult (∼11 m)	18 and 36 d	Aqueous	Fibrosis	[99]
Nonylphenol: Zebrafish (*Danio rerio*)					
4-Nonylphenol					
100 μg/L	2 dph	Up to 60 dph	Aqueous	Fibrosis	[26]
Pentylphenol, Octylphenol: Other Species					
Carp (*Cyprinus carpio*); 4-*tert*-pentyl phenol					
32, 100, 320, and 1000 μg/L	Sexually mature (<210 dph)	3 m	Aqueous (flow-through)	Fibrosis	[29]
Eelpout (*Zoarces viviparus*); 4-*tert*-octylphenol					
10, 50, or 100 μg/L	Adult	3 w	Aqueous (flow-through)	50 and 100 μg/L: increased interstitial fibrosis	[152]
Pesticides: Medaka (*Oryzias latipes*)					
Vinclozolin (Ronilan)					
1 and 5 mg/L	1 dph	100 d	Aqueous	Moderate lobular fibrosis in testes of males (golden strain)	[27]

(continued)

TABLE 6.3 (*Continued*)

Concentration	Stage	Duration	Route of Exposure	Histopathological Observations	Ref.
		Pesticides: Other Species			
Freshwater murrel (*Channa punctatus*); fenitrothion					
1.5 ppm	15–20 cm	120 d	Aqueous (semistatic)	After 120 d: collagenous (fibrous) capsules around degenerating germ cells	[182]
Snakehead murrel (*Channa striatus*)					
Carbaryl 10 and 20 ppm Endosulfan 0.00075 and 0.001 ppm	~15 cm	30 d	Aqueous (semistatic)	Increase in connective tissue of tunica albuginea	[183]
		Heavy Metals: Other Species			
Goldfish (*Carassius auratus*); cadmium					
10 mg/kg	Adult	Up to 615 d	Single and repeated intraperitoneal injections	Fibrosis	[85]

Compounds and compound classes observed to induce testicular fibrosis in conjunction with a defined concentration, developmental stage, duration, and form/route of exposure. Tables are ordered according to the main three laboratory fish species and other species and distinguish laboratory experiments from field observations. Included are field trials with "contaminated" water of unknown chemical composition. The most prominent findings are listed along with the respective reference. d, days; dpf, days post-fertilization; dph, days post-hatch; °C dph, accumulated daily degrees °C post-hatch; h, hatch; hpf, hours post-fertilization; hr, hours; n.d., not defined or specifiable; m, months; w, weeks; y, years.

TABLE 6.4 Germ Cell Effects in Male Fish

Concentration	Stage	Duration	Route of Exposure	Histopathological Observations	Ref.
17α-ethinylestradiol: Fathead Minnow (*Pimephales promelas*)					
0.2, 1, 4, 16, and 64 ng/L	<24 h	305 d	Aqueous (flow-through)	56 dph, 4 ng/L: testicular apoptosis	[68]
50 ng/L	Adult	15 d	Aqueous	Decrease in spermatocytes	[188]
6.0 ± 2.8 ng/L	Adult	113 d	Aqueous	Delayed spermatogenesis	[33]
17α-ethinylestradiol: Zebrafish (*Danio rerio*)					
10 and 25 ng/L	Adult	3, 6, 12, and 24 d	Aqueous	10 ng/L: after 24 d mainly spermatogonia, fewer spermatozoa	[28]
1 and 10 ng/L	2 dph	Up to 60 dph	Aqueous	1 ng/L: regressed testes, enhanced pyknosis in male germ cells	[26]
25 ng/L	Adult (2.5–4 cm)	21 d	Aqueous (flow-through)	Degeneration of spermatozoa, phagocytosis of sperm cells by Sertoli cells	[54]
17α-ethinylestradiol: Other Species					
Rare minnow (*Gobiocypris rarus*)					
0, 1, 5, and 25 ng/L	Adult (~7 m)	28 d	Aqueous (flow-through)	Inhibition of spermatogenesis	[165]
Guppy (*Poecilia reticulata*)					
10, 50, and 200 ng/L	Birth	3.5 m (to adulthood)	Aqueous (flow-through)	10 ng/L: fewer spermatocytes and spermatozeugmata 50 ng/L and 200 ng/L: fewer spermatids	[87]

(*continued*)

TABLE 6.4 *(Continued)*

Concentration	Stage	Duration	Route of Exposure	Histopathological Observations	Ref.
		17β-estradiol: Fathead Minnows (*Pimephales promelas*)			
2780 ng/L (10 nM)	160 d	14 d	Aqueous (flow-through)	Germ cell syncytia, sperm necrosis	[41]
0.0625, 0.1, 0.125, 0.25, 0.5, 1, 2, 10, 100, and 1000 nM	Adult	14 d	Aqueous	Degeneration of spermatozoa, loss of germ cells, germ cell syncytia	[55]
		17β-estradiol, Estrone: Medaka (*Oryzias latipes*)			
29.3, 55.7, 116, 227, and 463 ng/L E_2	Adult (4 m)	21 d	Aqueous	Degeneration of spermatozoa and spermatocytes in the highest concentration group	[4]
Pellets with unknown estrone dose	Adult	5 w	Repeated subcutaneous pellets (1/wk)	Second after last implant: inhibition of spermatogenesis	[105]
		17β-estradiol, Ultrasitosterol: Zebrafish (*Danio rerio*)			
10 and 100 μg/L oxidized ultrasitosterol preparation	Adult (4–6 months)	3 w	Aqueous (flow-through)	Relative increase in spermatocytes 10 μg/L only: increased amount of spermatozoa 100 μg/L only: increase in atretic oocytes, immature germ cells in testicular lumen	[189]

0.27 µg/L E_2	Adult (4–6 months)	3 w	Aqueous (flow-through)	Relative decrease in spermatogonia, relative increase in spermatocytes, immature germ cells in testicular lumen	[189]
0.27, 2.7, and 27 ng/L E_2	Adult	21 d	Aqueous (semistatic)	≥ 2.7 ng/L: regressed ovary, shift toward spermatogonia	[73]
1 nmol/kg and 1 µmol/kg E_2	Adult	10 d	Maternal exposure through injection	Defective spermatogenesis	[38]
17β-estradiol: Other Species					
Carp (*Cyprinus carpio*)					
10 and 100 µg/L	50 dph	90 d	Aqueous (flow-through)	Reduced number of PGCs, inhibited spermatogenesis	[30]
0.1 and 1 µg/L	Sexually mature (<210 dph)	3 m	Aqueous (flow-through)	Regressed testes	[29]
Eelpout (*Zoarces viviparus*)					
0.5 µg/L	Adult	3 w	Aqueous (flow-through)	No spermatocytes	[152]
Guppy (*Poecilia reticulata*)					
10 and 50 ng/L	Birth	3.5 m (to adulthood)	Aqueous (flow-through)	Fewer spermatocytes	[87]
Platyfish (*Xiphophorus maculatus*)					
96 µg/L	Adult	28 d	Aqueous	Reduced number of spermatozeugmata in testicular ducts, increased number of deformed spermatozeugmata	[71]

(*continued*)

TABLE 6.4 (*Continued*)

Concentration	Stage	Duration	Route of Exposure	Histopathological Observations	Ref.
Summer flounder (*Paralichthys dentatus*)					
0.1, 1.0, and 10.0 mg/kg	Juvenile (2 y)	4 w	Two injections, second after 2 w	Clusters of dead germ cells in cysts, residual sperm in efferent ducts, regressive spermatogenesis	[147]
Androgens: Fathead Minnow, Medaka, Zebrafish					
Fathead minnow (*Pimephales promelas*)					
Dihydrotestosterone 50 μg/L	Adult	15 d	Aqueous	Increase in spermatids	[188]
Methyltestosterone 0.12 mg/L; 1.7 mg/L	Reproductive stage	12 d	Aqueous	Degeneration and necrosis of germ cells in testis, acceleration of spermatogenesis	[42]
Medaka (*Oryzias latipes*, Qurt strain); 11-keto-testosterone					
0.01, 0.1, and 1.0 mg/L	1 w	96 h	Aqueous (semistatic)	Germ cell necrosis	[44]
Zebrafish (*Danio rerio*), testosterone					
0.28 μg/L	Adult (4–6 months)	3 w	Aqueous (flow-through)	Increased amount of spermatozoa, immature germ cells in testicular lumen	[189]

Antiandrogens: Fathead Minnow, Medaka					
Fathead minnow (*Pimephales promelas*); flutamide					
62.7 ± 6 and 651 ± 45 μg/L	Adult (6 m)	21 d	Aqueous (flow-through)	In 2 out of 6 tests, some cells with pycnotic nuclei located in seminiferous tubule epithelium, spermatocyte degeneration, and necrosis	[150]
Medaka (*Oryzias latipes*, Qurt strain); flutamide					
0.032, 0.1, 0.32, 1.0, and 3.2 mg/L	1 w	96 h	Aqueous (semistatic)	After 3 m: severe germ cell necrosis	[44]
Medaka (*Oryzias latipes*, golden strain); cyproterone acetate					
1 and 10 μg/L	1 dph	100 d	Aqueous	Decrease in mature spermatozoa	[27]
Anti-androgens: Other Species					
Guppy (*Poecilia reticulata*); flutamide					
1, 10, and 100 μg/g	Adult	30 d	Diet	Reduced amount of spermatozeugmata in testicular ducts, free spermatozoa in sperm ducts	[31]
Aromatase Inhibitors: Zebrafish					
Zebrafish (*Danio rerio*); tributyltin					
0.01–100 ng/L	0 and 30 dph	0 to 30, 30 to 60, 0 to 70 dph	Aqueous	Adults after 3–5 m: sex ratio biased toward male fish; 0.1 ng/L: some sperm lacked flagella; 10 ng/L: all sperm lacked flagella	[186]

(continued)

TABLE 6.4 *(Continued)*

Concentration	Stage	Duration	Route of Exposure	Histopathological Observations	Ref.
		Aromatase Inhibitors: Other Species			
Guppy (*Poecilia reticulata*); tributyltin					
11.2–22.3 ng/L	Adult males	12 d	Aqueous	Significant declines (by 40–75%) in total sperm counts	[22]
Mummichog (*Fundulus heteroclitus*); tributyltin					
0.5, 1, 2, 4, 8, and 10 μg/L	Adult (8–15 g)	2 w	Aqueous	Number of spermatozoa in seminal ducts slightly reduced, inhibition of germ cell proliferation	[190]
		Bisphenol A: Fathead Minnow, Medaka			
Fathead minnow (*Pimephales promelas*)					
0, 1, 16, 160, 640, and 1280 μg/L	Adult (122 d)	43, 71, and 164 d	Aqueous (flow-through)	Shift toward increased numbers of immature germ cells at > 1 μg/L ♂: significant inhibition of spermatogenesis after 164 d	[6]
0, 1, 16, 64, 160, and 640 μg/L	Reproductive stage	164 d	Aqueous (flow-through)	Increased number of immature germ cells at > 640 μg/L	[23]
Medaka (*Oryzias latipes*)					
0, 10, 50, 100, and 200 μg/L	1 dph	85–110 d	Aqueous	>10 μg/L: decreased number of sperm cells	[102]

Bisphenol A: Guppy (*Poecilia reticulata*)					
274–549 µg/L	Adult males	12 d	Aqueous	Significant declines (by 40–75%) in total sperm counts	[22]
5, 50, 500, and 5000 µg/L	Adult	30 d (5000 µg/L: 21 d)	Aqueous	Testes filled with spermatozeugmata, some of them ruptured (free spermatozoa)	[31]
Nonylphenol: Fathead Minnow, Medaka, Zebrafish					
Fathead minnow (*Pimephales promelas*)					
4-nonyphenol 0.09, 0.1, 0.33, 0.93, and 2.4 µg/L	Adult	42 d	Aqueous (flow-through)	≥ 0.33 µg/L: sperm necrosis, and germ cell syncytia	[43]
Medaka (*Oryzias latipes*)					
4-nonylphenol 24.8, 50.9, 101, and 184 µg/L	Adult (4 m)	21 d	Aqueous	Inhibition of spermatogenesis in the highest concentration group	[34]
100 ppb	Adult	6 w	Aqueous	6× increased rate of apoptosis of spermatocytes	
p-Nonylphenol 10, 50, and 100 µg/L	1 dph	3 m	Aqueous	Testicular regression	[57]
Zebrafish (*Danio rerio*)					
4-Nonylphenol 100 µg/L	2 dph	Up to 60 dph	Aqueous	Regressed testes, enhanced pyknosis in male germ cells	[26]
Nonylphenol 10^{-7} M	2–2.5 hpf	Up to 22–24 hpf	Aqueous	Significant alteration of somite-specific domain aggregation but not number of PGCs	[191]

(*continued*)

TABLE 6.4 (*Continued*)

Concentration	Stage	Duration	Route of Exposure	Histopathological Observations	Ref.
		Pentylphenol, Nonylphenol, Octylphenol: Other Species			
Carp (*Cyprinus carpio*)					
4-*tert*-Pentylphenol					
100, 320, and 1000 μg/L	50 dph	90 d	Aqueous (flow-through)	Reduced number of PGCs, inhibited spermatogenesis	[146]
100, 320, and 1000 μg/L	50 dph	90 d	Aqueous (flow-through)	Reduced number of PGCs, inhibited spermatogenesis	[30]
32, 100, 320, and 1000 μg/L	Sexually mature (<210 dph)	3 m	Aqueous (flow-through)	Regressed testes	[29]
140 μg/L	−3, 0, and 3 dph; 24 dph	To 110 dph	Aqueous	Reduced number of PGCs	[110]
Eelpout (*Zoarces viviparus*)					
4-*tert*-Octylphenol					
10, 50 or 100 μg/L 50 μg/L 4-*tert*-octylphenol plus ZM 189154 (anti-estrogen, 20 μg/g per week)	Adult	3 w	Aqueous (flow-through)	Impaired spermatogenesis, no cysts containing spermatocytes (instead, occupied with clusters of spermatogonia and/or Sertoli cells); very few or no spermatozoa (if present, phagocytosed by Sertoli cells lining the lobular walls or within lobules); accumulations of yellowish-brown pigmented (sometimes hypertrophied) cells in lumen of seminiferous lobules 50 μg/L plus ZM 189154: no spermatocytes	[152]

Guppy (*Poecilia reticulata*)					
Octylphenol					
25 µg/L	Adult	♂: 28 d ♀: 26–36 d	Aqueous	Inhibition of spermatogonial mitosis	[192]
4-*tert*-Octylphenol					
100, 300, and 900 µg/L	Adult	60 d (900 µg/L: 30 d)	Aqueous	Highest treatment: increased number of spermatozeugmata in enlarged sperm ducts; some of spermatozeugmata were ruptured, resulting in free spermatozoa; few spermatogenetic cysts at the periphery of these testes (conversion of early stages into mature sperm without replenishment of early stages)	[31]
Platyfish (*Xiphophorus maculatus*); nonylphenol					
80, 320, 960, and 1280 µg/L	Adult	28 d	Aqueous	Reduced amount of spermatozeugmata in testicular ducts, increased number of deformed spermatozeugmata, free sperm in testicular ducts	[71]

(continued)

TABLE 6.4 (*Continued*)

Concentration	Stage	Duration	Route of Exposure	Histopathological Observations	Ref.
Summer flounder (*Paralichthys dentatus*); octylphenol					
1.0, 10.0, and 100 mg/kg	Juvenile (2 y)	8 w	Two injections, second after 2 w	Highest treatment: numerous spermatogonia, no developing sperm cysts, testicular regression from stages IV and V to stage I After 8 w: testis development similar to controls, with ongoing stage IV and V spermatogenesis	[147]
Industrial Chemicals: Fathead Minnow, Zebrafish					
Fathead minnow (*Pimephales promelas*); benzophenone-2 (UV filter)					
0.002, 0.1, 1.2, 5.0 and 9.7 mg/L	Adult	15 d	Aqueous	$\geq$ 1.2 mg/L: inhibition of spermatogenesis	[188]
Zebrafish (*Danio rerio*); PCB-60					
1 μmol/kg	Adult	10 d	Maternal exposure through injection	Sexually mature offspring: necrosis in testis and defective spermatogenesis	[38]
Pesticides: Fathead Minnow, Medaka, Zebrafish					
Fathead minnow (*Pimephales promelas*)					
Prochloraz 0.03, 0.1, and 0.3 mg/L	Adult	21 d	Aqueous (flow-through)	0.03 mg/L: increase in spermatogonia 0.1 and 0.3 mg/L: increase in testicular stage	[193]
Fenarimol 0.1 and 1 mg/L	Adult	21 d		Increase in spermatogonia	

Medaka (*Oryzias latipes*); pentachlorophenol					
10, 20, 50, 100, and 200 μg/L	Adult	28 d	Aqueous (flow-through)	Inhibition of spermatogenesis (two highest treatment groups)	[171]
Zebrafish (*Danio rerio*); endosulfan					
10^{-7} M	2–2.5 hpf	up to 22–24 hpf	Aqueous	Alteration of somite specific domain aggregation but not number of PGCs	[191]
Pesticides: Other Species					
Bluegill sunfish (*Lepomis macrochirus*); endosulfan					
1 μg/L	Adult (2 y)	1, 2, 3, and 4 d; 1 and 2 w	Aqueous (semistatic)	1 d: smaller primary spermatogonia 2 d: separation of primary spermatocytes from seminiferous tubule walls 3 d: primary spermatogonia expelled to the lumen, damaged spermatids 4 d: clumping of spermatozoa, aggregation of spermatogonia and spermatocytes, damaged primary spermatogonia, fewer spermatozoa and spermatids 1 and 2 w: smaller primary spermatogonia	[123]

(continued)

TABLE 6.4 *(Continued)*

Concentration	Stage	Duration	Route of Exposure	Histopathological Observations	Ref.
Freshwater murrel (*Channa punctatus*)					
Fenitrothion 1.5 ppm	15–20 cm	Up to 120 d	Aqueous (semistatic)	After 30 d: primary spermatocytes under initial stages of necrosis and detaching from lobule boundary wall After 60 d: most of lobules devoid of intact spermatogenetic cells (mainly spermatocyte necrosis) After 90 d: lobules with masses of necrotic germ cells under necrosis, few intact spermatogonia and spermatocytes After 120 d: most of lobules devoid of intact germ cells, spermatid necrosis	[182]
Carbofuran 5 ppm	15–20 cm	Up to 120 d	Aqueous (semistatic)	After 30 d: secondary spermatogonia undergoing transformation into primary spermatocytes After 60 d: spermatocyte necrosis and forming of a ball-like structure in the lobular lumen After 90 d: few intact spermatogonia, degenerating primary spermatocytes as ball-like structures, few lobules with spermatids and sperm After 120 d: most spermatids undergoing necrosis	[182]

Nile tilapia (*Oreochromis niloticus*); paraquat					
0.5 mg/L at 17°C or 27°C	Adult	45 d	Aqueous (flow-through)	Lower visibility of spermatozoa	[156]
Redbelly tilapia (*Tilapia zillii*); metronidazole					
5, 10, and 20 mg/L	~75 g	2, 7, and 14 d	Aqueous (flow-through)	5 mg/L: no effects observed 10 and 20 mg/L: lumen of tubules devoid of spermatozoa; simultaneous administration of 200 g/kg vitamin C abolished effects	[157]
Summer flounder (*Paralichthys dentatus*); *o,p*′-DDT; *p,p*′-DDE					
15, 30, and 60 mg/kg	Juvenile (2 y)	8 w	Two injections, second after 2 w	DDT: testicular regression from stages IV and V to immaturity (stage I); DDE: no effects observed	[147]
Industrial Chemicals: Other Species					
Atlantic cod (*Gadus morhua*); Aroclor 1245					
1, 5, 10, 25, and 50 μg/g	Adult	$5^1/_2$ m	Diet	1 μg/g: inhibition of spermatogenesis 5 and 10 μg/L: necrosis of spermatogenic elements 25 and 50 μg/g: disorganization and degeneration of spermatogenic elements	[194]

(*continued*)

TABLE 6.4 (*Continued*)

Concentration	Stage	Duration	Route of Exposure	Histopathological Observations	Ref.
		Heavy Metals: Other Species			
Giant gourami (*Colisa fasciata*); nickel sulfate					
64 mg/L	Adult	96 h	Aqueous (semistatic)	Reduced spermatogenic activity, degenerating germ cells	[195]
Goldfish (*Carassius auratus*); cadmium					
10 mg/kg	Adult	Up to 615 d	Single and repeated intraperitoneal injections	Inhibition of spermatogenesis	[85]
Guppy (*Poecilia reticulata*)					
Cadmium					
225 ppm	adult	30 d	Aqueous	After 11d: reduced number of spermatogonial nest, primary and secondary spermatocytes After 21 d: vacuolized spermatogonial nest, primary and secondary spermatocytes, pycnosis in primary spermatocytes After 31 d: primary and secondary spermatocytes with distinct cytoplasmatic vacuolization, pycnosis in and reduced number of spermatocytes	[86]

Methylmercurychloride					
1, 1.8, 3.2, 5.6, and 10 μg/L	4 w	1 and 3 m	Aqueous	Inhibition of spermatogenesis, sperm degeneration/necrosis, absence of mature sperm	[32]
Rosy barb (*Puntius conchonius*)					
Zinc: ~ 11 mg/L Copper: ~ 0.2 mg/L Lead: ~ 0.12 mg/L	~5 cm	2, 3, and 4 m	Aqueous (semistatic)	Cu: transient arrest of spermatogenesis (disappearance of all stages between spermatogonia and sperms within 2 m, reappearance after 3 m)	[196]
Field: Fathead Minnow (*Pimephales promelas*)					
Experimental lake in Ontario, Canada					
5–6 ng/L EE_2	Adult	3 y	Aqueous	Arrest of spermatogenesis at spermatogonial stage	[7]
Field: Other Species					
Catfish (*Pangasianodon hypophthamus*); Mekong River water (containing heavy metals and POPs)					
n.d.	0 dph	Lifelong	Field	Only necrotic spermatogonia in fish with small testis, karyolysis of nuclei, swelling and breaking of spermatogonia; fish with normal testes: normal morphology	[163]

(continued)

TABLE 6.4 *(Continued)*

Concentration	Stage	Duration	Route of Exposure	Histopathological Observations	Ref.
Crucian carp (*Carassius carassius*); treated sewage effluent					
25%, 50%, and 100%	Adult (1–2 y)	28 d	Field	Spermatogenesis progressively reduced to total inhibition in fish exposed to 100%	[48]
Gudgeon (*Gobio gobio*); treated sewage effluent					
n.d.	Adult	Lifelong	Field	Delayed spermatogenesis	[177]
Mirror carp (*Cyprinus carpio*); treated sewage effluent					
25%, 50%, and 100%	Adult (2–3 y)	28 d	Field	Restriction of spermatogenesis, in some cases total inhibition with gonadal recrudescence and some very reduced lobules containing only spermatogonia	[178]

Compounds and compound classes observed to induce germ cell effects (e.g., changes in the number/localization of PGCs, absence of specific stages, increased necrosis/apoptosis, degeneration, and phagocytosis of specific germ cell stages, inhibition/acceleration of spermatogenesis) in conjunction with a defined concentration, developmental stage, duration, and form/route of exposure. Tables are ordered according to the main three laboratory fish species and other species and distinguish laboratory experiments from field observations. Included are field trials with "contaminated" water of unknown chemical composition. The most prominent findings are listed along with the respective reference. d, days; dpf, days post-fertilization; dph, days post-hatch; °C dph, accumulated daily degrees °C post-hatch; h, hatch; hpf, hours post-fertilization; hr, hours; n.d., not defined or specifiable; m, months; w, weeks; y, years.

TABLE 6.5 Somatic Cell Effects in Male Fish

Concentration	Stage	Duration	Route of Exposure	Histopathological Observations	Ref.
		17α-ethinylestradiol: Zebrafish (*Danio rerio*)			
25 ng/L	Adult (2.5–4 cm)	21 d	Aqueous (flow–through)	Increased number and phagocytosis of Sertoli cells	[54]
		17α-ethinylestradiol: Guppy (*Poecilia reticulata*)			
10, 50, and 200 ng/L	Birth	3.5 m (to adulthood)	Aqueous (flow-through)	200 ng/L: increased interstitial tissue, Sertoli cell hypertrophy	[87]
		17β-estradiol: Fathead Minnow (*Pimephales promelas*)			
0.0625, 0.1, 0.125, 0.25, 0.5, 1, 2, 10 100, and 1000 nM	Adult	14 d	Aqueous	Sertoli cell hypertrophy and hyperplasia	[55]
		17β-estradiol: Other Species			
Guppy (*Poecilia reticulata*)					
0.03 and 0.1 μg/L	Adult	60 d	Aqueous	Increased number of hypertrophied Sertoli cells	[31]
Platyfish (*Xiphophorus maculatus*)					
96 μg/L	Adult	28 d	Aqueous	Sertoli cell hypertrophy and hyperplasia	[71]
Summer flounder (*Paralichthys dentatus*)					
0.1, 1.0, and 10.0 mg/kg	Juvenile (2 y)	4 w	Two injections, second after 2 w	Increase in interstitial tissue	[147]

(continued)

TABLE 6.5 (*Continued*)

Concentration	Stage	Duration	Route of Exposure	Histopathological Observations	Ref.
		Androgens: Zebrafish (*Danio rerio*)			
17α-Methyldihydrotestosterone 1, 10, and 100 μg/L	Adult	21 d	Aqueous (Semistatic)	Sertoli cell hypertrophy and hyperplasia	[73]
		Anti-androgens: Guppy (*Poecilia reticulata*)			
Flutamide 1, 10, and 100 μg/g	Adult	30 d	Diet	Sertoli cell hypertrophy and hyperplasia	[31]
		Bisphenol A: Fathead Minnow (*Pimephales promelas*)			
0, 1, 16, 160, 640, and 1280 μg/L	Reproductive stage	43, 71, and 164 d	Aqueous (flow-through)	No effects observed	[6]
0, 1, 16, 64, 160, and 640 μg/L	Reproductive stage	164 d	Aqueous (flow-through)	Decreased number of Leydig cells at 160 and 640 μg/L	[23]
		4-Nonylphenol: Fathead Minnow (*Pimephales promelas*), Medaka (*Oryzias latipes*)			
0.09, 0.1, 0.33, 0.93, and 2.4 μg/L	Adult	42 d	Aqueous (flow-through)	≥ 0.33 μg/L: Sertoli cell hyperplasia	[43]
100 ppb	Adult	6 w	Aqueous	Six-fold increased rate of apoptosis of Sertoli and Leydig cells	[51]
		Nonylphenol, Octylphenol: Other Species			
Eelpout (*Zoarces viviparus*); 4-*tert*-octylphenol 10, 50, or 100 μg/L	Adult	3 w	Aqueous (flow-through)	No cysts containing spermatocytes (instead, occupied with clusters of spermatogonia and/or Sertoli cells)	[152]

Platyfish (*Xiphophorus maculatus*); nonylphenol					
80, 320, 960, and 1280 µg/L	Adult	28 d	Aqueous	Sertoli cell hypertrophy and hyperplasia	[71]
Summer flounder (*Paralichthys dentatus*); octylphenol					
1.0, 10.0, and 100 mg/kg	Juvenile (2 y)	8 w	Two injections, second after 2 w	Thickened interstitial tissue	[147]
Pesticides: Fathead Minnow (*Pimephales promelas*)					
Ketoconazole					
25 µg/L	Adult	21 d	Aqueous (flow-through)	Leydig cell proliferation, elevated mRNA transcripts for two key steroidogenic enzymes, cytochrome P450 side-chain cleavage (CYP11A) and cytochrome P450 c17α hydroxylase/17,20-lyase (CYP17)	[96]
Pesticides: Other Species					
Bluegill sunfish (*Lepomis macrochirus*); endosulfan					
1 µg/L	Adult (2 y)	1, 2, 3, and 4 d 1 and 2 w	Aqueous (semistatic)	3 d: some Sertoli cells appeared damaged, some Leydig cells and primary spermatogonia were in close proximity (possibly due to breakdown of seminiferous tubule wall) 1 and 2 w: smaller Sertoli cells	[123]
Freshwater murrel (*Channa punctatus*); fenitrothion					
1.5 ppm	15–20 cm	120 d	Aqueous (semistatic)	After 120 d: more or less involuted condition of interstitial (Leydig) cells	[182]

(continued)

TABLE 6.5 (*Continued*)

Concentration	Stage	Duration	Route of Exposure	Histopathological Observations	Ref.
Carbofuran					
5 ppm	15–20 cm	120 d	Aqueous (semistatic)	After 120 d: more or less involuted condition of interstitial (Leydig) cells	[182]
Redbelly tilapia (*Tilapia zillii*); metronidazole (antiparasitic, antibacterial)					
5, 10, and 20 mg/L	~75 g	2, 7, and 14 d	Aqueous (flow-through)	10 and 20 mg/L: few melanophores in interstitial tissue; simultaneous administration of 200 g/kg vitamin C abolished effect	[157]
Industrial Chemicals: Other Species					
Nile tilapia (*Oreochromis niloticus*); Aroclor 1254					
0.05 g/kg	10 dph	40 d	Diet	After 18 months: Leydig cell hyperplasia	[97]
Heavy Metals: Other Species					
Guppy (*Poecilia reticulata*)					
Cadmium					
225 ppm	Adult	30 d	Aqueous	After 11 days: interstitial cell necrosis After 21 and 31d: Sertoli cell hypertrophy	[86]
Methylmercury chloride					
1, 1.8, 3.2, 5.6, and 10 μg/L	4 w	1 and 3 m	Aqueous	Sertoli cell hypertrophy	[32]

Field: Other Species					
Carp (*Cyprinus carpio*)/goldfish (*Carassius auratus*); Hamilton Harbour, Lake Ontario, Canada					
n.d.	Adult	Lifelong	Field	Two different types of testicular tumours, one with spindle-shaped cells and a few Leydig cells, the other type of spermatogonial origin with interspersed Sertoli cells and, more rarely, Leydig cells	[174]
Catfish (*Pangasianodon hypophthamus*); Mekong River water (containing heavy metals and POPs)					
n.d.	0 dph	Lifelong	Field	In fish with small testis: Sertoli cell hypertrophy and vacuolization In fish with normal testes: normal morphology	[163]

Compounds and compound classes observed to induce somatic cell effects (e.g., Sertoli cells/Leydig cells/interstitial cells: hypertrophy, hyperplasia, apoptosis, necrosis) in conjunction with a defined concentration, developmental stage, duration, and form/route of exposure. Tables are ordered according to the main three laboratory fish species and other species and distinguish laboratory experiments from field observations. Included are field trials with "contaminated" water of unknown chemical composition. The most prominent findings are listed along with the respective reference, d, days; dpf, days post-fertilization; dph, days post-hatch; °C dph, accumulated daily degrees °C post-hatch; h, hatch; hpf, hours post–fertilization; hr, hours; n.d., not defined or specifiable; m, months; w, weeks; y, years.

TABLE 6.6 Testicular Inflammation and Macrophage Infiltration in Male Fish

Concentration	Stage	Duration	Route of Exposure	Histopathological Observations	Ref.
17α-ethinylestradiol: Zebrafish (*Danio rerio*)					
9.3 ng/L	0 dph	177 dph	Aqueous (flow-through)	Macrophage infiltration, inflammation	[101]
Bisphenol A: Fathead Minnow (*Pimephales promelas*)					
0, 1, 16, 160, 640, and 1280 μg/L	Reproductive stage	43, 71, and 164 d	Aqueous (flow-through)	No effects observed	[6]
0, 1, 16, 64, 160, and 640 μg/L	Reproductive stage	164 d	Aqueous (flow-through)	No effects observed	[23]
Bisphenol A: Carp (*Cyprinus carpio*)					
0, 1, 10, 100, and 1000 μg/L	1 y	14 d	Aqueous (flow-through)	1 μg/L only: granulocyte infiltrations in males	[151]
Heavy Metals: Other Species					
Goldfish (*Carassius auratus*); cadmium 10 mg/kg	Adult	Up to 615 d	Single and repeated intraperitoneal injections	Sharp increase in macrophages filled with numerous small granules	[85]
Guppy (*Poecilia reticulata*); methylmercurychloride 1, 1.8, 3.2, 5.6, and 10 μg/L	4 w	1 and 3 m	Aqueous	Testicular inflammation	[32]

Compounds and compound classes observed to induce testicular inflammation and macrophage infiltration in male fish in conjunction with a defined concentration, developmental stage, duration, and form/route of exposure. Tables are ordered according to the main three laboratory fish species and other species and distinguish laboratory experiments from field observations. Included are field trials with "contaminated" water of unknown chemical composition. The most prominent findings are listed along with the respective reference. d, days; dpf, days post-fertilization; dph, days post-hatch; °C dph accumulated daily degrees °C post-hatch; h, hatch; hpf, hours post-fertilization; hr, hours; n.d., not defined or specifiable; m, months; w, weeks; y, years.

TABLE 6.7 Overt Ovarian Changes in Female Fish

Concentration	Stage	Duration	Route of Exposure	Histopathological Observations	Ref.
17α-ethinylestradiol: Medaka, Zebrafish					
Medaka (*Oryzias latipes*), d-rR strain					
0.005–2.5 ng per egg	Egg stage	To adulthood	Nano-injection	0.5 and 2.5 ng/egg: small ovaries	[127]
Zebrafish (*Danio rerio*)					
5 ng/L	Egg stage	210 d (lifelong)	Aqueous (flow-through)	Arrest of development rather than producing functional females	[128]
17α-ethinylestradiol: Other Species					
Rare minnow (*Gobiocypris rarus*)					
0, 1, 5, and 25 ng/L	Adult (∼7 m)	28 d	Aqueous (flow-through)	Degenerated ovary	[165]
0.2, 1, 4, 16, and 64 ng/L (16 and 64 ng/L: toxic)	2 hpf	180 d	Aqueous (flow-through)	1 and 4 ng/L: degeneration of ovaries	[138]
Estrone, Estriol: Medaka (*Oryzias latipes*)					
0.01, 0.1, 1, and 10 μg/L Estriol	1 dph	85–110 d	Aqueous	Significantly more males	[102]
0.01, 0.1, 1, and 10 μg/L Estrone	1 dph	85–110 d	Aqueous	0.1 μg/L only: significantly more males	[102]
17β-estradiol: Other Species					
Baltic salmon (*Salmo salar*)					
0.94 or 9.4 μg/g	Fry	8, 13, 17, and 41 w	Diet	Lack of ovarian lamellae	[145]

(*continued*)

TABLE 6.7 (*Continued*)

Concentration	Stage	Duration	Route of Exposure	Histopathological Observations	Ref.
Androgens: Fathead Minnow, Zebrafish					
Fathead minnow (*Pimephales promelas*); methyldihydrotestosterone					
0, 0.10, 0.32, and 1.0 µg/L	Early blastula embryos	Up to 114 dph	Aqueous	Suppression of ovarian development	[124]
Zebrafish (*Danio rerio*)					
17α-methyltestosterone 26–1000 ng/L	20 dph	40 d	Aqueous	Complete masculinization (all treatments); 1000 ng/L: some fish with one gonad occupied with oocytes and one gonad consisted of testicular tissue	[130]
Trenbolone acetate (synthetic androgen)					
< 5, 9.7, 23, 117, and 2779 ng/L	Fertilized egg	59 d	Aqueous (flow-through)	≥9.7 ng/L: 100% masculinization	[133]
Androgens: Other Species					
Arctic char (*Salvelinus alpinus*); 17α-methyldihydrotestosterone					
0.5, 1, 3, 5, and 10 mg/L	Hatched	140°C dph	Aqueous (flow-through)	Increasing proportion of "not females"	[149]
0.5 mg/kg	140°C dph	600°C dph	Diet		

Chinook salmon (*Oncorhynchus tshawytscha*)					
17α-Methyltestosterone					
10 mg/L	3 days after 50% hatching	2 d	Aqueous	22% masculinization	[200]
17α-Methyldihydrotestosterone					
10 mg/L				91% masculinization (gonads indistinguishable from testes of normal XY males of the same age)	
European sea bass (*Dicentrarchus labrax*); 17α-methyltestosterone					
10 mg/kg	126 dpf	100 d	Diet	At 3 years of age: permanent masculinization, increased proportion of males with functional ducts producing sperm capable of fertilizing ova	[168]
	226 dpf	100 d			
	326 dpf	100 d			
	126–426 dpf	300 d		All males	
Nile tilapia (*Oreochromis niloticus*); 17α-methyltestosterone					
50 µg/g	8 dph	22 d	Diet	100% masculinization	[201]
Rainbow trout (*Oncorhynchus mykiss*)					
1,4,6-Androstatriene-3,17-dione 50 mg/kg	Onset of first feeding	900 to 950°C days	Diet	100% masculinization of a genetic all female population	[202]
11β-Hydroxyandrostenedione 1 mg/kg				100% masculinization of a genetic all female population	

(continued)

TABLE 6.7 *(Continued)*

Concentration	Stage	Duration	Route of Exposure	Histopathological Observations	Ref.
Three-spined stickleback (*Gasterosteus aculeatus*); 17α-methyltestosterone					
1 μg/L	0 dph	To 39–51 dph	Aqueous (semistatic)	Development of large branched cavities with unidentified origin	[136]
	0 dph	To 14 dph		No effects observed	
	14 dph	To 39–51 dph			
Tilapia (*Tilapia mossambica*); 17α-methyltestosterone					
50 μg/g	7 dph	19 d	Diet	No ovarian cavity, but testes with well-formed efferent ducts	[112]
1000 μg/g	7 dph	19 d		No effects observed in ovaries	
1000 μg/g	7 dph	44 d		Expanded efferent ducts	
P450-Aromatase Inhibitors: Zebrafish (*Danio rerio*)					
Fadrozole					
500 μg/g food	Fertilized egg	From 35 to 71 dpf	Diet	100% masculinization; gonads with testicular morphology; after termination of fadrozole treatment no recovery	[132]
10, 100, 1000 μg/g	Juvenile (15 d)	25 d	Diet (5% of fish body weight)	10 μg/L: 62.5% gonadal masculinization (sex reversal) 100 and 1000 μg/L: 100% gonadal masculinization	[203]

P450-Aromatase Inhibitors: Other Species					
Chinook salmon (*Oncorhynchus tshawytscha*); 4-(5,6,7,8-tetrahydroimidazo[1,5-a] pyridin-5-yl)benzonitrilemonohydrochloride					
10 mg/L	3 days after 50% hatching	2 d	Aqueous	23% masculinization of females, gonads indistinguishable from testes of normal XY males of the same age	[204]
Nile tilapia (*Oreochromis niloticus*); fadrozole					
0, 50, 75, and 100 mg/kg	9 dph	15 and 30 d	Diet	At 80 dph: proportion of males in all treatment groups significantly elevated (75 and 100 mg/kg for 30 d produced 100% males, no histological differences compared to male controls)	[170]
Bisphenol A: Fathead Minnow (*Pimephales promelas*)					
0, 1, 16, 160, 640, and 1280 μg/L	Reproductive stage	43, 71, and 164 d	Aqueous (flow-through)	No effects observed	[6]
0, 1, 16, 64, 160, and 640 μg/L	Reproductive stage	164 d	Aqueous (flow-through)	Increased intravascular and interstitial proteinaceous fluid at 640 μg/L	[23]
Bisphenol A: Other Species					
Carp (*Cyprinus carpio*)					
0, 1, 10, 100, and 1000 μg/L	1 y	14 d	Aqueous (flow-through)	Oocyte atresia in 57% of females exposed to 1000 μg/L	[151]
Octylphenol: Medaka (*Oryzias latipes*)					
10, 25, 50, and 100 μg/L	1 dph	6 m	Aqueous	Proliferation of the interstitial stroma surrounding oocytes	[99]

(continued)

TABLE 6.7 (*Continued*)

Concentration	Stage	Duration	Route of Exposure	Histopathological Observations	Ref.
		Pesticides: Fathead Minnow (*Pimephales promelas*), Zebrafish (*Danio rerio*)			
Vinclozolin 100 and 700 μg/L	Adult	21 d	Aqueous (flow-through)	Ovaries in a pre-vitellogenic state	[67]
Prochloraz 0, 16, 64, and 202 μg/L	24 hpf	60 d	Aqueous (flow-through)	202 μg/L: masculinization	[154]
		Pesticides: Other Species			
Bluegill sunfish (*Lepomis macrochirus*); diazinon					
60 μg/L	Adult	1, 2, 3, and 4 d; 1, 2, and 3 w	Aqueous (semistatic)	1 d: thinning of follicular lining 1 w: damage to ovigerous lamellae 2 w: thinning and necrosis of ovarian wall 3 w: extremely thin and broken ovarian wall, severe damage in ovigerous lamellae	[47]
60 μg/L	Adult	1, 2, 3, and 4 d; 1 and 2 w	Aqueous (semistatic)	3 d: ovigerous lamellae almost obliterated 1 and 2 w: ovarian damages increase	[46]
Catfish (*Heteropneustes fossilis*); carbofuran					
0.5 L and 2 mg/L	16–17 cm	30 d	Aqueous (semistatic)	Degeneration of follicular wall, damage in connective tissue	[205]

Snakehead murrel (*Channa striatus*)					
Carbaryl 10 and 20 ppm Endosulfan 0.00075 and 0.001 ppm	~15 cm	30 d	Aqueous (semistatic)	Presence of interfollicular spaces, increase in connective tissue of tunica albuginea	[183]
Heavy Metals: Other Species					
Giant gourami (*Colisa fasciata*); nickel sulfate					
64 mg/L	Adult	96 h	Aqueous (semistatic)	Occurrence of comparatively large interfollicular spaces (due to shrinkage of oocytes)	[195]
Striped gourami (*Colisa fasciata*); arsenic(III)oxide					
2 and 14 mg/L	Adult	15 and 30 d	Aqueous (semistatic)	Effects visible only at 14 mg/L after 30 d: prominent follicular spaces	[206]
Field: Other Species					
Common carp (*Cyprinus carpio*); Rio Grande basin (dioxins, PCBs, heavy metals, pesticides)					
n.d.	Adult	Lifelong	Field	Some fish with ovarian inflammation	[49]
Redear sunfish (*Lepomis microlophus*); Martin Lake (selenium contaminated)					
n.d.	Adult	Lifelong	Field	Ovarian connective tissue hypertrophy	[160]

(continued)

TABLE 6.7 *(Continued)*

Concentration	Stage	Duration	Route of Exposure	Histopathological Observations	Ref.
White suckers (*Catostomus commersoni*); Colorado River basin					
n.d.	Adult	Lifelong	Field	Asynchronous ovarian development	[161]
Whitefish (*Coregonus* sp.); Lake Thun					
n.d.	1–6 y; highest prevalence in 3-to 5-yold fish	n.d.	Field	35% affected gonads (281 out of 808): adhesions/fusions to peritoneal wall and lateral trunk musculature; asymmetry, atrophy, compartmentations, and constrictions of gonads Hermaphroditism: Consecutive sequence of testicular and ovarian lobes on the same gonad strand One gonad strand ovarian tissue and the other testicular tissue	[162]

Compounds and compound classes observed to induce overt ovarian changes (e.g, sex reversal, ovo-testis, undifferentiated gonad, immature gonad, lacking gonad, ovarian atrophy, ovarian dilatation via dilated interstitial spaces, ovarian cellular degeneration, interrupted or arrested ovarian/follicular development) in female fish in conjunction with a defined concentration, developmental stage, duration, and form/route of exposure. Tables are ordered according to the main three laboratory fish species and other species and distinguish laboratory experiments from field observations. Included are field trials with "contaminated" water of unknown chemical composition. The most prominent findings are listed along with the respective reference. d, days; dpf, days post-fertilization; dph, days post-hatch; °C dph, accumulated daily degrees °C post-hatch; h, hatch; hpf, hours post-fertilization; hr, hours; n.d., not defined or specifiable; m, months; w, weeks; y, years.

TABLE 6.8 Ovo–Testis in Female Fish

Concentration	Stage	Duration	Route of Exposure	Histopathlogical Observations	Ref.
17α-ethinylestradiol: Other Species					
Sheepshead minnow (*Cyprinodon variegates*); 17α-ethinylestradiol					
0.2, 2, 20, 200, 400, 800, 1600, and 3200 ng/L	Subadult	Up to 59 d	Aqueous (flow-through)	Spermatogonia in ovary	[69]
Androgens: Other Species					
Three-spined stickleback (*Gasterosteus aculeatus*); 17α-methyltestosterone					
1 μg/L	0 dph	To 39–51 dph	Aqueous (semistatic)	Ovo–testis	[136]
	0 dph	To 14 dph		Ovo–testis	
	14 dph	To 39–51 dph		No effects observed	
Bisphenol A: Fathead Minnow (*Pimephales promelas*)					
0, 1, 16, 160, 640, and 1280 μg/L	Reproductive stage	43, 71, and 164 d	Aqueous (flow-through)	No effects observed	[6]
0, 1, 16, 4, 160, and 640 μg/L	Reproductive stage	164 d	Aqueous (flow-through)	No effects observed	[23]

Compounds and compound classes observed to induce ovo–testis in female fish in conjunction with a defined concentration, developmental stage, duration, and form/route of exposure. Tables are ordered according to the main three laboratory fish species and other species and distinguish laboratory experiments from field observations. Included are field trials with "contaminated" water of unknown chemical composition. The most prominent findings are listed along with the respective reference. d, days; dpf, days post-fertilization; dph, days post-hatch; °C dph, accumulated daily degrees °C post-hatch; h, hatch; hpf, hours post-fertilization; hr, hours; n.d., not defined or specifiable; m, months; w, weeks; y, years.

TABLE 6.9 Ovarian Fibrosis in Female Fish

Concentration	Stage	Duration	Route of Exposure	Histopathological Observations	Ref.
		17α-ethinylestradiol: Zebrafish (*Danio rerio*)			
9.3 ng/L	0 dph	177 dph	Aqueous (flow-through)	Fibrosis	[101]
		Bisphenol A: Fathead Minnow (*Pimephales promelas*)			
0, 1, 16, 160, 640, and 1280 μg/L	Reproductive stage	43, 71, and 164 d	Aqueous (flow-through)	No effects observed	[6]
0, 1, 16, 64, 160, and 640 μg/L	Reproductive stage	164 d	Aqueous (flow-through)	No effects observed	[23]
		Pesticides: Other Species			
Snakehead murrel (*Channa striatus*) Carbaryl: 10 and 20 ppm Endosulfan: 0.00075 and 0.001 ppm	Juvenile (∼15 cm)	30 d	Aqueous (semistatic)	Increase in connective tissue of tunica albuginea	[183]
		Field: Other Secies			
Common carp (*Cyprinus carpio*); Rio Grande basin (dioxins, PCBs, heavy metals, pesticides)					
n.d.	Adult	Lifelong	Field	Fibrosis (1 out of 113) and inflammation	[49]
Redear sunfish (*Lepomis microlophus*); Martin Lake (selenium contaminated)					
n.d.	Adult	Lifelong	Field	Connective tissue hypertrophy	[160]

Compounds and compound classes observed to induce ovarian fibrosis in female fish in (conjunction with a defined concentration, developmental stage, duration, and form/route of exposure. Tables are ordered according to the main three laboratory fish species, and other species and distinguish laboratory experiments from field observations. Included are field trials with "contaminated" water of unknown chemical composition. The most prominent findings are listed along with the respective reference. (d, days; dpf, days post-fertilization; dph, days post-hatch; °C dph, accumulated daily degrees °C post-hatch; h, hatch; hpf, hours post-fertilization; hr, hours; n.d., not defined or specifiable; m, months; w, weeks; y, years.

TABLE 6.10 Germ Cell Effects in Female Fish

Concentration	Stage	Duration	Route of Exposure	Histopathological Observations	Ref.
		17α-ethinylestradiol: Fathead Minnow, Medaka, Zebrafish			
Fathead minnow (*Pimephales promelas*)					
50 ng/L	Adult	15 d	Aqueous	Increase in previtellogenic oocytes	[188]
Medaka (*Oryzias latipes*), d-rR strain					
0.005–2.5 ng per egg	Egg stage	To adulthood	Nano-injection	0.5 and 2.5 ng/egg: ovaries most in a pre- to early vitellogenic stage	[127]
32.6, 63.9, 116, 261, and 488 ng/L	Adult	21 d	Aqueous (flow-through)	488 ng/L: regressed ovary (many previtellogenic oocytes)	[1]
Zebrafish (*Danio rerio*)					
10 and 25 ng/L	Adult	3, 6, 12, and 24 d	Aqueous	Small regressed ovary without mature oocytes; recovery after 24 d	[28]
1 and 10 ng/L	2 dph	up to 60 dph	Aqueous	10 ng/L: suppression of ovarian development (only previtellogenic stages)	[26]
15.4 ng/L	20–60 dph; h to 60 dph	Up to 60 dph	Aqueous	All females with small ovaries containing few primary oocytes	[131]
		17β-estradiol: Fathead Minnow, Medaka, Zebrafish			
Fathead minnow (*Pimephales promelas*)					
0.0625, 0.1, 0.125, 0.25, 0.5, 1, 2, 10, 100, and 1000 nM	Adult	14 d	Aqueous	More immature follicles, fewer secondary and Graafian follicles	[55]

(continued)

TABLE 6.10 *(Continued)*

Concentration	Stage	Duration	Route of Exposure	Histopathological Observations	Ref.
Medaka (*Oryzias latipes*)					
0.939, 2.86, 8.66, 27.9, and 92.4 ng/L	12 hpf	101 d	Aqueous	≥ 92.4 ng/L: regressed ovary (many previtellogenic oocytes)	[166]
Zebrafish (*Danio rerio*)					
0.27, 2.7, and 27 ng/L	Adult	21 d	Aqueous (semistatic)	≥ 2.7 ng/L: regressed ovary	[73]
17β-estradiol: Other Species					
Guppy (*Poecilia reticulata*)					
0.85 μg/L	Adult	26–36 d	Aqueous	Inhibited yolk deposition, depletion of late stage oocytes	[192]
Androgens: Fathead Minnow (*Pimephales promelas*)					
Dihydrotestosterone					
50 μg/L	Adult	15 d	Aqueous	Decrease in late vitellogenic oocytes	[188]
Methyltestosterone					
0.12 and 1.7 mg/L	Reproductive stage	12 d	Aqueous	Inhibition of oogenesis	[42]
22.5, 46.8, 88.1, 188, and 380 ng/L	Adult	3 w	Aqueous (flow-through)	Degenerated oocytes (abnormal shape, fusion of yolk globule, yolk vesicle, and oil droplet; indicative of regression of oogenesis); increasing severity with increasing concentrations	[167]

Anti-androgens: Fathead Minnow, Medaka					
Fathead minnow (*Pimephales promelas*); Flutamide					
62.7 ± 6 and 651 ± 45 μg/L	Adult (6 m)	21 d	Aqueous (flow-through)	Increase in number of early-stage follicles	[150]
Medaka (*Oryzias latipes*, Qurt strain); Flutamide					
0.032, 0.1, 0.32, 1.0, and 3.2 mg/L	1 w	96 h	Aqueous (semistatic)	After 3 m: severe germ cell necrosis	[44]
Medaka (*Oryzias latipes*), golden strain; Cyproterone acetate					
1 and 10 μg/L	1 dph	100 d	Aqueous	Moderate inhibition of oogenesis	[27]
P450-Aromatase Inhibitors: Fathead Minnow (*Pimephales promelas*)					
Fadrozole					
2, 10, and 50 μg/L	Reproductive stage	21 d	Aqueous	Inhibition of oogenesis	[64]
P450-Aromatase Inhibitors: Coho salmon (*Oncorhynchus kisutch*)					
Fadrozole					
1 × 1 mg/kg, 1 × 10 mg/kg, 5 × 10 mg/kg	Adult	32 d	Weekly, intraperitoneal injection	Single injections: increase in follicle diameter	[216]
Bisphenol A: Fathead Minnow, Medaka, Zebrafish					
Fathead minnow (*Pimephales promelas*); Bisphenol A					
0, 1, 16, 160, 640, and 1280 μg/L	Reproductive stage	43, 71, and 164 d	Aqueous (flow-through)	No effects observed	[6]
0, 1, 16, 64, 160, and 640 μg/L	Reproductive stage	164 d	Aqueous (flow-through)	Increased numbers of early vitellogenic follicles at 640 μg/L	[23]

(*continued*)

TABLE 6.10 (*Continued*)

Concentration	Stage	Duration	Route of Exposure	Histopathological Observations	Ref.
Medaka (*Oryzias latipes*); Bisphenol A					
0, 10, 50, 100, and 200 µg/L	1 dph	85–110 d	Aqueous	Advanced stage of oogenesis in 200 µg/L group	[102]
Zebrafish (*Danio rerio*); tetrabromo-bisphenol A					
0.023, 0.047, 0.094, 0.188, 0.375, 0.750, and 1.5 µM	Adult	30 d	Aqueous (static renewal)	1.5 µM: inhibition of oogenesis	[212]
Nonylphenol: Medaka (*Oryzias latipes*)					
100 µg/L	3 dph	100 d	Aqueous	Inhibition of oogenesis	[39]
Industrial Chemicals: Fathead Minnow, Zebrafish					
Fathead minnow (*Pimephales promelas*); benzo-phenone-2 (UV filter)					
0.002, 0.1, 1.2, 5.0 and 9.7 mg/L	Adult	15 d	Aqueous	$\geq$ 1.2 mg/L: inhibition of oogenesis	[188]
Zebrafish (*Danio rerio*); mixture of 20 PCBs					
0.008, 0.08, and 0.4 µg/g of each congener	Adult	13 w	Diet	Reduced number of mature oocytes, only primary oocytes in the highest dose group	[40]
Industrial Chemicals: Other Species					
Baltic salmon (*Salmo salar*); polychlorinated naphthalenes					
0.1, 1, 2, and 10 µg/g	Fry	8, 13, 17, and 41 w	Diet	After 13 w, 10 µg/g: lower number of primary oocytes; no differences at any other time or concentration	[145]

Pesticides: Fathead Minnow, Zebrafish					
Fathead minnow (*Pimephales promelas*); vinclozolin					
100 and 700 μg/L	Adult	21 d	Aqueous (flow-through)	Retardation of oocyte maturation, ovaries in a pre-vitellogenic stage	[67]
Zebrafish (*Danio rerio*); 2,3,7,8-TCCD					
0.08, 0.32, and 0.80 ng per female and day	Adult (6 m)	4 w	Diet	Significantly reduced number and size of vitellogenic follicles (inhibition of transition)	[66]
Pesticides: Other Species					
Bluegill sunfish (*Lepomis macrochirus*); diazinon					
60 μg/L	Adult	1, 2, 3, and 4 d 1, 2 and 3 w	Aqueous (semistatic)	1 d: cytoplasm retraction (CR) 2 d: karyoplasmic clumping (KC), very few nucleoli in oocyte III, oocyte adhesion (OA), severe damage in most oocytes II, fewer effects on oocyte IV 3 d: more OA, severe CR, numerous atretic oocytes, cytoplasmic expulsion from oocyte IV 4 d: KC, CR, severe damage to the follicular epithelium of oocyte IV, anucleated oocyte IV due to loss nucleus, cytoplasmic necrosis 1 w: karyoplasm expulsion along with nucleoli from oocyte IV, OA between oocyte III and IV, empty oocytes, damage to ovigerous lamellae	[47]

(*continued*)

TABLE 6.10 (*Continued*)

Concentration	Stage	Duration	Route of Exposure	Histopathological Observations	Ref.
				2 w: more OA, greater karyoplasm retraction and CR, severe damage to oocyte III, anucleated oocyte IV with entire cytoplasm filled with yolk vacuoles 3 w: KC, CR, extreme cellular necrosis, large amount of OA, KR, oocyte necrosis	
60 μg/L	Adult	1, 2, 3, and 4 d; 1 and 2 w	Aqueous (semistatic)	1 d: cytoplasmic retraction (CR) and adhesion (AD) of follicles 2 d: karyoplasmic retraction (KR), anucleated oocytes (AN) 3 d: additional CR and KR 4 d: karyoplasmic clumping (KC), AN, fragmented and necrotic oocytes, KC in almost all follicles; ovarian damage increases	[46]
Catfish (*Heteropneustes fossilis*); carbofuran					
0.5, 1, and 2 mg/L	Juvenile (16–17 cm)	30 d	Aqueous (semistatic)	Degeneration of ooplasm, clumping of yolk globules, inhibition of oogenesis	[205]
False kelpfish (*Sebastiscus marmoratus*); tributyltin					
1, 10, and 100 ng/L	Juvenile (12–17 g)	50 d	Aqueous	10 ng/L: lack of vitellogenic stage follicles and higher proportions of primary growth stage follicles 100 ng/L: follicles entirely at the earliest (primary) stages of development 10 and 100 ng/L: increasing apoptosis	[217]

Freshwater murrel (*Channa punctatus*)					
Carbofuran 5 ppm Fenitrothion 1.5 ppm	Juvenile (15–20 cm)	120 d	Aqueous (semistatic)	With fenitrothion > carbofuran: inhibited/disturbed oogenesis. Fenitrothion only: significant decline in diameter of stage III oocytes and number of nucleoli in stage I oocytes	[218]
Nile tilapia (*Oreochromis niloticus*); paraquat					
0.5 mg/L at 17°C or 27°C	Adult	45 d	Aqueous (flow-through)	Significantly elevated proportion of late vitellogenic and mature stages, relative decrease in primary oocytes	[156]
Rosy barb (*Puntius conchonius*)					
Aldrin 0.0466 μg/L Methoxychlor 0.00746 μg/L Monocrotophos 53 μg/L	Juvenile (~5 cm)	2, 3, and 4 m	Aqueous (semistatic)	Increase in atretic oocytes, mainly former stage IV oocytes, which disappear simultaneously	[219]
Snakehead murrel (*Channa striatus*)					
Carbaryl 10 and 20 ppm Endosulfan 0.00075 and 0.001 ppm	Juvenile (~15 cm)	30 d	Aqueous (semistatic)	Lower number of oocytes, reduction in size and deformity in different stages of oocytes, damage to yolk vesicles	[183]

(*continued*)

TABLE 6.10 (*Continued*)

Concentration	Stage	Duration	Route of Exposure	Histopathological Observations	Ref.
		Heavy Metals: Other Species			
Catfish (*Clarias Batrachus*)					
Mercuric chloride 0.05 mg/L Methylmercuric chloride 0.04 mg/L Emisan 6 (organic mercurial fungicide) 0.5 mg/L	Adult	180 d	Aqueous (semistatic)	All treatments: impairment of ovarian recrudescence through inhibition of vitellogenesis	[215]
Goldfish (*Carassius auratus*); cadmium					
10 mg/kg	Adult	Up to 615 d	Single and repeated intraperitoneal injections	Inhibition of oogenesis	[85]
Striped gourami (*Colisa fasciatus*); arsenic(III) oxide					
2 and 14 mg/L	Adult	15 and 30 d	Aqueous (semistatic)	Effects visible only at 14 mg/L after 30 d: decreased development of oocytes II and III, reduced number and diameter of nucleoli	[206]
		Field: Fathead Minnow (*Pimephales promelas*)			
Michigan wastewater treatment plant effluent					
n.d.	Adult	3 w	Field	No effects observed	[158]
		Field: Other Species			
Common carp (*Cyprinus carpio*)/black bass (*Micropterus* spp.)/channel catfish (*Ictalurus punctatus*)					
Rio Grande basin (dioxins, PCBs, heavy metals, pesticides)		[illegible]	Field	Some fish with follicular	[49]

Crucian carp (*Carassius carassius*); treated sewage effluent					
25%, 50%, and 100%	Adult (1–2 y)	28 d	Field	Oocyte degeneration, inhibition of oogenesis	[48]
Gudgeon (*Gobio gobio*); treated sewage effluent					
n.d.	Adult	Lifelong	Field	Delayed oocyte maturation	[177]
Mirror carp (*Cyprinus carpio*); treated sewage effluent					
25%, 50%, and 100%	Adult (2–3 y)	28 d	Field	Stroma follicular cell degeneration	[178]
Redear sunfish (*Lepomis microlophus*); Martin Lake (selenium contaminated)					
n.d.	Adult	Lifelong	Field	Ovarian necrosis, abnormally shaped follicles, asynchronous oocyte development, overall reduction in number of developing oocytes	[160]
White suckers (*Catostomus commersoni*); Colorado River basin					
n.d.	Adult	Lifelong	Field	Delayed follicular maturation, asynchronous ovarian development	[161]
Bleached kraft pulp mill effluent					
n.d.	Adult	Lifelong	Field	Elevated ovarian follicular apoptosis	[53]

Compounds and compound classes observed to induce germ cell effects (e.g., number/localization of PGCs, absence of specific follicular stages, necrosis, apoptosis or degeneration of follicles, inhibition of vitellogenesis, changes of oocyte diameter, inhibition or acceleration of oogenesis) in female fish in conjunction with a defined concentration, developmental stage, duration, and form/route of exposure. Tables are ordered according to the main three laboratory fish species, and other species and distinguish laboratory experiments from field observations. Included are field trials with “contaminated” water of unknown chemical composition. The most prominent findings are listed along with the respective reference. d, days; dpf days post-fertilization; dph, days post-hatch; °C dph, accumulated daily degrees °C post-hatch; h, hatch; hpf, hours post-fertilization; hr, hours; n.d, not defined or specifiable m, months; w, weeks; y, years.

TABLE 6.11 Germ Cell Atresia in Female Fish

Concentration	Stage	Duration	Route of Exposure	Histopathological Observations	Ref.
		17α-ethinylestradiol: Medaka, Zebrafish			
Medaka (*Oryzias latipes*), d-rR strain					
0.005–2.5 ng per egg	Egg stage	To adulthood	Nano-injection	0.5 and 2.5 ng/egg: atretic oocytes	[127]
Zebrafish (*Danio rerio*)					
10 and 25 ng/L	Adult	3, 6, 12, and 24 d	Aqueous	Increased follicular atresia; recovery after 24 d	[28]
9.3 ng/L	0 dph	177 dph	Aqueous (flow-through)	Follicular atresia in all individuals displaying gonads with ovarian morphology	[101]
		17α-ethinylestradiol: Other Species			
Sheepshead minnow (*Cyprinodon variegates*)					
0.2, 2, 20, 200, 400, 800, 1600, and 3200 ng/L	Subadult	Up to 59 d	Aqueous (flow-through)	Atretic oocytes	[69]
		17β-estradiol, Ultrasitosterol: Fathead Minnow, Zebrafish			
Fathead minnows (*Pimephales promelas*); 17β-estradiol					
2780 ng/L (10 nM)	160 d	14 d	Aqueous (flow-through)	Ovary follicular atresia	[41]
0.0625, 0.1, 0.125, 0.25, 0.5, 1, 2, 10, 100, and 1000 nM	Adult	14 d	Aqueous	Increased follicular atresia	[55]

Zebrafish (*Danio rerio*)					
Oxidized ultrasitosterol preparation					
10 and 100 μg/L	Adult (4–6 m)	3 w	Aqueous (flow-through)	100 μg/L only: increase in atretic oocytes	[189]
17β-Estradiol					
0.27 μg/L	Adult (4–6 m)	3 w	Aqueous (flow-through)	Increase in atretic oocytes (80%)	[189]
		17β-estradiol: Other Species			
Guppy (*Poecilia reticulata*)					
0.85 μg/L	Adult	26–36 d	Aqueous	Atretic oocytes	[192]
		Androgens: Fathead Minnow, Medaka			
Fathead minnow (*Pimephales promelas*)					
Methyltestosterone					
0.12 and 1.7 mg/L	Reproductive stage	12 d	Aqueous	Atretic follicles	[42]
17α-Methyltestosterone					
0.1, 1, 5, and 50 μg/L	Adult (10–15 m)	3 w	Aqueous (flow-through)	≥0.1 μg/L: atretic follicles (increasing along with concentration up to 24%)	[148]
Medaka (*Oryzias latipes*), Qurt strain; 11-keto-testosterone					
0.01, 0.1, and 1.0 mg/L	1 w	96 h	Aqueous (semistatic)	Severe ovarian atresia	[44]
		Anti-androgens: Fathead Minnow, Medaka			
Fathead minnow (*Pimephales promelas*); Flutamide					
62.7 ± 6 and 651 ± 45 μg/L	Adult (6 m)	21 d	Aqueous (flow-through)	Increase in oocyte atresia (high treatment)	[150]

(continued)

TABLE 6.11 (*Continued*)

Concentration	Stage	Duration	Route of Exposure	Histopathological Observations	Ref.
Medaka (*Oryzias latipes*), Qurt strain; Flutamide					
0.032, 0.1, 0.32, 1.0, and 3.2 mg/L	1 w	96 h	Aqueous (semi-static)	After 3 m: ovarian atresia	[44]
P450-Aromatase Inhibitors: Fathead Minnow, Medaka					
Fathead minnow (*Pimephales promelas*); fadrozole					
2, 10, and 50 μg/L	Reproductive stage	21 d	Aqueous	Follicular atresia	[64]
Medaka (*Oryzias latipes*); quercetin					
100 ppb	Adult	6 w	Aqueous	Significant increase in atretic follicles	[51]
P450-Aromatase Inhibitors: Other Species					
Coho salmon (*Oncorhynchus kisutch*); fadrozole					
1 × 1 mg/kg 1 × 10 mg/kg 5 × 10 mg/kg (weekly)	Adult	32 d	Intraperitoneal injection	Multiple injections: significant increase in atretic oocytes	[216]
Bisphenol A: Fathead Minnows, Zebrafish					
Fathead minnow (*Pimephales promelas*); bisphenol A					
0, 1, 16, 160, 640, and 1280 μg/L	Reproductive stage	43, 71, and 164 d	Aqueous (flow-through)	No effects observed	[6]
0, 1, 16, 64, 160, and 640 μg/L	Reproductive stage	164 d	Aqueous (flow-through)	No effects observed	[23]

[illegible] (*Danio rerio*), tetrabromo-bisphenol A					
0.023, 0.047, 0.094, 0.188, 0.375, 0.750, and 1.5 μM	Adult	30 d	Aqueous (static renewal)	Atretic oocytes in all treatment groups	[212]
		Bisphenol A: Other Species			
Carp (*Cyprinus carpio*) 0, 1, 10, 100, and 1000 μg/L	1 y	14 d	Aqueous (flow-through)	Oocyte atresia in 57% of females exposed to 1000 μg/L	[151]
		Nonylphenol, Octylphenol: Medaka, Zebrafish			
Medaka (*Oryzias latipes*)					
Octylphenol 100 μg/L	1 dph	1, 2, and 3 m	Aqueous	Atresia	[99]
p-Nonylphenol 10, 50, and 100 μg/L	1 dph	3 m	Aqueous	Increased follicular atresia	[57]
Zebrafish (*Danio rerio*); 4-nonylphenol 100 μg/L	2 dph	Up to 60 dph	Aqueous	Increased follicular atresia	[26]
		Industrial Chemicals: Fathead Minnow, Zebrafish			
Fathead minnow (*Pimephales promelas*)					
Methoxychlor 0.55 and 3.56 μg/L	Reproductive stage	21 d	Aqueous	Increased follicular atresia in high concentration group (20%)	[42]
Benzophenone-2 (UV filter) 0.002, 0.1, 1.2, 5.0, and 9.7 mg/L	Adult	15 d	Aqueous	$\geq$ 1.2 mg/L: atretic oocytes	[188]

(continued)

TABLE 6.11 *(Continued)*

Concentration	Stage	Duration	Route of Exposure	Histopathological Observations	Ref.
Zebrafish (*Danio rerio*), (AB strain); 2,3,7,8-tetrachlorodibenzo-*p*-dioxin					
0.08, 0.32, and 0.80 ng per female and day	Adult (6 m)	4 w	Diet	Follicular atresia	[66]
Zebrafish (*Danio rerio*); 3 PCB congeners					
1 μmol/kg	Adult	10 d	Maternal exposure through injection	Sexually mature offspring PCB-60 and PCB-104: ovarian atresia	[38]
Industrial Chemicals: Other Species					
Nile tilapia (*Oreochromis niloticus*); Aroclor 1254					
0.05 g/kg	10 dph	40 d	Diet	After 18 months: atretic oocytes	[97]
Pesticides: Fathead Minnow, Medaka					
Fathead minnow (*Pimephales promelas*)					
Prochloraz 0.03, 0.1, and 0.3 mg/L Fenarimol 0.1 and 1 mg/L	Adult	21 d	Aqueous (flow-through)	Increase in oocyte atresia	[193]
Medaka (*Oryzias latipes*); Pentachlorophenol					
10, 20, 50, 100, and 200 μg/L	Adult	28 d	Aqueous (flow-through)	≥ 50 μg/L: ovarian atresia	[171]
Medaka (*Oryzias latipes*), golden strain; vinclozolin (Ronilan)					
1 and 5 mg/L	1 dph	100 d	Aqueous	Moderate ovarian atresia	[27]

Pesticides: Other Species					
Blue spotted tilapia (*Tilapia leucosticta*); lebaycid					
7 ppm; 5 single "shots" during exposure time, 24 h each	Adult	14 d	Aqueous (semistatic)	90% of ovaries with total atresia (filled with atretic follicles and no mature eggs)	[224]
Bluegill sunfish (*Lepomis macrochirus*); diazinon					
60 μg/L	Adult	1, 2, 3, and 4 d	Aqueous (semistatic)	2 d: several atretic oocytes 3 d: numerous atretic oocytes, cytoplasmic expulsion from oocyte IV	[47]
		1, 2, and 3 w		1 w: numerous atretic oocytes 3 w: large amount of atretic oocytes	
60 μg/L	Adult	1, 2, 3, and 4 d	Aqueous (semistatic)	2 d: ovarian atresia 4 d: increasing ovarian atresia	[46]
Freshwater murrel (*Channa punctatus*)					
Carbofuran 5 ppm Fenitrothion 1.5 ppm	15–20 cm	120 d	Aqueous (semistatic)	With fenitrothion > carbofuran: Atresia of previtellogenic and vitellogenic oocytes Fenitrothion only: 100% atretic oocytes after 120 d	[218]
Goldfish (*Carassius auratus*); atrazin					
100 and 1000 μg/L	Adult	21 d	Aqueous (flow-through)	Significantly higher proportion of oocyte atresia (20–25%)	[155]
Rosy barb (*Puntius conchonius*)					
Aldrin 0.0466 μg/L	~5 cm	2, 3 and 4 m	Aqueous (semistatic)	Increase in atretic oocytes, mainly former stage IV oocytes, which disappear simultaneously	[219]

(continued)

TABLE 6.11 (*Continued*)

Concentration	Stage	Duration	Route of Exposure	Histopathological Observations	Ref.
Methoxychlor 0.00746 μg/L Monocrotophos 53 μg/L					
Snakehead murrel (*Channa striatus*) Carbaryl 10 and 20 ppm Endosulfan 0.00075 and 0.001 ppm	~15 cm	30 d	Aqueous (semistatic)	Increase in atretic oocytes	[183]
Heavy Metals: Other Species					
Catfish (*Clarias batrachus*) Mercuric chloride 0.05 mg/L Methylmercuric chloride 0.04 mg/L Emisan 6 (organic mercurial fungicide) 0.5 mg/L	Adult	180 d	Aqueous (semistatic)	All treatments: atretic follicles	[215]
Giant Gourami (*Colisa fasciata*); nickel sulfate 64 mg/L	Adult	96 h	Aqueous (semistatic)	Atretic oocytes	[195]
Rosy barb (*Puntius conchonius*) Zinc: ~ 11 mg/L Copper: ~ 0.2 mg/L Lead: ~0.12 mg/L	~5 cm	2, 3, and 4 m	Aqueous (semistatic)	Increase in atretic oocytes	[196]

Striped gourami (*Colisa fasciatus*); arsenic(III)oxide					
2 and 14 mg/L	Adult	15 and 30 d	Aqueous (semistatic)	Effects visible only at 14 mg/L after 30 d: increased number of atretic follicles	[206]
Field: Other Species					
Common carp (*Cyprinus carpio*)/black bass (*Micropterus* spp.)/channel catfish (*Ictalurus punctatus*) Rio Grande basin (dioxins, PCBs, heavy metals, pesticides)					
n.d.	Adult	Lifelong	Field	Some fish with atretic follicles	[49]
Crucian carp (*Carassius carassius*); treated sewage effluent					
25%, 50%, and 100%	Adult (1–2 y)	28 d	Field	Atretic oocytes	[48]
Stoneloach (*Barbatula barbatula*); treated sewage effluent					
n.d.	Adult	Lifelong	Field	Elevated proportion of atresia at one sampling time point	[177]
Mirror carp (*Cyprinus carpio*); treated sewage effluent					
25%, 50%, and 100%	Adult (2–3 y)	28 d	Field	Follicular atresia	[178]
Redear sunfish (*Lepomis microlophus*); Martin Lake (selenium contaminated)					
n.d.	Adult	Lifelong	Field	Atretic follicles	[160]
Influence of pH					
Fathead minnows (*Pimephales promelas*)					
pH 5; pH 6; pH 8	0.6 g	5 m	Aqueous (flow-through)	Elevated ovarian atresia in pH 5 and pH 6 treatments	[225]

Compounds and compound classes observed to induce germ cell atresia in female fish in conjunction with a defined concentration, developmental stage, duration, and form/route of exposure. Tables are ordered according to the main three laboratory fish species, and other species and distinguish laboratory experiments from field observations. Included are also field trials with "contaminated" water of unknown chemical composition. The most prominent findings are listed along with the respective reference. d, days; dpf, days post-fertilization; dph, days post-hatch; °C dph, accumulated daily degrees °C post-hatch; h, hatch; hpf, hours post-fertilization; hr, hours; n.d., not defined or specifiable; m, months; w, weeks; y, years.

TABLE 6.12 Ovarian Inflammation and Macrophage Infiltration in Female Fish

Concentration	Stage	Duration	Route of Exposure	Histopathological Observations	Ref.
Bisphenol A: Fathead Minnow (Pimephales promelas)					
0, 1, 16, 160, 640, and 1280 μg/L	Reproductive stage	43, 71, and 164 d	Aqueous (flow-through)	No effects observed	[6]
0, 1, 16, 64, 160, and 640 μg/L	Reproductive stage	164 d	Aqueous (flow-through)	No effects observed	[23]
Field: Other Species					
Mirror carp (*Cyprinus carpio*), treated ewage effluent					
25%, 50%, and 100%	Adult (2–3 y)	28 d	Field	Macrophage infiltration	[178]
Common carp (*Cyprinus carpio*), black bass (*Micropterus* spp.), channel catfish (*Ictalurus punctatus*)					
Rio Grande basin (dioxins, PCBs, heavy metals, pesticides)					
n.d.	Adult	Lifelong	Field	Some fish with ovarian inflammation	[49]

Compounds and compound classes observed to induce ovarian inflammation and macrophage infiltration in female fish in conjunction with a defined concentration, developmental stage, duration, and form/route of exposure. Tables are ordered according to the main three laboratory fish species, and other species and distinguish laboratory experiments from field observations. Included are field trials with "contaminated" water of unknown chemical composition. The most prominent findings are listed along with the respective reference. d, days; dpf, days post-fertilization; dph, days post-hatch; °C dph, accumulated daily degrees °C post-hatch; h, hatch; hpf, hours post-fertilization; hr, hours; n.d. not defined or specifiable; m, months; w, weeks; y, years.

6.2.1 Effects in Male Fish Associated with Exposure to Specific Compounds or Compound Classes

6.2.1.1 Testis Architecture, Integrity, and Gross Appearance. Compounds affecting testicular architecture, integrity, or gross appearance were taken from the literature by scanning the published literature for terms characterizing gonadotype or testes integrity:

Gonadotype: sex reversal, real intersex (large proportion of gonadal tissue of the other gender), undifferentiated gonad, immature gonad, lacking gonad

Testes integrity: ovarian cavity, female ducts, two mesentarial attachments, testicular atrophy, large interstitial spaces, loss of structure, degeneration

The tables demonstrate clearly that the majority of experiments reporting testicular architecture, integrity, or gross appearance have been carried out with true estrogens or androgens. Of these experiments, whether run as static, static renewal, or flow-through experiments, nearly all employed moderate to high concentrations (>1 ng/L) of these steroids compared to concentrations generally found in the aquatic environment (i.e., excluding hot spots immediately downstream of sewage treatment plants [121,122]), and the effects reported were observed uniformly in the high-dose groups (10 to >1000 ng/L). Similarly, exposures to steroid mimics (e.g., pentyl-, nonyl-, or octyl-phenol and bisphenol A) resulted in no effects on the testis unless exposures were extremely high (>500 μg/L) for a moderate-exposure duration or high (>100 μg/L) in chronic experiments. The latter also holds true for the effect of pesticides, possibly with the exception of the report by Dutta et al. [123], who noted significant damage to connective tissue and disrupted and missing seminiferous tubule walls in adult male bluegill sunfish exposed to 1 μg/L endosulfan for up to 14 days under semistatic conditions. As the latter experiment is the only one reporting findings at such low concentrations, and due to the fact that only low numbers of fish ($n = 5$) and replicates ($n = 1$) were used for each experimental unit, these findings may represent false positive findings (see Chapter 9).

In contrast, no reports were noted for effects of anti-androgens and aromatase inhibitors in males of fish species other than the fathead minnow. Moreover, no effects of "industrial chemicals," heavy metals, or pH were noted for testicular architecture, integrity, or gross appearance in any species.

Overall, the available findings strongly suggest that EDC-mediated adverse effects on testis architecture can be observed only at extremely high compound concentrations. One must therefore assume that such effects, even when observed under field conditions, are either specific to a location experiencing untypically high environmental concentrations of EDCs or to factors not causally related to EDC exposure.

6.2.1.2 Testis–Ova The occurrence of oocytes interspersed within normal functional testicular tissue is often referred to as *testis–ova*. However, *ovo-testis* is used synonymously for *testis–ova* in some publications. For clarity, *ovo-testis* is denoted specifically in this book as ovarian tissue in females containing spermatogenic cysts or primary spermatogonia. It is interesting to note that none of the aromatase-inhibitors, industrial chemicals, or heavy metals tested resulted in significantly increased incidences of testis–ova in male fish. The latter is of importance, as many field studies do report testis–ova in wild fish (see Table 6.2 field studies). The group of compounds that appear to induce significantly higher incidences of testis–ova again is comprised of the steroids (estrogens, androgens), anti-androgens, phenols, bisphenol A (BPA), and pesticides. However, as noted in Section 6.2.1.1, the concentrations required to induce a veritable number of males with testis–ova with the latter compounds is extremely high.

Unfortunately, not all authors report actual incidences of testis–ova observed or when reported used an extremely low number of individual fish to arrive at the reported incidence (e.g., 5 of 9 males caught in a lake treated with 5 to 6 ng/L 17α-ethinylestradiol (EE_2) for three years [7]). The question thus arises whether the testis–ova observed in field experiments is really representative for the entire population of the site of investigation or is the result of extreme local levels of EDC exposure. It is noteworthy, for example, that despite lifelong exposure of zebrafish to 5 ng/L EE_2 [128], only a very low incidence of testis–ova was observed under experimental conditions. Conversely, despite a 164-day exposure of fathead minnows to high concentrations (640 μg/L) of BPA, no testis–ova was observed [23]. Moreover, treatment of medaka with high concentrations of BPA for only 21 days resulted in an concentration-independent incidence of testis–ova [3]. These observations strongly suggest that a very elaborate field study design, employing a high number of replicates or independent samples of numerous well-described and possibly genetically and ecotype-characterized fish populations [162], is required to really elucidate the causality of testis–ova in a given fish population with local EDC exposure. Indeed, the latter statement is supported by recent field experiments from areas heavily contaminated with mixtures of persistent organic pollutants, pesticides, and sewage treatment water [163,164] and despite this heavy contamination no testis–ova was found, whereas in the sub-alpine lake Thun, considered nearly pristine, a 1.1% incidence of testis–ova was reported in an indigenous coregonid population also presenting with 40% (151 out of 225) gonadal alterations [162].

6.2.1.3 Fibrosis Testicular fibrosis, in the worst case a nearly complete replacement of gametogenic and supportive testicular cells by connective tissue, appears to be associated primarily with exposure to high concentrations of estrogens over an intermediate-to-extended period of time. Indeed, no fibrosis was observed in any fish species exposed to androgens, anti-androgens, aromatase inhibitors, or industrial chemicals, nor were reports found that demonstrate testicular fibrosis in fish from field surveys, with the exception of Palace et al. [33] and the ensuing follow-up report by Kidd et al. [7] of a fathead minnow population in Canadian Experimental Lake 260 dosed with 6 ng/L 17α-ethinylestradiol (EE_2) for three years. Beyond estrogens, moderate concentrations (i.e., <100 μg/L of phenols bisphenol A (BPA), nonylphenol, octylphenol, 4-*tert*-pentyl- or octylphenol) and high concentrations of pesticides and heavy metals (Cd) applied for an intermediate-to-extended period of time also appear to promote testicular fibrosis. In the latter context it appears noteworthy that despite high concentrations and chronic exposure, no testicular fibrosis was observed in fathead minnows exposed to BPA [6,23], while even low BPA concentrations (>10 μg/L) and a much shorter exposure duration led to appreciable testicular fibrosis in medaka [102,103]. These observations possibly suggest that certain fish species (e.g., medaka) may be more susceptible to certain EDC- induced testicular fibrosis. Moreover, the numerous reports on phenols inducing testicular fibrosis at moderate concentrations raises a the question as to whether these compounds also induce enhanced apoptosis and necrosis of testicular gametogenic and support (somatic) cells with a subsequent inflammatory reaction. If the latter were the case, testicular fibrosis as observed with these compounds could be seen as the consequence of the precedent tissue damage rather than the direct or indirect effect of endocrine-disruptive activity. Unfortunately, and as discussed in Section 6.2.1.6, the lack of reports on phenols-induced inflammation prevents more in-depth corroboration of the foregoing interpretation. An exception to this is the lack of an inflammatory response to chronic BPA treatment in fathead minnows [6,23]. However, as heavy metals and even EE_2 treatment were reported to produce apoptosis and necrosis as well as an inflammatory response and testicular fibrosis, the cytotoxic rather than the endocrine activity–mediated response of testicular cells may be the most important mechanism leading to testicular fibrosis.

6.2.1.4 Germ Cell Effects The effects on germ cells reported comprise following observations: number and/or localization of PGCs, absence of germ cell stages, necrosis or apoptosis, degeneration, germ cell pyknosis, germ cell phagocytosis, inhibition or acceleration of spermatogenesis, spermatogenesis arrest, and others. It is highly noteworthy that despite the numerous endpoints listed as germ cell effects, in essence there are only a few field studies reporting the arrest of spermatogenesis at the spermatogonial stage [7,48,177,178]. The Kidd et al. study employed high concentrations (5 to 6 ng/L) of 17α-ethinylestradiol (EE_2) in the experimental lake area of Canada. As noted earlier, the arrest of spermatogenesis was not the only pathological finding, but testicular fibrosis [33] and pathology associated with high concentrations of interstitial vitellogenin were also rampant in nearly all of the male fathead minnows studied. The studies of treated sewage by Diniz et al. and Douxfils et al. [48,177,178] observed effects at 100% sewage treatment water (no effluent dilution).

Unfortunately, none of the latter studies provided for extensive concurrent analytical data such that the "estrogenic effect" reported could be related to steroid or other EDC concentrations in the effluents used. Indeed, only Diniz et al. [178] provided some analytical information (>5 μg/L alkylphenols, 0.55 μg/L bisphenol A, and 1 ng/L 17β-estradiol) while neglecting the analysis of EE_2, estriol or estrone. Moreover, Douxfils et al. [177] found the same effects in gudgeon above and below the effluent discharge of the sewage treatment plants investigated, thus effectively questioning the causal influence of the effluents in the effects observed in male and female fish. The data above, however, raise a question as to whether the effects on germ cells observed (see below) result primarily from the fact that this was a laboratory experiment dealing with extraordinarily contaminated sewage treatment effluent observed at extremely high concentrations of EDCs. Indeed, germ cell apoptosis and necrosis and abrogated spermatogenesis are observed primarily at high concentrations of estrogens, indicating that the organism inherent endocrine regulation of normal development is overrun to the extent that either cells become nonvariable or lack the necessary proper endocrine stimulus to progress to the next stage during spermatogenesis. A similar picture can be drawn for androgens; however, in this case only concentration in the high μg/L to mg/L range will provide necrotic and apoptotic sperm cells. Lower concentrations will promote spermatogenic progression; thereby decreasing the number of early stages and promoting the number of late-stage germ cells.

Anti-androgens, on the other hand, have the effect of reducing the amount of systemically and locally available androgens, thereby abrogating the required growth and maturation stimulus to germ cells. The consequence of this is not necessarily an inhibition of spermatogenesis but rather, a degeneration of sperm cells due to lack of maturation stimuli.

Alkylphenols generally promote inhibition of spermatogenesis in fish when applied at extremely high concentrations for prolonged periods of time. The underlying mechanism may be due in part to this estrogen-like activity, albeit competition with androgens for the respective receptor (anti-androgenic activity [184]) cannot be excluded. The latter may explain why even at lower concentrations, sperm necrosis is reported. These findings cannot, however, be generalized to all alkylphenols, especially as bisphenol A (BPA) was reported to have a significantly reduced number of sperm cells at >10 and >1 μg/L in medaka [102] and fathead minnows [185], respectively. This would imply that BPA was much more potent than any of the other alkylphenols tested. Some doubt with regard to the latter findings must be raised, as a more recent and experimentally much more profound and robust study could not confirm inhibition of spermatogenesis at these low concentrations in fathead minnows, but rather found the earliest signs of effects at the highest concentration of 640 μg/L [23]. As discussed in Chapter 9, many of the discrepancies within the effects reported in the current literature stem primarily from a faulty experimental design (i.e., insufficient independent replicates, number of animals, faulty detection methods, statistical analyses, etc.). Using a weight-of-evidence approach, it remains to be stated that inhibitory effects of alkylphenols on spermatogenesis in fish certainly appear to be reproducible when applied at high concentrations whereas the evidence for these effects at lower or lowest concentrations is highly questionable.

A similar observation can be made for pesticides, where germ cell effects are generally noted at concentrations in the high μg/L to low mg/L range. An exception to this is tributyltin, whose aromatase inhibitory activity [22,186,187] may have contributed to an imbalance of the estrogens and androgens required for germ cell proliferation. Similar to the observations for BPA, a species-sensitivity range of approximately 1000-fold appears to be associated with tributyltin, as mummichog exposed to μg/L concentrations showed barely any effect on germ cells, while zebrafish [186] and guppies [22] exposed to pg/L to ng/L tributyltin present with nearly completely inhibited sperm motility, sperm lacking flagella, or reduced sperm count.

6.2.1.5 Somatic Cell Effects Somatic cell effects include any change in morphology or number of Sertoli, Leydig, and interstitial cells including overt apoptosis or necrosis. Probably due to the more demanding and detailed pathological assessment, relatively few publications actually report somatic cell effects in [37] fish testes. While treatment of fish with estrogens and androgens resulted primarily in Sertoli cells hypertrophy and hyperplasia with no effects reported on Leydig cells, treatment with alkylphenols resulted in interstitial cell hyperplasia, Sertoli cell hypertrophy and hyperplasia, as well as increased Sertoli and Leydig cell apoptosis. However, all of the effects observed, irrespective of the compound type, were associated with high concentrations applied during the respective experiments. This may, beyond the fact that somatic cell effects were not sought specifically, in part explain that no somatic cell effects were reported from field studies, other than the one in conjunction with testicular tumors in carp [174] and a field study of catfish in the Mekong delta [163].

If indeed, only relatively high concentrations of EDCs lead to overt (visible) changes in somatic cells, the most prominent change (e.g., Sertoli cell hypertrophy), would clearly be secondary to primary germ cell necrosis or apoptosis [54]. This reaction would then represent a normal physiological and functional compensatory response of somatic cells to any challenge, not necessarily restricted to EDCs. This interpretation is corroborated by the observation that catfish with small testes presented with necrotic spermatogonia and vacuolized and hypertrophied Sertoli cells, whereas catfish with normal testes also had normal Sertoli cell morphology, indicating a generally abnormal health status, probably associated with an increased body burden of heavy metals of catfish with small testes [163]. Heavy metal exposure was demonstrated to increase testicular inflammation [32,85,197], thus emphasizing the more generalized necrotic effect of heavy metals rather than an EDC-mediated effect in fish testes (see Section 6.2.1.6). The latter would also suggest that low or moderate concentrations of EDCs would not have any visible effect on somatic cells, but rather, by virtue of their adverse interaction with the hypothalamic–gonadal axis and steroid synthesis in Leydig and Sertoli cells, would result in changes observable in downstream effector cells of Leydig and Sertoli cells (e.g., spermatogenic maturation and/or viability). The latter assumption is corroborated by the recent findings of Ankley et al. [96] in that the proliferation of Leydig cells in fathead minnows exposed to ketaconazole were seen as potential adaptive or compensatory responses to impaired steroidogenic capacity.

6.2.1.6 Testicular Cell Inflammation and Macrophage Infiltration The paucity of publications specifically investigating inflammatory responses in the testis does not allow for a proper and solid interpretation or derivation of generally applicable findings. The fibrosis observed in fathead minnows exposed to 17α-ethinylestradiol (EE_2) in a lake of the Canadian experimental lakes area [33], as well as the generally observed fibrotic response in various fish exposed to EE_2 (see Sections 6.1.6 and 6.2.1.3) would suggest that some inflammatory response may have been present, albeit not recognized or reported [7,33]. The more recent publication on EE_2- induced macrophage infiltration, inflammation, and fibrosis by Schafers et al. [101] supports the interpretation above. In this case the inflammatory changes were always observed in conjunction with overt germ cell degeneration.

Similarly, heavy metal–induced germ cell necrosis and testicular fibrosis must be seen in conjunction with the inflammatory response reported (see below). Overwhelming cell necrosis will provide for a stimulus for macrophage invasion into the testis followed by fibroblast infiltration, solely with the aim of maintaining the structural integrity of the testes. In contrast the granulocyte infiltration reported in testes of carp exposed to bisphenol A (BPA) at 1 μg/L exposure concentrations [151] appears questionable, due to experimental flaws as well as to the fact that exposure of fathead minnows to >1000-fold-higher BPA concentrations [23,185] for a much longer duration resulted in no determinable inflammatory response.

6.2.2 Effects in Female Fish Associated with Exposure to Specific Compounds or Compound Classes

6.2.2.1 Ovary Architecture, Integrity, and Gross Appearance The overt pathological effects mediated by EDCs or EDC-associated compounds were grouped roughly into:

Gonadotype: sex reversal, real intersex (large proportion of gonadal tissue of the other gender), undifferentiated gonad, immature gonad, lacking gonad

Ovarian integrity: atrophy, interstitial spaces, and degeneration, including necrosis and apoptosis

High concentrations of estrogens induce primarily degeneration or developmental arrest of ovaries. High concentrations of estriol and estrone are able to induce a significantly higher number of male medaka upon exposure of fish at 1 dph for up to 110 days [102]. In contrast, no reports were found regarding effects of anti-androgens, albeit effects of these compounds would be expected in ovaries, specifically in the oocyte surrounding steroidogenic theca and granulosa cells, at least at very high concentrations.

Low to moderately low concentrations of androgens are capable of suppressing ovarian growth or of inducing complete masculinization in fathead minnows and zebrafish, whereas significantly higher concentrations are required to induce a similar response in other fish species (e.g., salmonids, sea bass) Similarly, and commensurate with their specific activity of converting testosterone to 17β-estradiol, exposure to aromatase inhibitors results in masculinization of fish (testis formation), although only when fish were exposed to very high (mg/L or μg/g) concentrations.

Alkylphenols, including bisphenol A and other industrial chemicals, do not seem to have any effect on fish ovarian tissue, with the exception of 4-*tert*-octylphenol, which induced a proliferation of the oocyte interstitial stroma in medaka when applied to 1-dph fish at concentrations of <100 μg/L for up to six months [99]. Whether the effect observed represents an endocrine-mediated effect or a more generalized toxic effect to the compound is difficult to assess from the data presented.

Pesticides and heavy metals both appear to induce primarily ovarian damage, including connective tissue damage and necrosis or shrinking of oocytes. The latter effects are generally observed at extremely high (mg/L) concentrations only. The exceptions are prochloraz and vinclozolin; the anti-androgenic vinclozolin or its metabolites do not bind competitively to the fathead minnow androgen receptor but are capable of antagonizing AR receptor function and binding to the estrogen receptor [198]. Similarly prochloraz is known to act as an androgen receptor antagonist but also to reduce steroid synthesis significantly via interaction with cytochrome P450 c17α hydroxylase/17,20-lyase (CYP17) activity [96,199]. Thus, an imbalance between androgen and estrogen synthesis in zebrafish during development could explain the higher degree of masculinization observed in prochloraz exposed 24-hpf zebrafish [154].

Overall, there are a paucity of reports from field studies on the effects in ovaries. Although one report observed ovarian inflammation and follicular atresia as a reaction of carp to exposure with a mixture of compounds, the report lacks sufficient descriptive pathology that world enable more in-depth understanding of the incidences and severity of the findings and associate these findings with the assumed exposure [49]. Indeed, at least one other report finds degenerated or malformed ovaries, including hermaphrodism in coregonids from a pristine pre-alpine lake [162]. The examples above demonstrate that results obtained from laboratory experiments, usually employing high concentrations of a compound, are rarely compared with concentrations found in field experiments and that incidences of pathological changes may very well be due to inherited or ecological traits of a respective fish population.

6.2.2.2 Ovo–Testis For clarity, ovo-testis is denoted specifically in this book as ovarian tissue in females containing spermatogenic cysts or primary spermatogonia. This is a rather rare observation and could be explained primarily by the androgenic activity of EDCs (e.g., 17α-methyltestosterone) [136]. However, as subadult sheepshead minnow exposed to 17α-ethinylestradiol also present with ovo-testis, a general imbalance of steroids during sexual determination may be considered the most likely explanation. Indeed, ovo–testis was observed primarily in studies employing fish not yet having completed sexually determination [136]. The latter interpretation is supported by the fact that 14-dph stickleback fish exposed to 17α-methyltestosterone for up to 51 days do not present with ovo-testis. Conversely, it is not surprising that adult fathead minnows exposed over a long period to high bisphenol A concentrations, a putative estrogen mimic, do not present with ovo–testis.

6.2.2.3 Fibrosis Commensurate with the atretic effect on oocytes (see Section 6.2.2.5), high concentrations of 17α-ethinylestradiol will also induce a general degeneration of the ovarian structure (see Section 6.2.2.1) and thus provoke infiltration of fibroblasts in order that the structural integrity of the ovaries can be maintained. Prolonged exposure thus would be expected to result in ovarian fibrosis [101]. The latter interpretation is also supported by field studies in which ovarian inflammation, connective tissue hypertrophy, and fibrosis are reported in conjunction with exposure of fish to high concentrations of heavy metals or mixtures of xenobiotics, including PCBs, pesticides, heavy metals, and dioxins.

6.2.2.4 Germ Cell Effects Contrary to the effects described for male germ cells (see Section 6.2.1.4), the effects on female germ cells reported comprise observations on the ovarian follicles, including the oocytes and associated supporting granulosa and theca cells. The effects include, for example, the number and/or localization of PGCs, the absence of certain follicular stages, necrosis or apoptosis, degeneration, inhibition of vitellogenesis, oocyte diameter, inhibition or acceleration of oogenesis [regressed ovary, delayed oogenesis (see Section 8.3)], and impairment of ovarian recrudescence.

Noteworthy is the effect of 17α-ethinylestradiol and 17β-estradiol, in that rather than promoting or enhancing early maturation of follicles, high concentrations of these steroids provide for a regression or inhibited maturational development of the ovary proper. In addition, a lower number of primary oocytes or inhibited follicular development is reported. The latter observations suggest that high concentrations of estrogens will interfere with pituitary-gonadal axis signaling via negative feedback inhibition, possibly via interaction with the estrogen receptor β [207,208]. Estrogens apparently had a much higher binding affinity for the estrogen receptor β of roach than for the corresponding estrogen receptor α [207]. Whether the effects observed are the result of the interaction of membrane-bound ERs, belonging to the G protein–coupled receptor family, which is distinct and different from nuclear ERs [209,210], or to the estrogenic transduction pathway involving genomic (i.e., nuclear) control, in which ERs act as transcription factors within the cell nucleus, cannot be ascertained without additional research. However, most recent research related to *in vitro* experiments with croaker and zebrafish oocytes suggests that the reduced maturation of follicles is associated with excessive estrogen binding to the membrane bound G protein–coupled estrogen receptor [211].

The promotion of oogenesis by bisphenol A (BPA) [23,102,212] and nonylphenol [39] at moderate and very high concentrations of these compounds, respectively, may be explained by the potential of these compounds to bind with low affinity to the fish estrogen receptors α and β. On the other hand, the lack of any observed [185] or marginal [23] follicular effects of BPA demonstrate that only the most exhaustive experimental design and pathological analysis will provide for the detection of minimal effects at the highest BPA concentration.

In contrast, the nonaromatizable dihydrotestosterone and the aromatase inhibitor fadrozole, when given at high concentrations (μg/L), may interfere with estrogen-mediated vitellogenin synthesis in the liver and incorporation into oocytes in fathead minnows [64,188]. Contrary to the activity of dihydrotestosterone, the oogenesis inhibitory effects of extreme concentrations (mg/L) of methyltestosterone in fathead minnows [42] may be explained in part by the fact that methyltestosterone is aromatizable to 17β-estradiol and thus represents more of an estrogenic than an androgenic effect. The latter may also hold true for the degeneration and regression of oocytes observed in fathead minnows exposed to ng/L methyltestosterone concentrations for 3 weeks [167].

The germ cell necrotic and maturation-inhibiting effects of the anti-androgen flutamide in fathead minnows and medaka, on the other hand, cannot readily be explained. However, direct intervention of flutamide on membrane-bound G protein-coupled testosterone receptor [213] as well as negative feedback loop nuclear receptors of the pituitary–gonadal axis, resulting in a down-regulation of ovarian ERα [214], could serve as a possible explanation of the effects observed. By analogy, the inhibition of aromatase by fadrozole will lead to a down-regulation of endogenous estrogen synthesis and thus estrogen levels. Thus, a reduction of estrogen receptors or estrogen levels will result in inhibition of oogenesis. Whether competitive binding of extreme concentrations of benzophenone-2 or PCBs for the estrogen and androgen receptors account for the inhibition of oogenesis observed at the highest concentrations and feed doses applied cannot be ascertained, as concomitant receptor-binding assays or measurements of endogenous steroid levels were not carried out.

Of all other compounds, including heavy metals, pesticides, and mixtures demonstrated to have effects on ovarian germ cells, only the inhibition of oogenesis by tributyltin and the impairment of recrudescence can be explained mechanistically. Indeed, tributyltin is a known aromatase inhibitor [186], even at very low concentrations (ng/L), thus effectively blocking estrogen synthesis. Thus, the oogenesis inhibiting effects of tributyltin are comparable to the effects of flutamide.

In contrast, the reported inhibition of vitellogenesis by mercuric chloride and mercurial fungicides at high concentrations (mg/L) and subsequent to this, impaired recrudescence in catfish [215], appears a more coincidental than causally related finding. Exposures to mg/L concentrations for up to 180 days will result in general systemic toxicity, including abrogated protein and steroid synthesis. The latter findings must thus be seen in context with unspecific high-concentration toxicity rather than specific endocrine-mediated effects.

Field observations report primarily inhibition of oogenesis, delayed oocyte maturation and some oocyte necrosis and apoptosis [48,177,178]. Most of the latter observations are not solitary effects but come in conjunction with other relevant endpoints [e.g., follicular atresia (see Section 6.2.2.5), macrophage infiltration (see Section 6.2.2.6)] and appear closely associated with treated sewage effluent and thus compare well with the high-dose effects found with estrogens in laboratory experiments. Indeed, as sewage treatment effluents reportedly contain high natural estrogen concentrations (originating primarily from marginally diluted sewage plant effluents receiving human and domesticated farm animal excrements) [122], the effects observed in the field are probably a result of high concentrations of natural estrogens.

6.2.2.5 Germ Cell Atresia Germ cell or follicular atresia (*corpora atretica;* see Section 4.3.1) may serve to regulate de-recruitment and/or resorption of malfunctional younger-stage oocyte in order to preserve normal ovarian physiology during the breeding season. A more important role is attributed to atresia at the end of the breeding season, especially in fish with asynchronous ovaries (i.e., asynchronous spawning) and defined seasonal breeding patterns (e.g., medaka and fathead minnows), whereby atresia could be seen as a cleaning-up process [62,220]. Indeed, atresia appears to be carried out via an apoptotic and not a necrotic process, thus clearly enabling the organism to remove excess or damaged follicles via a controlled process. However, atresia is also observed in conjunction with stress and other environmental insults (e.g., poor nutrition, physical factors, overpopulation, or stressors), that could decrease gonadotropin levels such as xenobiotics [61–63]. Thus, an imbalance of gonadotropin levels systemically or of steroids either locally or systemically may promote an early onset or increased incidences of follicular atresia. Consequently, it is not surprising to find that very high concentrations of 17α-ethinylestradiol (EE_2) (>10 ng/L) will provide for follicular atresia in various fish species [28,69,101,127], especially when exposed for a long period. As observed earlier, the 17β-estradiol (E_2) concentrations required to achieve effects comparable to those observed for EE_2 exposures are orders of magnitude higher, suggesting that the higher lipophilicity as well as the lowered metabolic conversion of EE_2 than of the corresponding natural E_2 may serve for a much higher relative systemically active concentration.

Suppression of available E_2 and thus an imbalance of endogenous steroids *in situ* may also explain the atresia observed in conjunction with extreme concentrations (μg/L to mg/L) of androgens [42,44,148] or aromatase inhibitors [51,64,216,221].

Flutamide, on the other hand, was demonstrated to increase endogenous 11–ketotestosterone levels in female fathead minnows [150], thus providing effect an similar to that for 11–ketotestosterone given directly [44].

Somewhat surprising is the fact that low-to-moderate bisphenol A (BPA) and nonylphenol concentrations (approximately 100 μg/L) have been reported by various authors to induce follicular atresia in a number of species. Although these levels are still approximately a factor 100 to 1000 above those measured in the environment [222,223], these data suggest that some of the alkylphenols, with the exception of BPA [6,23], could provide for an adverse effect on normal oocyte maturation probably by virtue of their weak estrogen receptor-binding properties and thus competitive binding at the membrane-bound estrogen receptor *in situ* rather than via interaction with nuclear steroid receptors and thus systemic interference. A similar explanation may be assumed to hold true for other industrial chemicals (e.g., benzophenone-2, methoxychlor, PCB congeners, dioxins, Aroclor 1254) demonstrated to increase follicular atresia at high exposure concentrations. A common characteristic of the latter compounds is their high lipophilicity as well as persistence in the environment. The latter characteristics may be important for specific inclusion of these compounds into developing oocytes along with other lipids, and thus may serve to produce some of the transgenerational effects reported [212,223].

In contrast to the more specific endocrine activities of xenobiotics and natural steroids described above, the follicular atresia reported in conjunction with pesticide exposure appears to be the result of unspecific toxicity secondary to a generalized adverse effect on the whole organism The latter may also hold true for the atresia observed in fish exposed to heavy metals, although in the case of mercury and mercurial fungicides [215], inhibition of protein synthesis as well as competition for energy due to increased energy consumption resulting from metallothioneine synthesis may contribute to "starvation" of the developing follicles and thus their subsequent atresia.

The combination of a multitude of interactions, directly endocrine as well as indirectly general toxicity related, may be responsible for follicular atresia observed in field experiments. However, in most cases the effects observed appear to be related to high concentrations of a specific class of compounds (e.g., steroids [48,177,178], heavy metals [160], PCBs, and/or dioxins [49]) rather than the result of additive or synergistic activity of a multitude of compounds at very low concentrations in a mixture.

6.2.2.6 Germ Cell Inflammation and Macrophage Infiltration

Ovarian inflammation and macrophage infiltration are the only reported effects associated with female fish exposure to xenobiotics in the field. The latter is all the more surprising as overt ovarian fibrosis has been causally related to steroid exposures (see Section 6.2.2.3). If, indeed, fibrosis is secondary to ovarian cell necrosis or apoptosis (Sections 6.2.2.1 and 6.2.2.4), rather than the result of a mitogenic stimulus for mesenchymal tissue growth, a higher number of experiments would have been expected to report inflammatory reactions in ovaries. The question thus remains to be elucidated whether the paucity of reported inflammatory changes in fish ovaries post-exposure to xenobiotics at high concentrations is due to a true lack of inflammatory change or is a matter of limited pathological recognition of this pathological process by the investigators involved.

6.3 POPULATION EFFECTS

All of the effects observed in male and female fish upon exposure to xenobiotics or mixtures thereof may be limited in their importance if they cannot be extrapolated from the effects observed at the level of the individual within a population to the consequences these effects may have for the cohort or even the population within the ecosystem as a whole. The overt toxicity and consequently lethality [7] subsequent to organ pathology noted in exposure scenarios employing extremely high concentrations of compounds, albeit relevant for the immediate survival of a population, are rare and unusual occurrences in the field setting. Thus, nearly all of the laboratory experiments describing pathological effects in fish gonads following exposure to extreme concentrations of xenobiotics may provide some information regarding the compound- or exposure-scenario-specific effects possibly encountered, but are compromised with regard to the ability to extrapolate these findings from the high (or extreme) concentration scenario in the laboratory to the low-dose situation in the field as well as to provide insight as to how much of a pathological or biochemical change is required to really produce quantifiable and population-relevant effects. Indeed, of all the parameters determined, only a few would appear to be of relevance for the population:

- Reduced number or lack of reproductively capable adults (impairment of F_0 fecundity and/or fertility)
- Impaired mating behavior or brood maintenance of adults and thus lowered hatching success of the subsequent generation
- Transgenerational effects regarding fecundity and/or fertility or impaired mating and brood maintenance behavior of succeeding generations

Impaired fecundity, at least at the pathological level, could be a consequence of functional effects (e.g., decreased or inhibited gametogenesis, enhanced gamete atresia) or structural effects (e.g., absence of gametogenic cells, replacement of gametogenic cells by mesenchymal tissue, absent or deformed gonadal ducts). Reduced fertility, on the other hand, would not be observable at the level of histopathology, possibly with the exception of spermatids with lacking or deformed tails. Even when observed in the laboratory or in the field, the problem remains of causally relating a given percentage change in a specific effect with a change in fecundity or fertility and thus a population effect. For example, Seki et al. [1] demonstrated that in medaka exposed to high concentrations of 17α-ethinylestradiol (EE_2), the physiological and histological measurements were approximately eightfold more responsive than overt reproductive effects to EE_2 exposure. The latter, especially, holds true for laboratory experiments employing low-to-moderate xenobiotic concentrations, thereby aiming to reproduce environmentally relevant situations as closely as possible.

As mentioned earlier and exemplified in Chapter 9, the experimental design (number of replicates employed per test group) will determine whether or not a given pathological effect can be detected significantly [23]. Thus, high demands are placed on knowledge of natural variability of a given parameter within a fish species [226] prior to exposing the fish to a given xenobiotic and then inferring environmental relevance or population (fecundity or fertility) effects from the changes observed. The latter has been especially relevant with regard to the discrepant interpretations of potential effects of bisphenol A (BPA) on fathead minnows, whereby a limited and suboptimal experimental design [6] led to the finding of significant effects of BPA on fathead minnow spermatogenesis while a more rigorous design [23] led to the realization that no effect occurred at the concentrations previously reported as being effective. The significant findings of BPA on spermatogenesis by Sohoni et al. [6] were somewhat contradicted by their own concurrent fecundity and hatchability trials, which demonstrated no BPA-specific effect. The lack of BPA-induced reproductive and transgenerational effects, even when employing very high concentrations, was corroborated by Kang et al. [3] in medaka. This not only highlights the problems associated with extrapolating low-dose pathological effects to population-relevant parameters, but stresses the high demands placed on the experimental design (see Chapter 9) resulting from natural species variability in fecundity and fertility variables [226] to allow detection of any significant xenobiotic-induced change, especially those relevant for a population [227]. Paull et al. [228] determined that a minimum number of six replicate zebrafish colonies, each with six fish, assessed daily for a period of 4 days was required per treatment to detect a 40% change in egg output (fecundity), whereas a minimum of 32 individual male zebrafish per treatment was required to detect a 40% change in sperm quality (fertility) (e.g., percentage motility).

6.3.1 Fecundity and Fertility

Although fecundity can be quantified separately, fertility can only be determined via "quality" parameters of eggs or sperm produced. The latter are rarely determined and thus the most common measure of fertility is the hatching success of fertilized eggs.

Hatching in itself thus also incorporates the fertility of sperm and eggs as well as embryo mortality. Irrespective of the parameters used for measurement of population effects, it is extremely vital to understand the natural variability as well as the limitations of determinations (qualitative or quantitative) of each of the parameters for the fish species in question, as discussed in Chapter 9 and suggested by Mihaich et al. [23] and Paull et al. [228]. It is thus not surprising that most of the experiments to date reporting on effects of xenobiotics on fecundity and fertility lack the rigor in experimental design required to maintain the significant effects associated with xenobiotic exposure. In consequence and similar to overt effects on gonad pathology, inhibition of fecundity and fertility is often observed at extreme experimental concentrations (extreme xenobiotic concentrations, short duration of exposure) used. For example, the number of eggs and spawnings were significantly affected in fathead minnows exposed to 4-nonylphenol (4-NP) for three weeks at $57.7 \pm 3\,\mu g/L$ [229], concentrations at least 20-fold higher than those encountered in the environment. Similarly reduced fecundity was reported by Kang et al. [4] in medaka exposed for 21 days to 463 ng/L 17β-estradiol, a concentration up to two orders of magnitude higher than what is found directly below sewage treatment plant influents into recipient streams or surface waters [122,230]. Along the same lines, exposure of medaka to 0.101, 0.202, 0.397, 0.787, or 1.56 mg/L of the anti-androgen flutamide for 3 weeks [169] resulted in significantly reduced fecundity and fertility of paired medaka at 1.56 mg/L only, compared with those of the controls.

Treatment of paired medaka with methyltestosterone concentrations of 22.5, 46.8, 88.1, 188, and 380 ng/L for 3 weeks [167] resulted in a significant reduction in fecundity and fertility of paired medaka in the 46.8-, 88.1-, 188-, and 380-ng/L methyltestosterone groups compared with those of controls. However, as flutamide and methyltestosterone concentrations rarely exceed 1 µg/L and 0.5 ng/L, respectively [231], the latter results obtained with concentrations up to three orders of magnitude higher than realistically observable in the environment are of little help with regard to risk assessment for fish populations. The latter also holds true for present information on ultravioletfilters and other industrial and personal care products. Although a 15-day exposure of fathead minnows to 0.002, 0.1, 1.2, 5.0, and 9.7 mg/L benzophenone-2 resulted in a decrease in egg production at 5 mg/L and a complete cessation of spawning activity at 9.7 mg/L [188], the concentrations employed are again three orders of magnitude higher than those observed in the environment. Moreover, as the lowest concentrations (0.002, 0.1 mg/L) had no effect at all on any of the parameters determined, including fecundity, it may be assumed that the weak endocrine disruptive activity of benzophenone-2 is observed at extreme concentrations only and does not allow extrapolation of the high-dose short-term exposure scenario to a lower concentration but prolonged exposure duration scenario, at least for a population risk assessment.

In contrast, Yokota et al. [104] using a two-generation approach with the medaka parental (F_0) generation exposed from within 24 hpf, demonstrated that none of the 4-NP exposure concentrations (4.2, 8.2, and 17.7 µg/L) employed changed fecundity in the F_0 or the F_1 generation, despite the fact that the sex ratio of the F_1 generation was skewed toward more female and significantly more fry mortality was observed at the highest concentrations used. The latter is strong testimony that environmentally relevant concentrations, at least of 4-NP, may provide some pathological (testis-ova) and sex-determination changes, but not necessarily for population-relevant effects. Indeed, no effects were observed at any of the concentrations tested on hatching success, post-hatch mortality, or growth in the F_1 generation. The latter study is, however, somewhat contradicted by Schwaiger et al. [223], who demonstrated that adult rainbow trout exposed for 4 months prior to spawning to 4-NP concentrations of 1 and 10 μg/L resulted in significantly reduced hatching rates of the spawns produced by the 10 μg/L adults, albeit no change in sex ratio or histopathological changes in the offspring were observed. However, analysis of sex steroid levels revealed a twofold increase in 17β-estradiol (E_2) in the plasma of male offspring and a 13-fold elevation of testosterone in the plasma of female progeny, suggesting a hormonal imbalance in the filial generation. This discrepancy in the observation of effects may be due in part to species-sensitivity differences; however, experimental design artifacts and thus limitations in detection of significant findings as well as the possibility of false positive findings due to limitations in the use of independent replicates are also highly likely.

Lifelong exposure of zebrafish to 5 ng/L 17α-ethinylestradiol (EE_2), on the other hand, caused a 56% reduction in fecundity and complete population failure with no fertilization in the F_1 generation [128]. Similar results were obtained by Schaefers et al. [101], who demonstrated that partial life-cycle exposure of a parental (F_1) zebrafish generation [i.e., until completion of gonad differentiation (0 to 75 dpf)], impaired juvenile growth, time to sexual maturity, adult fecundity (egg production per female per day), and adult fertilization success at 1.1 ng/L EE_2 and higher whereas lifelong exposure of the F_1 generation (i.e., until 177 dpf) resulted in the same effects, albeit with a steeper slope of the dose-response curve. Moreover, full life-cycle exposure of the filial (F_2) generation (i.e., until 162 dpf) impaired growth, delayed the onset of spawning, and reduced fecundity and fertilization success at 2 ng/L EE_2. Similar findings were reported by Hirai et al. [232] in that exposure to 150 ng/L E_2 until the end of the medaka larval period produced sex-transformed female medaka (XY females) with high reproductive ability, similar to that of normal females (XX females), albeit longer exposure to E_2 inhibited the sexual maturation of sex-transformed females and therefore in the long run reduced overall fecundity. The latter data compare well with those obtained in fathead minnows, whereby 48-hpf eggs exposed to EE_2 at nominal concentrations of 0.32 and 0.96 ng/L and measured concentrations of 3.5, 9.6, and 23 ng/L until 150 dph (i.e., through maturation and reproduction) demonstrated a significantly decreased egg fertilization and sex ratio (skewed toward females) at 0.32 ng/L [233]. These findings are supported by the more recent field experiment carried out in the experimental lakes area in northwestern Ontario, Canada [7], where continuous dosing of the experimental lake with 5 to 6 ng/L EE_2 led to impacts on gonadal development of fathead minnows, as evidenced by intersex in males and altered oogenesis in females, and, ultimately, near-extinction of this species from the lake. The question remains, though whether a concentration of 0.32 ng/L, (i.e., a more realistic and environmentally relevant EE_2 concentration) would also have led to "feminization" of fathead minnow males, reduced fecundity and fertility, and ultimately to the extinction of the fathead population in this lake.

The data above (and as indicated earlier [1]) emphasize that EE_2 has a much higher potency than E_2 and that even the "low" concentrations of EE_2 used in the laboratory and field experiments actually represent high, if not in some cases extreme, concentrations rarely found in the real environment [121]. Indeed, when calculating estrogenic potentials, (i.e., summing the present steroids and immediate steroid analogs in sewage treatment effluents), concentrations of E_2 equivalents generally do not exceed 0.5 ng/L [122,231].

The fact that experimental data shown have veritable population-relevant effects at 0.32 ng/L [233] and that overt population effects were observed at concentrations of 5 to 6 ng/L when applied over three years [7], raises the question about the window of observation. Indeed, subtle changes in population structure would not be likely to be observable or quantifiable, due to experimental design restrictions in the laboratory ([228] and Chapter 9) and even more so in the field. However, if such subtle changes did occur, they would become manifest only after a prolonged period of continuous exposure to xenobiotics (e.g., after 10 to 20 years). This interpretation is somewhat supported by the recent predictive population model used to investigate population dynamics of fathead minnow exposed to the androgen receptor agonist 17β-trenbolone [234]. This model uniquely combines a Leslie population projection matrix and a logistic equation to translate changes in fecundity of the fathead minnow in a short-term laboratory toxicity test to alterations in population growth rate, thus effectively requiring a life table for the organism of interest, a measure of carrying capacity for the given population, and an estimation of the effect of a stressor on vital rates. Use of this model demonstrated that fathead minnow populations occurring at carrying capacity and subsequently exposed to 27 ng/L of 17β-trenbolone exhibited a 51% projected decrease in average population size after two years of exposure. Populations at carrying capacity exposed to concentrations of ≥266 ng/L exhibited a 93% projected decrease in average population size after two years of exposure, again projecting a similar high-concentration exposure population response as observed with 5 to 6 ng/L EE_2 in the experimental lakes area in northwestern Ontario, Canada [7].

The foregoing finding for industrial chemicals, pharmaceuticals, pesticides, personal-use products, and natural and synthetic steroids pales in comparison with the effects observed with antifouling biocide organotins [235]. In female false kelpfish, an exposure for 50 days to 10 ng/L tributyltin (TBT), an aromatase inhibitor, led to lack of vitellogenic-stage follicles and higher proportions of primary-growth-stage follicles [217]. Although Zhang et al. did not determine fecundity or fertility, the results obtained suggested impaired fecundity in females exposed to environmentally relevant concentrations of TBT over a prolonged period. More important, impaired reproductive function was observed in zebrafish exposed to environmentally realistic levels (0.01 to 100 ng/L) of TBT (from 0 to 30, 30 to 60, and 0 to 70 dph) examined three to five months post cessation of exposure for sex ratio and sperm motility of the adults. Fish exposed for 70 days to 0.1 ng/L TBT, a concentration presently below the detection limit in water [187,236], showed a male-biased population that produced a high incidence of sperm lacking flagella, while at 1 ng/L, the motility of sperm was significantly lower than that of control fish, and at 10 ng/L all sperm lacked flagella. The latter findings emphasize the problem of TBT in the environment, as this persistent and lipophilic compound can be found in marine and freshwater body sediments at concentrations up to 7.2 μg/L TBT [22,187,237] and therefore directly, via uptake into fish eggs, or indirectly, via maternal transfer to the eggs, affect the sexual development and reproductive functionality of all subsequent generations and thus of entire fish populations.

6.3.2 Breeding and Reproductive Behavior of Fishes

Limited information is available regarding the xenobiotic-induced change in reproductive behavior and breeding success of fish. However, a review by Jones and Reynolds [238] does suggest that xenobiotics have effects on courtship [99], including decreased or increased frequencies of displays, increased courtship duration, and performance of male-like behavior by masculinized females [117,239]. Studies of parental care have found decreased nest-building activity, decreased offspring defense, and changes in division of parental care between the sexes. However, despite these indications there is yet a paucity of data that causally relate adversely EDC exposure-affected reproductive behavioral and breeding success to population structure and dynamics [240]. An exception to the latter may be a study by Nash et al. [128], who demonstrated that lifelong exposure to 5 ng/L 17α-ethinylestradiol (EE_2) of an F_1 zebrafish generation caused a 56% reduction in fecundity and complete population failure with no fertilization. Infertility in the F_1 generation was due to disturbed sexual differentiation, with males having no functional testes and either undifferentiated or intersex gonads. However, despite the fact that F_1 males lacked functional testes, they showed male-pattern reproductive behavior, inducing the spawning act and competing with healthy males, thereby affecting the breeding dynamics and reproductive success in this group-spawning fish. In conjunction with this study and the reported effects on fecundity observed with 0.32 ng/L EE_2 in fathead minnows [233], the question thus remains to be answered whether behavioral effects would also occur at the lowest environmental EE_2 concentrations and thus whether or not these could really contribute to the adverse effects on the population.

6.3.3 Transgenerational Effects

Transgenerational effects denote any adverse effect in the filial generation of parents exposed to xenobiotics, including EDCs. Within this context, adverse effects also include reduced hatchability, deformation, reduced growth, and survival of offspring as well as the typical adverse effects associated with exposure to EDCs. For example, Gray and co-workers [99] exposed male and female medaka to nominal concentrations of 10, 25, 50, and 100 μg/L 4-*tert*-octylphenol from one day to six months post-hatch and carried out pair trials, always mating an exposed male or female with a corresponding unexposed partner. Males exposed to nominal concentrations of 25 and 50 μg/L showed a reduction in some courtship activity and overall reproductive success, and eggs produced by mating of exposed males and females demonstrated various developmental problems, including circulatory system difficulties, incomplete eye development (i.e., anisophthalmia), and failure to inflate swim-bladders upon hatch. The latter results were confirmed by Kuiper et al. [212], who showed that exposure of adult zebrafish for 30 days to waterborne tetrabromobisphenol A (TBBPA) in nominal concentrations ranging from 0 (control) to 1.5 μM (815.9 μg/L), followed by exposure of the offspring to the same concentrations in early life stages up to 47 days post-hatching, resulted in reduced fecundity of adults exposed to TBBPA concentrations of 0.047 μM (25,6 μg/L) and higher, and decreased hatching of TBBPA-exposed larvae concentrations higher than 0.375 μM (204 μg/L).

Reduced hatching capabilities were associated with malformation and pericardial fluid accumulation. Similarly, Schwaiger et al. [223] reported on hormonal imbalance in offspring of parental rainbow trout exposed for four months to 10 μg/L 4-nonylphenol prior to spawning but not during egg development. The data above effectively suggest that sperm and eggs of the adult fish were affected by 4-*tert*-octylphenol, TBBPA, and 4-nonylphenol, as noted previously by Yang and co-workers in zebrafish [241]. In contrast, a comparable study carried out with medaka and high concentrations of bisphenol A (BPA) [3] demonstrated no reduction of fecundity and fertility in any of the treatment groups and no reduction in survival and growth or imbalance of the sex ratio of the offspring. Similarly, no effects of BPA were observed on the development and hatching rate of the filial generation of fathead minnows [23] exposed to high concentrations similar to those used by Kang et al., but expanding the exposure of the parental generation to 164 days. Moreover, exposure of the viviparous guppies as adults via the water or as embryos via the mother to 26 μg/L octylphenol [192] had no adverse effect on the development of the filial generation, albeit having an effect on the gonads of the adult fish. The data above, as well as recent reports on pentachlorophenol (PCP)-induced decreased hatching rates and time to hatch of offspring of parental medaka exposed to over 200 μg/L PCP for 28 days [171], therefore highlight that no inferences can be made from compound classes. Rather, each compound will have to be investigated in its own right with regard to EDC activity in fish, always considering species-specific sensitivities.

High concentrations of natural estrogens [e.g., estrone (307 and 781 ng/L)] do not appear to induce transgenerational effects in fathead minnows [242], although the parental generation presented with reduced fecundity and spawning frequency of females and mortality of males. Similar, albeit more differentiated effects were observed in adult zebrafish exposed for 21 days and offspring for another 42 days to 17β-estradiol (E_2) and the estrogen receptor antagonist tamoxifen (TMX) [227]. With E_2, the most sensitive endpoints were feminization of offspring (at 0.1 nM 27.2 ng/L) and increased vitellogenin production in males (at 0.32 nM = 87 ng/L). At 1 nM (272.4 ng/L), decreased F_1 survival, increased F_1 body length and weight, VTG-related edema and kidney lesions, and inhibited spermatogenesis were observed. Comparable observations were made by Brown and co-workers in adult rainbow trout exposed for 56 days to either 0.8, 8.3, or 65 ng/L 17α-ethinylestradiol [243], finding that F_1 embryo survival was significantly reduced at 19 dpf compared with the corresponding control. Thus reconfirming that natural as well as synthetic estrogens provide for transgenerational effects only at high concentrations rarely observed in the environment [121]. The latter is also corroborated, in that exposure to TMX resulted in specific effects, including oocyte atresia with granulosa cell transformation and disturbed spermatogenesis in adult fish at an extreme test concentration only (87 nM = 323.2 μg/L) and decreased hatching, survival, and body weight and length as well as decreased feminization in the filial F_1 generation.

Overall, it appears at present that transgenerational effects are restricted primarily to reduced survival as well as to developmental deficiencies in the surviving offspring. Whether these observations can be found only in situations where high to extreme concentrations of EDCs have been employed with the parental generation cannot be ascertained at present. More in-depth understanding of transgenerational effects would certainly be beneficial for the development of more advanced population models and thus environmental risk assessment.

REFERENCES

1. Seki, M., et al., Effect of ethinylestradiol on the reproduction and induction of vitellogenin and testis-ova in medaka (*Oryzias latipes*). *Environmental Toxicology and Chemistry*, 2002, **21**(8):1692–1698.
2. Shioda, T. and M. Wakabayashi, Effect of certain chemicals on the reproduction of medaka (*Oryzias latipes*). *Chemosphere*, 2000, **40**(3):239–243.
3. Kang, I.J., et al., Effects of bisphenol A on the reproduction of Japanese medaka (*Oryzias latipes*). *Environmental Toxicology and Chemistry*, 2002, **21**(11):2394–2400.
4. Kang, I.J., et al., Effect of 17β-estradiol on the reproduction of Japanese medaka (*Oryzias latipes*). *Chemosphere*, 2002, **47**(1):71–80.
5. Wester, W., et al., Identification of endocrine disruptive effects in the aquatic environment: A partial life cycle assay in zebrafish, Rijksinstituut voor Volksgezondheid en Milieu, Bilthoven, The Netherlands, 2003 p. 112.
6. Sohoni, P., et al., Reproductive effects of long-term exposure to bisphenol A in the fathead minnow (*Pimephales promelas*). *Environmental Science and Technology*, 2001, **35**(14): 2917–2925.
7. Kidd, K.A., et al., Collapse of a fish population after exposure to a synthetic estrogen. *Proceedings of the National Acadamy of Sciences of the USA*, 2007, **104**(21):8897–8901.
8. Ayscough, N.J., et al., Review of human pharmaceuticals in the environment: Technical report. Environmental Agency, Bristol, UK, 2000 p. 390.
9. Belfroid, A.C., et al., Analysis and occurrence of estrogenic hormones and their glucuronides in surface water and waste water in The Netherlands. *Science of the Total Environment*, 1999, **225**:101–108.
10. Desbrow, C., et al., Identification of estrogenic chemicals in STW effluent. 1. Chemical fractionation and in vitro biological screening. *Environmental Science and Technology*, 1998, **32**(11):1549–1558.
11. Lee, H.B. and T.E. Peart, Determination of 17 beta-estradiol and its metabolites in sewage effluent by solid-phase extraction and gas chromatography/mass spectrometry. *Journal of the AOAC International*, 1998, **81**(6):1209–1216.
12. Pickering, A.D. and J.P. Sumpter COMPREHENDing endocrine disrupters in aquatic environments. *Environmental Science and Technology*, 2003, **37**(17):331A–336A.

13. Rodgers-Gray, T.P., et al., Long-term temporal changes in the estrogenic composition of treated sewage effluent and its biological effects on fish. *Environmental Science and Technology*, 2000, **34**:1521–1528.

14. Braga, O., et al., Fate of steroid estrogens in Australian inland and coastal wastewater treatment plants. *Environmental Science and Technology*, 2005, **39**(9):3351–3358.

15. Braga, O., et al., Steroid estrogens in primary and tertiary wastewater treatment plants. *Water Science Technology*, 2005, **52**(8):273–278.

16. Chimchirian, R.F., R.P. Suri, and H. Fu, Free synthetic and natural estrogen hormones in influent and effluent of three municipal wastewater treatment plants. *Water Environmental Research*, 2007, **79**(9):969–974.

17. Cicek, N., et al., Removal of selected natural and synthetic estrogenic compounds in a Canadian full-scale municipal wastewater treatment plant. *Water Environmental Research*, 2007, **79**(7):795–800.

18. Hinteman, T., et al., Field study using two immunoassays for the determination of estradiol and ethinylestradiol in the aquatic environment. *Water Research*, 2006, **40** (12):2287–2294.

19. Johnson, A., et al., Estrogen content and relative performance of Japanese and British sewage treatment plants and their potential impact on endocrine disruption. *Environmental Science*, 2007, **14**(6):319–329.

20. Johnson, A.C., et al., Comparing steroid estrogen, and nonylphenol content across a range of European sewage plants with different treatment and management practices. *Water Research*, 2005, **39**(1):47–58.

21. Fernandez, M.P., M.G. Ikonomou, and I. Buchanan, An assessment of estrogenic organic contaminants in Canadian wastewaters. *Science of the Total Environment*, 2007, **373**(1): 250–269.

22. Haubruge, E., F. Petit, and M.J. Gage, Reduced sperm counts in guppies (*Poecilia reticulata*) following exposure to low levels of tributyltin and bisphenol A. *Proceedings of the Royal Society of London Part B: Biological Sciences*, 2000, **267**(1459):2333–2337.

23. Mihaich, E., et al, Fathead minnow (*Pimephales promelas*) reproductive and gonadal histopathology study with bisphenol A. *SETAC Europe 17th Annual Meeting*, Porto, Portugal; 2007

24. Jobling, S., et al., Altered sexual maturation and gamete production in wild roach (*Rutilus rutilus*) living in rivers that receive treated sewage effluents. *Biology of Reproduction*, 2002, **66**:272–281.

25. Jobling, S., et al., Widespread sexual disruption in wild fish. *Environmental Science and Technology*, 1998, **32**:2498–2506.

26. Weber, L.P., R.L. Hill, Jr, and D.M. Janz, Developmental estrogenic exposure in zebrafish (*Danio rerio*): II. Histological evaluation of gametogenesis and organ toxicity. *Aquatic Toxicology*, 2003, **63**(4):431–446.

27. Kiparissis, Y., et al., Effects of the anti-androgens, vinclozolin and cyproterone acetate on gonadal development in the Japanese medaka (*Oryzias latipes*). *Aquatic Toxicology*, 2003, **63**(4):391–403.

28. van den Belt, K., et al., Effects of ethynylestradiol on the reproductive physiology in zebrafish (*Danio rerio*): Time dependency and reversibility. *Environmental Toxicology and Chemistry*, 2002, **21**(4):767–775.

29. Gimeno, S., et al., Demasculinisation of sexually mature male common carp, *Cyprinus carpio*, exposed to 4-*tert*-pentylphenol during spermatogenesis. *Aquatic Toxicology*, 1998 **43**(2–3):93–109.
30. Gimeno, S., et al., Feminisation of young males of the common carp, *Cyprinus carpio*, exposed to 4-*tert*-pentylphenol during sexual differentiation. *Aquatic Toxicology*, 1998, **43**(2–3):77–92.
31. Kinnberg, K. and G. Toft, Effects of estrogenic and antiestrogenic compounds on the testis structure of the adult guppy (*Poecilia reticulata*). *Ecotoxicology and Environmental Safety*, 2003, **54**(1):16–24.
32. Wester, P.W. and H.H. Canton, Histopathological effects in *Poecilia reticulata* (guppy) exposed to methyl mercury chloride. *Toxicological Pathology*, 1992, **20**(1):81–92.
33. Palace, V.P., et al., Induction of vitellogenin and histologic effects in wild fathead minnows from a lake experimentally treated with the synthetic estrogen, ethynylestradiol. *Water Quality Research Journal of Canada*, 2002, **37**(3):637–650.
34. Kang, I.J., et al., Effects of 4-nonylphenol on reproduction of Japanese medaka, *Oryzias latipes*. *Environmental Toxicology and Chemistry*, 2003, **22**(10):2438–2445.
35. Wester, P.W. and J.G. Vos, Toxicological pathology in laboratory fish: An evaluation with two species and various environmental contaminants. *Ecotoxicology*, 1994, **2**(1):21–44.
36. Wester, P.W., et al., Aquatic toxicology: Opportunities for enhancement through histopathology. *Environmental Toxicology and Pharmacology*, 2003, **22**(4):908–913.
37. van der Ven, L.T.M., P.W. Wester, and J.G. Vos, Histopathology as a tool for the evaluation of endocrine disruption in zebrafish (*Danio rerio*). *Environmental Toxicology and Chemistry*, 2003, **22**(4):908–913.
38. Olsson, P.-E. et al., Effects of maternal exposure to estrogen and PCB on different life stages of zebrafish (*Danio rerio*). *Ambio*, 1999, **28**(1):100–106.
39. Janz, D.M., Effects of chronic nonylphenol exposure on medaka (*Oryzias latipes*) gonadal histology. University of Saskatchewan, Saskatoon, Saskatchewan, Canada, 2004.
40. Oern, S., et al., The impact on reproduction of an orally administered mixture of selected PCBs in zebrafish (*Danio rerio*). *Archives of Environmental Contamination and Toxicology*, 1998, **35**(1):52–57.
41. Wolf, J.C., et al., Qualitative and quantitative histomorphologic assessment of fathead minnow *Pimephales promelas* gonads as an endpoint for evaluating endocrine-active compounds: A pilot methodology study. *Toxicologic Pathology*, 2004, **32**(5):600–612.
42. Ankley, G.T., et al., Description and evaluation of a short-term reproduction test with the fathead minnow (*Pimephales promelas*). *Environmental Toxicology and Chemistry*, 2001, **20**(6):1276–1290.
43. Miles-Richardson, S.R., et al., Effects of waterborne exposure to 4-nonylphenol and nonylphenol ethoxylate on secondary sex characteristics and gonads of fathead minnows (*Pimephales promelas*). *Environmental Research, Section A*, 1999, **80**(2):S122–S137
44. Leon, A., et al., Androgen disruption of early development in Qurt strain medaka (*Oryzias latipes*). *Aquatic Toxicology*, 2007, **82**(3):195–203.
45. Wallig, M.A., Morphologic manifestations of toxic cell injury, in *Handbook of Toxicologic Pathology*, W.M. Haschek, C.G. Rousseaux, and M.A. Wallig, Editors. Academic Press, San Diego, CA, 2002 39–66.
46. Maxwell, L.B. and H.M. Dutta, Diazinon-induced endocrine disruption in bluegill sunfish, *Lepomis macrochirus*. *Ecotoxicology and Environmental Safety*, 2005, **60**(1):21–27.

47. Dutta, H.M. and L.B. Maxwell, Histological examination of sublethal effects of diazinon on ovary of bluegill, *Lepomis macrochirus*. *Environmental Pollution*, 2003, **121**(1):95–102.
48. Diniz, M.S., et al., Estrogenic effects in crucian carp (*Carassius carassius*) exposed to treated sewage effluent. *Ecotoxicology and Environmental Safety*, 2005, **62**(3):427–435.
49. Schmitt, C.J., et al., Environmental contaminants and biomarker responses in fish from the Rio Grande and its U.S. tributaries: Spatial and temporal trends. *Science of the Total Environment*, 2005, **350**(1–3):161–193.
50. Blazquez, M., et al., Structural and functional aspects of early exposure to estradiol 17-β and 17α-ethynylestradiol on the gonads of the gonochoristic teleost *Dicentrarchus labrax*. *Fish Physiology and Biochemistry*, 1998, **18**(1):37–47.
51. Weber, L.P., et al., Increased cellular apoptosis after chronic aqueous exposure to nonylphenol and quercetin in adult medaka (*Oryzias latipes*). *Comparative Biochemistry and Physiology, Part C: Toxicology and Pharmacology*, 2002, **131**(1):51–59.
52. Janz, D.M., et al., Recovery of ovary size, follicle cell apoptosis, and HSP70 expression in fish exposed to bleached pulp mill effluent. *Canadian Journal of Fisheries and Aquatic Sciences*, 2001, **58**(3):620–625.
53. Janz, D.M., et al., Elevated ovarian follicular apoptosis and heat shock protein-70 expression in white sucker exposed to bleached kraft pulp mill effluent. *Toxicology and Applied Pharmacology*, 1997, **147**(2):391–398.
54. Islinger, M., et al., Effects of 17α-ethinylestradiol on the expression of three estrogen-responsive genes and cellular ultrastructure of liver and testes in male zebrafish. *Aquatic Toxicology*, 2003, **62**(2):85–103.
55. Miles-Richardson, S.R., et al., Effects of waterborne exposure of 17β-estradiol on secondary sex characteristics and gonads of fathead minnows (*Pimephales promelas*). *Aquatic Toxicology*, 1999, **47**(2):129–145.
56. Tilly, J.J. and G.I. Perez, Mechanisms and genes of physiological cell death: A new direction for toxicological risk assessments? in *Comprehensive Toxicology*, I.G. Sipes, C.A. McQueen, and A.J. Gandolfi, Editors. Elsevier Press, Oxford, UK, 1997, pp. 379–385.
57. Gray, M.A., A.J. Niimi, and C.D. Metcalfe, Factors affecting the development of testis-ova in medaka, *Oryzias latipes*, exposed to octylphenol. *Environmental Toxicology and Chemistry*, 1999, **18**(8):1835–1842.
58. Miles-Richardson, S.R., *Effects of Estrogenic Compounds on the Fathead Minnow (Pimephales promelas)*. PhD thesis, Michigan State University, East Lansing, MI, 1997.
59. Nagahama, Y., The functional morphology of teleost gonads, in *Fish Physiology*, W.S. Hoar, D.J. Randall, and E.M. Donaldson, Editors. Academic Press, San Diego, CA, 1983, pp. 223–275.
60. Tyler, C.R. and J.P. Sumpter, Oocyte growth and development in teleosts. *Reviews in Fish Biology and Fisheries*, 1996, **6**(3):287–318.
61. Patino, R. and J.M. Redding, Reproductive systems, in *The Handbook of Experimental Animals: The Laboratory Fish*, G.K. Ostrander, Editor. Academic Press, San Diego, CA, 2000, pp. 489–500.
62. Wallace, R.A. and K. Selman, Cellular dynamic aspects of oocyte growth in teleosts. *American Journal of Zoology*, 1981, **21**:325–343.
63. Papoulias, D.M., D.B. Noltie, and D.E. Tillitt, An in vivo model fish system to test chemical effects on sexual differentiation and development: Exposure to ethinyl estradiol. *Aquatic Toxicology*, 1999, **48**(1):37–50.

64. Ankley, G.T., et al., Evaluation of the aromatase inhibitor fadrozole in a short-term reproduction assay with the fathead minnow (*Pimephales promelas*). *Toxicological Sciences*, 2002, **67**:121–130.

65. Wood, A.W. and G. Van der Kraak, Inhibition of apoptosis in vitellogenic ovarian follicles of rainbow trout (*Oncorynchus mykiss*) by salmon gonadotropin, epidermal growth factor, and 17β-estradiol. *Molecular Reproduction and Development*, 2002, **61**(4):511–518.

66. Heiden, T.K., M.J. Carvan 3rd, and R.J. Hutz, Inhibition of follicular development, vitellogenesis, and serum 17β-estradiol concentrations in zebrafish following chronic, sublethal dietary exposure to 2,3,7,8-tetrachlorodibenzo-*p*-dioxin. *Toxicological Sciences*, 2006, **90**(2):490–499.

67. Makynen, E.A., et al., Effects of the mammalian antiandrogen vinclozolin on development and reproduction of the fathead minnow (*Pimephales promelas*). *Aquatic Toxicology*, 2000, **48**(4):461–475.

68. Laenge, R., et al., Effects of the synthetic estrogen 17α-ethynylestradiol on the life-cycle of the fathead minnow (*Pimephales promelas*). *Environmental Toxicology and Chemistry*, 2001, **20**(6):1216–1227.

69. Zillioux, E.J., et al., The sheepshead minnow as an in vivo model for endocrine disruption in marine teleosts: a partial life-cycle test with 17α-ethynylestradiol. *Environmental Toxicology and Chemistry*, 2001, **20**(9):1968–1978.

70. Tyler, C.R., et al., An in vivo testing system for endocrine disruptors in fish early life stages using induction of vitellogenin. *Environmental Toxicology and Chemistry*, 1999, **18**(2):337–347.

71. Kinnberg, K., et al., Effects of nonylphenol and 17β-estradiol on vitellogenin synthesis and testis morphology in male platyfish *Xiphophorus maculatus*. *The Journal of Experimental Biology*, 2000, **203**(2):171–181.

72. van der Ven, L.T.M. and P.W. Wester, Normal histology and effects of hormonal active agents of reproductive organs of the zebrafish (*Danio rerio*). *OECD Technical Workshop on Gonadal Histology of Small Laboratory Fish as a Tool for Endocrine Disruption Screening and Testing*, Bilthoven, the Netherlands, 2002.

73. van der Ven, L.T.M. and P.W. Wester, Toxicological pathology atlas of small laboratory fish with special reference to normal histology and effects of endocrine disruption in the zebrafish (*Danio rerio*). http://www.rivm.nl/fishtoxpat 2003

74. Dietrich, D.R., et al., Optimization of tissue preparation for the quantification of germinal cells in histologic sections of fathead minnow gonads. *SETAC Europe 13th Annual Meeting*, Hamburg, Germany 2003.

75. Grier, H.J., Comparative organization of Sertoli cells including the Sertoli cell barrier, in *The Sertoli Cell*, L.D. Russell and M.D. Griswold, Editors. Cache River Press, Clearwater, FL, 1993, pp. 704–739.

76. Grier, H.J., Cellular organization of the testis and spermatogenesis in fishes. *American Zoologist*, 1981, **21**:345–357.

77. Grier, H.J., J.M. Fitzsimons, and J.R. Linton, Structure and ultrastructure of the testis and sperm formation in goodeid teleosts. *Journal of Morphology*, 1978, **156**(3):419–435.

78. Grier, H.J., Aspects of germinal cyst and sperm development in *Poecilia latipinna* (Teleostei: Poeciliidae). *Journal of Morphology*, 1975, **146**:229–250.

79. Pudney, J., Spermatogenesis in nonmammalian vertebrates. *Microscopy Research and Technique*, 1995, **32**(6):459–497.

80. U.S. EPA, A short-term test method for assessing the reproductive toxicity of endocrine-disrupting chemicals using the fathead minnow (*Pimephales promelas*). U.S. Environmental Protection Agency, Washington, DC, 2002, pp. 141–144.

81. Kinnberg, K., B. Korsgaard, and P. Bjerregaard, Concentration-dependent effects of nonylphenol on testis structure in adult platyfish *Xiphophorus maculatus. Marine Environmental Research*, 2000, **50**(1–5):169–173.

82. Christiansen, T., et al., γ-Glutamyl transpeptidase as a possible marker of Sertoli cells in fish testes for studies of xenoestrogens. *Marine Environmental Research*, 2000, **50**:213–216.

83. Loir, M., et al., Cell–cell interactions in the testis of teleosts and elasmobranchs. *Microscopy Research and Technique*, 1995, **32**(6):533–552.

84. Russell, L.D. and M.D. Griswold, *The Sertoli Cell.* Cache River Press, Clearwater, FL, 1993, p. 747.

85. Tafanelli, R. and R.C. Summerfelt, Cadmium-induced histopathological changes in goldfish, in *The Pathology of Fishes*, W.E. Ribelin and G. Migaki, Editors. University of Wisconsin Press, Madison WI, 1975, pp. 613–645.

86. Sehgal, R. and A.K. Pandey, Effect of cadmium chloride on testicular activities in guppy *Lebistes reticulatus. Comparative Physiology and Ecology*, 1984, **9**:225–230.

87. Nielsen, L. and E. Baatrup, Quantitative studies on the effects of environmental estrogens on the testis of the guppy, *Poecilia reticulata. Aquatic Toxicology*, 2006, **80**(2):140–148.

88. Miura, T., et al., Estradiol-17β stimulates the renewal of spermatogonial stem cells in males. *Biochemical and Biophysical Research Communications*, 1999, **264**:230–234.

89. Miura, T. and C.I. Miura, Molecular control mechanisms of fish spermatogenesis. *Fish Physiology and Biochemistry*, 2003, **28**(4):181–186.

90. Miura, T., et al., Hormonal induction of all stages of spermatogenesis in vitro in the male Japanese eel, *Anguilla japonica. Proceedings of the National Academy of Sciences of the USA*, 1991, **88**(13):5774–5778.

91. Miwa, S., L. Yan, and Swanson P., Localization of two gonadotropin receptors in the salmon gonad by in vitro ligand autoradiography. *Biology of Reproduction*, 1994, **50**:629–642.

92. Nagahama, Y., Endocrine regulation of gametogenesis in fish. *International Journal of Developmental Biology*, 1994, **38**:217–229.

93. Nagahama, Y., Gonadal steroid hormones: major regulators of gonadal sex differentiation and gametogenesis in fish. *Sixth International Symposium on the Reproductive Physiology of Fish*, Bergen, Norway 1999.

94. Nagahama, Y., H. Kagawa, and G. Young, Cellular sources of sex steroids in teleost gonads. *Canadian Journal of Fisheries and Aquatic Sciences*, 1982, **39**:56–64.

95. Nobrega, R.H. and I. Quagio-Grassiotto, Morphofunctional changes in Leydig cells throughout the continuous spermatogenesis of the freshwater teleost fish, *Serrasalmus spilopleura* (Characiformes, Characidae): An ultrastructural and enzyme study. *Cell Tissue Research*, 2007, **329**(2):339–349.

96. Ankley, G.T., et al., Ketoconazole in the fathead minnow (*Pimephales promelas*): Reproductive toxicity and biological compensation. *Environmental Toxicology and Chemistry*, 2007, **26**(6):1214–1223.

97. Coimbra, A.M. and M.A. Reis-Henriques, Tilapia larvae Aroclor 1254 exposure: Effects on gonads and circulating thyroid hormones during adulthood. *Bulletin of Environmental Contamination and Toxicology*, 2007, **79**(5):488–493.

98. Weber, L.P., et al., Increased kidney, liver, and testicular cell death after chronic exposure to 17alpha-ethinylestradiol in medaka (*Oryzias latipes*). *Environmental Toxicology and Chemistry*, 2004, **23**(3):792–797.

99. Gray, M.A., K.L. Teather, and C.D. Metcalfe, Reproductive success and behaviour of Japanese medaka (*Oryzias latipes*) exposed to 4-*tert*-octylphenol. *Environmental Toxicology and Chemistry*, 1999, **18**(11):2587–2594.

100. Haschek, W.M., C.G. Rousseaux, and M.A. Wallig, *Handbook of Toxicologic Pathology*, Vol 2, 2nd ed. Academic Press, San Diego, CA, 2002, p. 973.

101. Schafers, C., et al., Concentration- and time-dependent effects of the synthetic estrogen, 17alpha-ethinylestradiol, on reproductive capabilities of the zebrafish, *Danio rerio*. *Journal of Toxicology and Environmental Health A*, 2007, **70**(9):768–779.

102. Metcalfe, C.D., et al., Estrogenic potency of chemicals detected in sewage treatment plant effluents as determined by in vivo assays with Japanese medaka (*Oryzias latipes*). *Environmental Toxicology and Chemistry*, 2001, **20**(2):297–308.

103. Yokota, H., et al., Effect of bisphenol A on the early life stage in Japanese medaka (*Oryzias latipes*). *Environmental Toxicology and Chemistry*, 2000, **19**(7):1925–1930.

104. Yokota, H., et al., Life-cycle toxicity of 4-nonylphenol to medaka (*Oryzias latipes*). *Environmental Toxicology and Chemistry*, 2001, **20**(11):2552–2560.

105. Egami, N., Production of testis–ova in adult males of *Oryzias latipes*: I. Testis-ova in fish receiving estrogens. *Japanese Journal of Zoology*, 1955, **11**:353–365.

106. Wing, L.Y., et al., Expression and mitogenic effect of fibroblast growth factor-9 in human endiometriotic implant is regulated by aberrant production of estrogens. *Journal of Clinical Endocrinology and Metabolism*, 2003, **88**(11):5547–5554.

107. Cline, J.M., et al., Assessment of hormonally active agents in the repoductive tract of female nonhuman primates. *Toxicological Pathology*, 2001, **29**(1):84–90.

108. Shimizu, I., Impact of oestrogens on the progression of liver disease. *Liver International*, 2003, **23**(1):63–69.

109. Xu, J.W., et al., Effects of estradiol on liver estrogen receptor-alpha and its mRNA expression in hepatic fibrosis. *World Journal of Gastroenterology*, 2004, **10**(2):250–254.

110. Gimeno, S., et al., Disruption of sexual differentiation in genetic male common carp (*Cyprinus carpio*) exposed to an alkylphenol during different life stages. *Environmental Science and Technology*, 1997, **31**(10):2884–2890.

111. van Aerle, R., et al., Windows of sensitivity for the estrogenic effects of ethinylestradiol in early life-stages of fathead minnow, *Pimephales promelas*. *Ecotoxicology*, 2002, **11** (6):423–434.

112. Nakamura, M. and H. Takahashi, Gonadal sex differentation in *Tilapia mossambica* with special regard to the time of oestrogen treatment effective in inducing feminization of genetic fishes. *Bulletin of the Faculty of Fisheries,* Hokkaido University, 1973, **24**:1–23.

113. van Aerle, R. et al., Sexual disruption in a second species of wild cyprinid fish (the gudgeon, *Gobio gobio*) in United Kingdom freshwaters. *Environmental Toxicology and Chemistry*, 2001, **20**(12):2841–2847.

114. Rodgers-Gray, T.P., et al., Exposure of juvenile roach (*Rutilus rutilus*) to treated sewage effluent induces dose-dependent and persistent disruption in gonadal duct development. *Environmental Science and Technology*, 2001, **35**(3):462–470.

115. Hemming, J.M., et al., Assessment of the estrogenicity and toxicity of a domestic wastewater effluent flowing through a constructed wetland system using biomarkers in

male fathead minnows (*Pimephales promelas* Rafinesque, 1820). *Environmental Toxicology and Chemistry*, 2001, **20**(10):2268–2275.

116. Mellanen, P., et al., Expression of the vitellogenin gene in the liver of juvenile whitefish (*Coregonus lavaretus* L. *s.l.*) exposed to effluents from pulp and paper mills. *Ecotoxicology and Environmental Safety*, 1999, **43**:133–137.

117. Ellis, R.J., et al., In vivo and in vitro assessment of the androgenic-potential of a pulp and paper mill effluent. *Environmental Toxicology and Chemistry*, 2003, **22**(7):786–798.

118. Larsson, D.G.J., H. Hällman, and L. Förlin, More male embryos near a pulp mill. *Environmental Toxicology and Chemistry*, 2000, **19**(12):2911–2917.

119. Denton, T.E., et al., Masculinisation of female mosquitofish by exposure to plant sterols and *Mycobacterium smegmatis*. *Bulletin for Environmental Contamination and Toxicology*, 1985, **35**(5):627–632.

120. Drèze, V., et al., Effects of 4-nonylphenol on sex differentiation and puberty in mosquitofish (*Gambusia holbrooki*). *Ecotoxicology*, 200, **9**(1–2):93–103.

121. Larsson, D.G.J., et al., Ethinyloestradiol: An undesired fish contraceptive? *Aquatic Toxicology*, 1999, **45**(2–3):91–97.

122. O'Brien, E. and D.R. Dietrich, Hindsight rather than foresight: Reality versus the EU draft guideline on pharmaceuticals in the environment. *Trends in Biotechnology*, 2004, **22**(7):326–330.

123. Dutta, H.M., D. Misquitta, and S. Khan, The effects of endosulfan on the testes of bluegill fish, *Lepomis macrochirus*: A histopathological study. *Archives Environmental Contamination Toxicology*, 2006, **51**(1):149–156.

124. Bogers, R., et al., Development of chronic tests for endocrine active chemicals, Part 2: An extended fish early-life stage test with an androgenic chemical in the fathead minnow (*Pimephales promelas*). *Aquatic Toxicology*, 2006, **80**(2):119–130.

125. Panter, G.H., et al., Development of chronic tests for endocrine active chemicals. Part 1. An extended fish early-life stage test for oestrogenic active chemicals in the fathead minnow (*Pimephales promelas*). *Aquatic Toxicology*, 2006, **77**(3):279–290.

126. Scholz, S. and H.O. Gutzeit, 17α-Ethinylestradiol affects reproduction, sexual differentiation and aromatase gene expression of the medaka (*Oryzias latipes*). *Aquatic Toxicology*, 2000, **50**(4):363–373.

127. Papoulias, D.M., D.B. Noltie, and D.E. Tillitt, An in vivo model fish system to test chemical effects on sexual differentiation and development: Exposure to ethinyl estradiol. *Aquatic Toxicology*, 2000, **48**(1):37–50.

128. Nash, J.P., et al., Long-term exposure to environmental concentrations of the pharmaceutical ethynylestradiol causes reproductive failure in fish. *Environmental Health Perspectives*, 2004, **112**(17):1725–1733.

129. Hill, L., Jr. and D.M. Janz, Developmental estrogenic exposure in zebrafish (*Danio rerio*): I. Effects on sex ratio and breeding success. *Aquatic Toxicology*, 2003, **63**(4):417–429.

130. Oern, S., et al., Gonad development and vitellogenin production in zebrafish (*Danio rerio*) exposed to ethinylestradiol and methyltestosterone. *Aquatic Toxicology*, 2003, **65**(4): 397–411.

131. Andersen, L., et al., Effects of exposure to 17alpha-ethinylestradiol during early development on sexual differentiation and induction of vitellogenin in zebrafish (*Danio rerio*). *Comparative Biochemistry and Physiology Part C: Toxicology and Pharmacology*, 2003, **134**(3):365–374.

132. Fenske, M. and H. Segner, Aromatase modulation alters gonadal differentiation in developing zebrafish (*Danio rerio*). *Aquatic Toxicology*, 2004, **67**(2):105–126.

133. Holbech, H., et al., Detection of endocrine disrupters: evaluation of a fish sexual development test (FSDT). *Comparative Biochemistry and Physiology Part C: Toxicology and Pharmacology*, 2006, **144**(1):57–66.

134. Allen, Y., et al., The extent of oestrogenic contamination in the UK estuarine and marine environments: Further surveys of flounder. *Science of the Total Environment*, 1999, **223**(1–3):5–20.

135. Zha, J., et al., Comparison of ethinylestradiol and nonylphenol effects on reproduction of Chinese rare minnows (*Gobiocypris rarus*). *Ecotoxicology and Environmental Safety*, 2008.

136. Hahlbeck, E., R. Griffiths, and B.-E. Bengtsson, The juvenile three-spined stickleback (*Gasterosteus aculeatus* L.) as a model organism for endocrine disruption: I. Sexual differentiation. *Aquatic Toxicology*, 2004, **70**(4):287–310.

137. Maunder, R.J., et al., Impaired reproduction in three-spined sticklebacks exposed to ethinyl estradiol as juveniles. *Biology of Reproduction*, 2007, **77**(6):999–1006.

138. Zha, J., et al., Assessment of 17alpha-ethinylestradiol effects and underlying mechanisms in a continuous, multigeneration exposure of the Chinese rare minnow (*Gobiocypris rarus*). *Toxicology and Applied Pharmacology*, 2008, **226**(3):298–308.

139. Iwamatsu, T., Convenient method for sex reversal in a freshwater teleost, the medaka. *Journal of Experimental Zoology*, 1999, **283**(2):210–214.

140. Koger, C.S., S.J. Teh, and D.E. Hinton, Determining the sensitive developmental stages of intersex induction in medaka (*Oryzias latipes*) exposed to 17β-estradiol or testosterone. *Marine Environmental Research*, 2000, **50**(1–5):201–206.

141. Nimrod, A.C. and W.H. Benson, Reproduction and development of Japanese medaka following an early life stage exposure to estrogens. *Aquatic Toxicology*, 1998, **44**(1–2):141–156.

142. Hirai, N., et al., Feminization of Japanese medaka (*Oryzias latipes*) exposed to 17beta-estradiol: Formation of testis-ova and sex-transformation during early-ontogeny. *Aquatic Toxicology*, 2006, **77**(1):78–86.

143. Yamamoto, T., A further study of induction of functional sex reversal in genotypic males of the medaka (*Oryzias latipes*) and progenies of sex reversals. *Genetics*, 1959, **44**:739–757.

144. Brion, F., et al., Impacts of 17beta-estradiol, including environmentally relevant concentrations, on reproduction after exposure during embryo-larval-, juvenile- and adult-life stages in zebrafish (*Danio rerio*). *Aquatic Toxicology*, 2004, **68**(3):193–217.

145. Akerblom, N., et al., Impact of polychlorinated naphthalenes (PCNs) in juvenile Baltic salmon, *Salmo salar*: Evaluation of estrogenic effects, development, and CYP1A induction. *Archives of Environmental Contamination and Toxicology*, 2000, **38**(2):225–233.

146. Gimeno, S., et al., Feminization of male carp. *Nature*, 1996, **384**(6606):221–222.

147. Zaroogian, G., et al., Effect of 17beta-estradiol, *o,p*-DDT, octylphenol and *p,p*-DDE on gonadal development and liver and kidney pathology in juvenile male summer flounder (*Paralichthys dentatus*). *Aquatic Toxicology*, 2001, **54**(1–2):101–112.

148. Pawlowski. S., et al., Androgenic and estrogenic effects of the synthetic androgen 17α-methyltestosterone on sexual development and reproductive performance in the fathead minnow (*Pimephales promelas*) determined using the gonadal recrudescence assay. *Aquatic Toxicology*, 2004, **68**(3):277–291.

149. Chiasson, M. and T.J. Benfey, Gonadal differentiation and hormonal sex reversal in Arctic charr (*Salvelinus alpinus*). *Journal of Experimental Zoology, Part A: Ecological Genetics and Physiology*, 2007, **307**(9):527–534.

150. Jensen, K.M., et al., Characterization of responses to the antiandrogen flutamide in a short-term reproduction assay with the fathead minnow. *Aquatic Toxicology*, 2004, **70**(2):99–110.

151. Mandich, A., et al., In vivo exposure of carp to graded concentrations of bisphenol A. *General and Comparative Endocrinology*, 2007, **153**(1–3):15–24.

152. Rasmussen, T.H., et al., Anti-estrogen prevents xenoestrogen-induced testicular pathology of eelpout (*Zoarces viviparus*). *Aquatic Toxicology*, 2005, **72**(3):177–194.

153. Metcalfe, T.L., et al., Gonadal development and endocrine responses in Japanese medaka (*Oryzias latipes*) exposed to *o,p*-DDT in water or through maternal transfer. *Environmental Toxicology and Chemistry*, 2000, **19**(7):1893–1900.

154. Kinnberg, K., et al., Effects of the fungicide prochloraz on the sexual development of zebrafish (*Danio rerio*). *Comparative Biochemistry and Physiology: Part C, Toxicology and Pharmacology*, 2007, **145**(2):165–170.

155. Spano, L., et al., Effects of atrazine on sex steroid dynamics, plasma vitellogenin concentration and gonad development in adult goldfish (*Carassius auratus*). *Aquatic Toxicology*, 2004, **66**(4):369–379.

156. Figueiredo-Fernandes, A., et al., The effect of paraquat on hepatic EROD activity, liver, and gonadal histology in males and females of Nile tilapia, *Oreochromis niloticus*, exposed at different temperatures. *Archives of Environmental Contamination and Toxicology*, 2006, **51**(4):626–632.

157. Khalil, W.K., et al., A sub-acute study of metronidazole toxicity assessed in Egyptian *Tilapia zillii*. *Jaurnal of Applied Toxicology*, 2007, **27**(4):380–390.

158. Nichols, K.M., et al., Effects of exposure to municipal wastewater in situ on the reproductive physiology of the fathead minnow (*Pimephales promelas*). *Environmental Toxicology and Chemistry*, 1999, **18**(9):2001–2012.

159. Sole, M., et al., Feminization of wild carp, *Cyprinus carpio*, in a polluted environment: Plasma steroid hormones, gonadal morphology and xenobiotic metabolizing system. *Comparative Biochemistry and Physiology, Part C: Toxicology and Pharmacology*, 2003, **136**:145–156.

160. Sorensen, E. M. and P. Thomas, Selenium accumulation, reproductive status, and histopathological changes in environmentally exposed redear sunfish. *Archives of Toxicology*, 1988, **61**(4):324–329.

161. Woodling, J.D., et al., Intersex and other reproductive disruption of fish in wastewater effluent dominated Colorado streams. *Comparative Biochemistry and Physiology, Part C: Toxicology and Pharmacology*, 2006, **144**(1):10–15.

162. Bernet, D., et al., Frequent and unexplained gonadal abnormalities in whitefish (central alpine *Coregonus* sp.) from an alpine oligotrophic lake in Switzerland. *Diseases of Aquatic Organisms*, 2004, **61**(1–2):137–148.

163. Yamaguchi, S., et al., Effects of lead, molybdenum, rubidium, arsenic and organochlorines on spermatogenesis in fish: Monitoring at Mekong Delta area and in vitro experiment. *Aquatic Toxicology*, 2007, **83**(1):43–51.

164. Hinck, J.E., et al., Chemical contaminants, health indicators, and reproductive biomarker responses in fish from the Colorado River and its tributaries. *Science of the Total Environment*, 2007, **378**(3):376–402.

165. Zha, J., et al., Histological alternation and vitellogenin induction in adult rare minnow (*Gobiocypris rarus*) after exposure to ethynylestradiol and nonylphenol. *Chemosphere*, 2007, **66**(3):488–495.

166. Seki, M., et al., Fish full life-cycle testing for 17β-estradiol on medaka (*Oryzias latipes*). *Environmental Toxicology and Chemistry*, 2005, **24**(5):1259–1266.

167. Kang, I.J., et al., The effects of methyltestosterone on the sexual development and reproduction of adult medaka (*Oryzias latipes*). *Aquatic Toxicology*, 2008.

168. Blazquez, M., et al., Development of sex control techniques for European sea bass (*Dicentrarchus labrax* L.) aquaculture: Effects of dietary 17α-methyltestosterone prior to sex differentiation. *Aquaculture*, 1995, **135**(4):329–342.

169. Kang, I.J., et al., Anti-androgen flutamide affects gonadal development and reproduction in medaka (*Oryzias latipes*). *Marine Environmental Research* 2006, **62**(Suppl):S253–S257.

170. Afonso, L.O.B., G.J. Wassermann, and R. Terezinha de Oliveira, Sex reversal in Nile tilapia (*Oreochromis niloticus*) using a nonsteroidal aromatase inhibitor. *Journal of Experimental Zoology*, 2001, **290**(1):177–181.

171. Zha, J., Z. Wang, and D. Schlenk, Effects of pentachlorophenol on the reproduction of Japanese medaka (*Oryzias latipes*). *Chemico-Biological Interactions*, 2006, **161**(1):26–36.

172. Vigano, L., et al., First observation of intersex cyprinids in the Po River (Italy). *Science of the Total Environment*, 2001, **269**(1–3): 189–194.

173. Vethaak, A.D., et al., Estrogenic effects in fish in The Netherlands: Some preliminary results. *Toxicology*, 2002, **181–182**: 147–150.

174. Leatherland, J.F. and R.A. Sonstegard, Structure of normal testis and testicular tumors in cyprinids from Lake Ontario. *Cancer Research*, 1978, **38**(10):3164–3173.

175. Gercken, J. and H. Sordyl, Intersex in feral marine and freshwater fish from northeastern Germany. *Marine Environmental Research*, 2002 **54** (3–5):651–655.

176. Hashimoto, S., et al., Elevated serum vitellogenin levels and gonadal abnormalities in wild male flounder (*Pleuronectes yokohamae*) from Tokyo Bay, Japan. *Marine Environmental Research*, 2000, **49**(1):37–53.

177. Douxfils, J., et al., Do sewage treatment plant discharges substantially impair fish reproduction in polluted rivers? *Science of the Total Environment*, 2007, **372**:497–514.

178. Diniz, M.S., I. Peres, and J.C. Pihan, Comparative study of the estrogenic responses of mirror carp (*Cyprinus carpio*) exposed to treated municipal sewage effluent (Lisbon) during two periods in different seasons. *Science of the Total Environment*, 2005, **349**(1–3): 129–139.

179. Barnhoorn, I.E., et al., Histological evidence of intersex in feral sharptooth catfish (*Clarias gariepinus*) from an estrogen-polluted water source in Gauteng, South Africa. *Environmental Toxicology*, 2004, **19**(6):603–608.

180. Harshbarger, J.C., M.J. Coffey, and M.Y. Young, Intersexes in Mississippi River shovelnose sturgeon sampled below Saint Louis, Missouri, USA. *Marine Environmental Research*, 2000, **50**(1–5)247–250.

181. Kavanagh, R.J., et al., Endocrine disruption and altered gonadal development in white perch (*Morone americana*) from the lower Great Lakes region. *Environmental Health Perspectives*, 2004, **112**(8):898–902.

182. Saxena, P.K. and K. Mani, Effect of safe concentrations of some pesticides on testicular recrudescence in the freshwater murrel, *Channa punctatus* (Bl.): A morphological study. *Ecotoxicology and Environmental Safety*, 1987, **4**:56–63.

183. Kulshrestha, S.K. and N. Arora, Impairments induced by sublethal doses of two pesticides in the ovaries of a freshwater teleost *Channa striatus* Bloch. *Toxicology Letters*, 1984, **20**(1):93–98.

184. Sohoni, P. and J. P. Sumpter, Several environmental oestrogens are also anti-androgens. *Journal of Endocrinology*, 1998, **158**(3):327–339.

185. Sohoni, P., P.A. Lefevre, J. Ashby, and J.P. Sumpter, Possible androgenic/anti-androgenic activity of the insecticide fenitrothion. *Journal of Applied Toxicology*, 2001, **21**(3):173–178.

186. McAllister, B.G. and D.E. Kime, Early life exposure to environmental levels of the aromatase inhibitor tributyltin causes masculinisation and irreversible sperm damage in zebrafish (*Danio rerio*). *Aquatic Toxicology*, 2003, **65**(3):309–316.

187. Fent, K., B.R. Woodin, and J.J. Stegeman, Effects of triphenyltin and other organotins on hepatic monooxygenase system in fish. *Comparative Biochemistry and Physiology, Part C: Toxicology and Pharmacology*, 1998, **121**(1–2): 277–288.

188. Weisbrod, C.J., et al., Effects of the UV filter benzophenone-2 on reproduction in fish. *Toxicology and Applied Pharmacology*, 2007, **225**(3):255–266.

189. Christianson-Heiska, I., et al., Endocrine modulating actions of a phytosterol mixture and its oxidation products in zebrafish (*Danio rerio*). *Comparative Biochemistry and Physiology Part C: Toxicology and Pharmacology*, 2007, **145**(4):518–527.

190. Mochida, K., et al., Molecular and histological evaluation of tributyltin toxicity on spermatogenesis in a marine fish, the mummichog (*Fundulus heteroclitus*). *Aquatic Toxicology*, 2007, **83**(1):73–83.

191. Willey, J.B. and P. H. Krone, Effects of endosulfan and nonylphenol on the primordial germ cell population in pre-larval zebrafish embryos. *Aquatic Toxicology*, 2001 **54**(1–2)113.

192. Kinnberg, K., B. Korsgaard, and P. Bjerregaard, Effects of octylphenol and 17β-estradiol on the gonads of guppies (*Poecilia reticulata*) exposed as adults via the water or as embryos via the mother. *Comparative Biochemistry and Physiology, Part C: Toxicology and Pharmacology*, 2003, **134**(1):45–55.

193. Ankley, G.T., et al., Effects of two fungicides with multiple modes of action on reproductive endocrine function in the fathead minnow (*Pimephales promelas*). *Toxicolical Science*, 2005, **86**(2):300–308.

194. Freeman, H.C., G. Sangalang, and B. Flemming, The sublethal effects of a polychlorinated biphenyl (Aroclor 1254) diet on the Atlantic cod (*Gadus morhua*). *Science of the Total Environment*, 1982, **24**(1):1–11.

195. Nath, K. and N. Kumar, Gonadal histopathology following nickel intoxication in the giant gaurami *Colisa fasciatus* (Bloch and Schneider), a freshwater tropical perch. *Bulletin of Environmental Contamination and Toxicology*, 1990, **45**(2):299–304.

196. Kumar, S. and S.C. Pant, Comparative effects of the sublethal poisoning of zinc, copper and lead on the gonads of the teleost *Puntius conchonius* Ham. *Toxicology Letters*, 1984, **23**(2):189–194.

197. Wester, P.W., Histopathological effects of environmental pollutants β-HCH and methyl mercury on reproductive organs in freshwater fish. *Comparative Biochemistry and Physiology, Part C: Toxicology and Pharmacology*, 1991, **100C**(1–2): 237–239.

198. Molina-Molina, J.-M., et al., Steroid receptor profiling of vinclozolin and its primary metabolites. *Toxicology and Applied Pharmacology*, 2006, **216**(1):44.

199. Blystone, C.R., et al., Prochloraz inhibits testosterone production at dosages below those that affect androgen-dependent organ weights or the onset of puberty in the male Sprague Dawley rat. *Toxicological Science*, 2007, **97**(1):65–74.

200. Piferrer, F., et al., Induction of sterility in coho salmon (*Oncorhynchus kisutch*) by androgen immersion before first feeding. *Aquaculture*, 1994, **119**(4):409.

201. Bhandari, R.K., et al., Suppression of steroidogenic enzyme expression during androgen-induced sex reversal in Nile tilapia (*Oreochromis niloticus*). *General and Comparative Endocrinology*, 2006, **145**(1):20–24.

202. Guiguen, Y., et al., Steroids and gonadal sex differentiation in the rainbow trout, *Oncorhynchus mykiss*. *Proceedings of the Sixth International Symposium on the Reproductive Physiology of Fish*, Institute of Marine Research, Bergen, Norway, 2000.

203. Uchida, D., et al., An aromatase inhibitor or high water temperature induce oocyte apoptosis and depletion of P450 aromatase activity in the gonad of genetic female zebrafish during sex reversal. *Comparative Biochemistry and Physiology, Part B: Biochemistry and Molecular Biology*, 2004, **137**(1):11–20.

204. Piferrer, F., et al., Brief treatment with an aromatase inhibitor during sex differentiation causes chromosomally female salmon to develop as normal functional males. *Journal of Experimental Zoology*, 1994, **270**(3):255–262.

205. Chatterjee, S., A.B. Dutta, and R. Ghosh, Impact of carbofuran in the oocyte maturation of catfish, *Heteropenustes fossilis* (Bloch). *Archives of Environmental Contamination and Toxicology*, 1997, **32**(4):426–430.

206. Shukla, J.P. and K. Pandey, Impaired spermatogenesis in arsenic treated freshwater fish, *Colisa fasciatus* (Bl. and Sch.). *Toxicology Letters*, 1984, **21**(2):191–195.

207. Katsu, Y., et al., Functional associations between two estrogen receptors, environmental estrogens, and sexual disruption in the roach (*Rutilus rutilus*). *Environmental Science and Technology*, 2007, **41**(9):3368–3374.

208. Nagler, J.J., et al., The complete nuclear estrogen receptor family in the rainbow trout: Discovery of the novel ERalpha2 and both ERbeta isoforms. *Gene*, 2007, **392**(1–2):164–173.

209. Revankar, C.M., et al., A transmembrane intracellular estrogen receptor mediates rapid cell signaling. *Science*, 2005, **307**(5715):1625–1630.

210. Thomas, P., et al., Identity of an estrogen membrane receptor coupled to a G protein in human breast cancer cells. *Endocrinology*, 2005, **146**(2):624–632.

211. Pang, Y., J. Dong, and P. Thomas, Estrogen signaling characteristics of Atlantic croaker GPR30 and evidence it is involved in maintenance of oocyte meiotic arrest. *Endocrinology*, 2008.

212. Kuiper, R., et al., Toxicity of tetrabromobisphenol A (TBBPA) in zebrafish (*Danio rerio*) in a partial life-cycle test. *Archives of Toxicology*, 2007, **81**(1):1.

213. Thomas, P., et al., Progestin, estrogen and androgen G-protein coupled receptors in fish gonads. *Steroids*, 2006, **71**(4):310–316.

214. Filby, A.L., et al., Gene expression profiles revealing the mechanisms of anti-androgen- and estrogen-induced feminization in fish. *Aquatic Toxicology*, 2007, **81**(2):219–231.

215. Kirubagaran, R. and K.P. Joy, Toxic effects of mercuric chloride,methylmercuric chloride, and emisan 6 (an organic mercurial fungicide) on ovarian recrudescence in the catfish *Clarias batrachus* (L.). *Bulletin of Environmental Contamination and Toxicology*, 1988, **41**(6):902–909.

216. Afonso, L.O.B., et al., Effects of the aromatase inhibitor fadrozole on plasma sex steroid and oocyte maturation in female coho salmon (*Oncorhynchus kisutch*) during vitellogenesis. *Fish Physiology and Biochemistry*, 1999, **20**:231–241.

217. Zhang, J., et al., Effect of tributyltin on the development of ovary in female cuvier (*Sebastiscus marmoratus*). *Aquatic Toxicology*, 2007, **83**(3):174–179.

218. Mani, K. and P. K. Saxena, Effect of safe concentrations of some pesticides on ovarian recrudescence in the freshwater murrel, *Channa punctatus* (BI.): A quantitative study. *Ecotoxicology and Environmental Safety*, 1985, **9**:241–249.

219. Kumar, S. and S.C. Pant, Comparative sublethal ovarian pathology of some pesticides in the teleost, *Puntius conchonius* Hamilton. *Bulletin of Environmental Contamination and Toxicology*, 1988, **41**:227–232.

220. Janz, D.M. and G.J. Van Der Kraak, Suppression of apoptosis by gonadotropin, 17β-estradiol, and epidermal growth factor in rainbow trout preovulatory ovarian follicles. *General and Comparative Endocrinology*, 1997, **105**(2):186–193.

221. Afonso, L.O.B., G.K. Iwama, and E.M. Donaldson, Effects of the aromatase inhibitor fadrozole on plasma sex steroid secretion and ovulation rate in female coho salmon, *Oncorhynchus kisutch*, close to final maturation. *General and Comparative Endocrinology*, 1999, **113**(2):221–229.

222. Staples, C.A., et al., Bisphenol A concentrations in receiving waters near US manufacturing and processing facilities. *Chemosphere*, 2000, **40**(5):521–525.

223. Schwaiger, J., et al., How estrogenic is nonylphenol? A transgenerational study using rainbow trout (*Oncorhynchus mykiss*) as a test organism. *Aquatic Toxicology*, 2002, **59**(3–4):177–189.

224. Kling, D., Total atresia of the ovaries of *Tilapia leucosticta* (Cichlidae) after intoxication with the insecticide Lebaycid. *Experientia*, 1981, **37**(1):73–74.

225. McCormick, J.H., G.N. Stokes, and R.O. Hermanutz, Oocyte atresia and reproductive success in fathead minnows (*Pimephales promelas*) exposed to acidified hardwater environments. *Archives of Environmental Contamination and Toxicology*, 1989, **18**(1–2):207–214.

226. Koger, C.S., S.J. Teh, and D.E. Hinton, Variations of light and temperature regimes and resulting effects on reproductive parameters in medaka (*Oryzias latipes*). *Biology of Reproduction*, 1999, **61**:1287–1293.

227. van der Ven, L.T., et al., Effects of the estrogen agonist 17beta-estradiol and antagonist tamoxifen in a partial life-cycle assay with zebrafish (*Danio rerio*). *Environmental Toxicology and Chemistry*, 2007, **26**(1):92–99.

228. Paull, G.C., et al., Variability in measures of reproductive success in laboratory-kept colonies of zebrafish and implications for studies addressing population-level effects of environmental chemicals. *Aquatic Toxicology*, 2008, **87**(2):115–126.

229. Harries, J.E., et al., Developmental of a reproductive performance test for endocrine disrupting chemicals using pair-breeding fathead minnows (*Pimephales promelas*). *Environmental Science and Technology*, 2000, **34**:3003–3011.

230. Robinson, C.D., et al., Effects of sewage effluent and ethynyl oestradiol upon molecular markers of oestrogenic exposure, maturation and reproductive success in the sand goby (*Pomatoschistus minutus*, Pallas). *Aquatic Toxicology*, 2003, **62**(2):119–134.

231. Kolpin, D.W., et al., Pharmaceuticals, hormones, and other organic wastewater contaminants in U.S. streams, 1999–2000: A national reconnaissance. *Environmental Science and Technology*, 2002, **36**(6):1202–1211.

232. Hirai, N., et al., Feminization of Japanese medaka (*Oryzias latipes*) exposed to 17beta-estradiol: Effect of exposure period on spawning performance in sex-transformed females. *Aquatic Toxicology*, 2006, **79**(3):288–295.

233. Parrott, J.L. and B.R. Blunt, Life-cycle exposure of fathead minnows (*Pimephales promelas*) to an ethinylestradiol concentration below 1 ng/L reduces egg fertilization success and demasculinizes males. *Environmental Toxicology and Chemistry*, 2005, **20**(2):131–141.

234. Miller, D.H. and G.T. Ankley, Modeling impacts on populations: Fathead minnow (*Pimephales promelas*) exposure to the endocrine disruptor 17beta-trenbolone as a case study. *Ecotoxicology and Environmental Safety*, 2004, **59**(1):1–9.

235. Schwaiger, J., et al., A prolonged toxicity study on the effects of sublethal concentrations of bis(tri-*n*-butyltin)oxide (TBTO): Histopathological and histochemical findings in rainbow trout (*Oncorhynchus mykiss*). *Aquatic Toxicology*, 1992, **23**(1):31.

236. Vos, J.G., et al., Health effects of endocrine-disrupting chemicals on wildlife, with special reference to the European situation. *Critical Reviews in Toxicology*, 2000, **30**(1):71–133.

237. Grinwis, G.C.M., et al., Short-term toxicity of bis(tri-*n*-butyltin)oxide in flounder (*Platichthys flesus*): pathology and immune function. *Aquatic Toxicology*, 1998, **42**(1):15.

238. Jones, J.C. and J.D. Reynolds, Effects of pollution on reproductive behaviour of fishes. *Reviews in Fish Biology and Fisheries*, 1997, **7**(4):463–491.

239. Baatrup, E. and M. Junge, Antiandrogenic pesticides disrupt sexual characteristics in the adult male guppy (*Poecilia reticulata*). *Environmental Health Perspectives*, 2001, **109**(10):1063–1070.

240. Clotfelter, E.D., A.M. Bell, and K.R. Levering, The role of animal behaviour in the study of endocrine-disrupting chemicals. *Animal Behaviour*, 2004, **68**(4):665.

241. Yang, F.-X., Y. Xu, and Y. Hui, Reproductive effects of prenatal exposure to nonylphenol on zebrafish (*Danio rerio*). *Comparative Biochemistry and Physiology, Part C*, 2006, **142**:77–84.

242. Thorpe, K.L., et al., Reproductive effects of exposure to oestrone in the fathead minnow. *Fish Physiology and Biochemistry*, 2003, **28**(4):451–452.

243. Brown, K.H., I.R. Schultz, and J.J. Nagler, Reduced embryonic survival in rainbow trout resulting from paternal exposure to the environmental estrogen 17alpha-ethynylestradiol during late sexual maturation. *Reproduction*, 2007, **134**(5):659–666.

CHAPTER 7

Determination of Effects of Exogenous Hormones and Endocrine-like Active Compounds

7.1 HISTOLOGICAL PROCESSING: MICRODISSECTION VERSUS WHOLE-FISH SECTIONING

When conducting experiments and histological examinations on small laboratory fish it is important to consider the size of the fish species utilized and the endpoints under evaluation prior to the fixation and embedding of the animal. Whole-fish sectioning (Figure 7.1) can be considered appropriate for smaller species such as medaka and zebrafish. Whole-body sectioning facilitates the analysis of multiple pathologies and of organs in their anatomical context, as well as providing a general impression of the health status of the experimental animal. Van der Ven et al. [1] concluded that the relatively small effort required to embed whole zebrafish has the potential to result in a greater integrated understanding of the effects produced, providing particular advantages when examining multiple organ effects. Additionally, whole-body sectioning can potentially avoid artifacts such as the rupture of spermatogenic cysts, which are easily induced when removing and handling vulnerable organs such as testes. An additional advantage of whole-body sectioning, (e.g., multiple cross sections; Figure 7.2A–C) is that spermatic ducts or oviducts are preserved and can be appraised histopathologically *in situ* [2–4], whereas this is highly unlikely in the microdissection scenario unless the ducts are specifically microdissected along with the ovary or testes. It is also possible that whole-body sectioning can reduce the number of experimental animals needed, as it can reveal information equivalent to that produced by other techniques, such as enzyme-linked immunosorbent assays: for example the accumulation of vitellogenin in interstitial spaces identified via immunohistochemistry, and basophilia in the liver, also associated with the production of vitellogenin as demonstrated by van der Ven et al. [1].

Histological Analysis of Endocrine-Disruptive Effects in Small Laboratory Fish
By Daniel R. Dietrich and Heiko O. Krieger

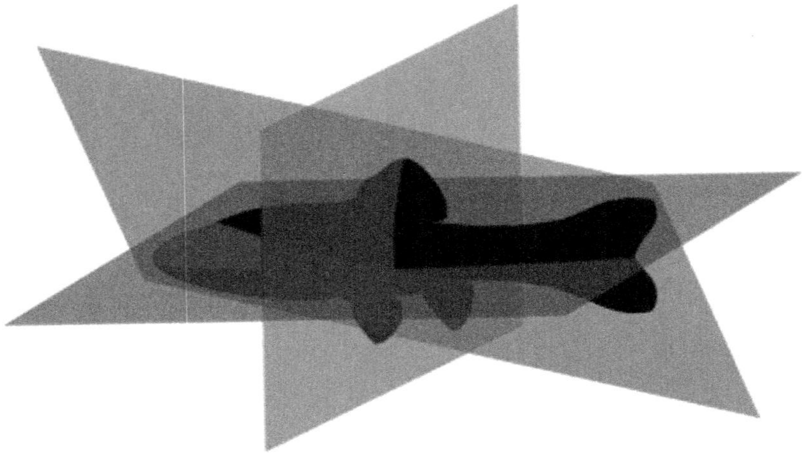

FIGURE 7.1 Different section planes through a whole-fish preparation. Green, transverse section; blue, horizontal section; red, transverse oblique section. Longitudinal section not shown. (See insert for color representation.)

There are, however, several reasons why whole-fish sectioning is less ideal for larger fish species such as the fathead minnow. These include the fact that fathead minnow are often too large for histocassettes, even with the head and tail removed. Fathead minnow gonads are elongated lobular structures oriented parallel to the axis of the fish. It is not possible for the entire length of a gonad to be recovered consistently on a single slide using longitudinal sectioning of whole fish (Figure 7.2B); thus, multiple cross sections would be necessary to provide for a histopathological impression representative for the testis. Additionally, microdissection of fathead minnow gonadal tissue (Figure 7.3) after excision avoids microtoming artifacts that can be caused by the presence of firmer tissues (such as bone and cartilage) in the whole-body sections (whether longitudinal/horizontal or cross-sectional; Figure 7.2A–C).

Finally, gonad excision allows for macroscopic observation *in situ* and the possibility of obtaining gonad weights [5] and thus of deriving additional endpoints e.g., gonadosomatic indices (GSIs). However, it is important to highlight the need for expert and precise excision of small fish gonads if microdissection and individual organ histology are required. Wolf et al. [5] concluded that the use of a dissecting microscope was essential for the removal of fathead minnow gonads, to minimize damage to these delicate tissues. Additionally, removal of the swimbladder prior to excision of the gonads facilitates the microdissection procedure, and damage to the tissue can be avoided by dissecting the gonads in a caudal-to-cranial direction while applying very gentle traction to the spermatic ducts or oviducts. An alternative is to fix the reproductive organs *in situ* (i.e., upon opening the abdominal cavity) and use fixative solution prior to the excision and prolonged fixation of the organs in the tissue cassettes. This may prove to be a rather cumbersome procedure, however, difficult to carry through when a large number of fish need to be processed following the termination of a study.

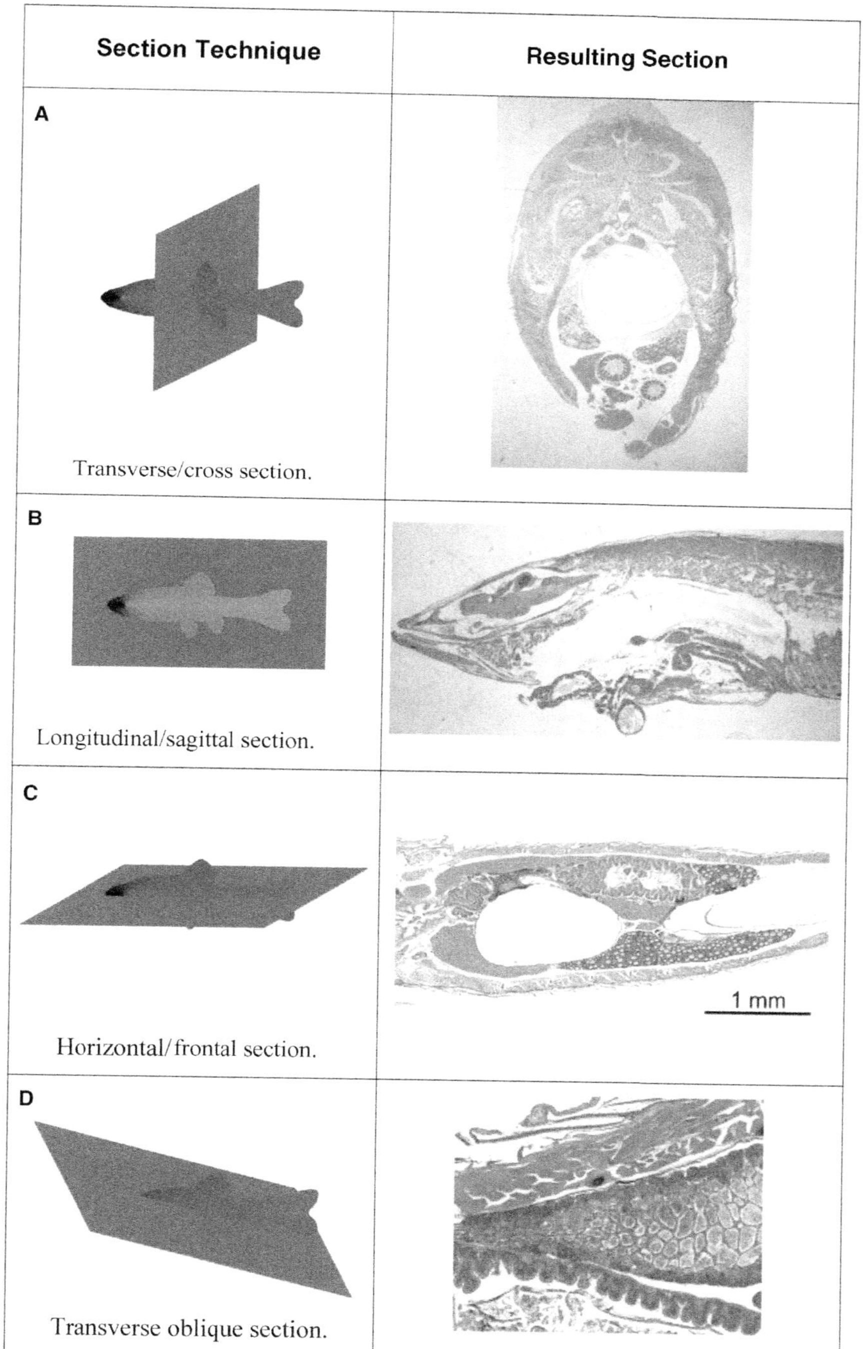

FIGURE 7.2 Whole-fish preparation: section-plane examples. (See insert for color representation.)

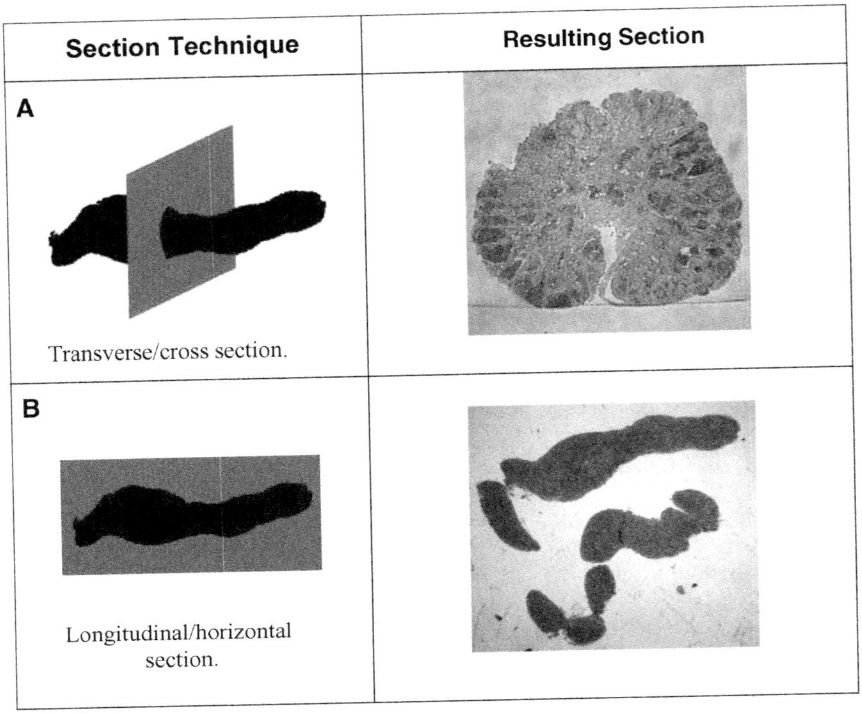

FIGURE 7.3 Single-organ preparation: section-plane examples. (See insert for color representation.)

7.2 OPTIMAL TISSUE PREPARATION AND HISTOLOGICAL TECHNIQUES

It is essential to use the most appropriate tissue fixation, embedding, sectioning, and staining techniques for the production of optimal tissue samples for later light-microscopic examinations. Additionally, for optimal cell-type identification and qualitative or quantitative evaluation, it is necessary to produce the best possible section thickness (i.e., thin sections) that facilitates single cell layers. Wolf et al. [5] conducted a histopathological tissue preparation and staining procedure comparison study to ascertain optimal procedures for gonadal histopathology in the fathead minnow. They evaluated various combinations of fixation (i.e., Bouin's solution, 10% neutral buffered formalin, or glutaraldehyde), embedding medium [paraffin or glycolmethacrylate (GMA)], and staining [hematoxylin and eosin (H&E) or Masson's trichrome] methods. It was concluded that for histopathological assessment of testicular tissue in fathead minnows the combination of glutaraldehyde fixation, GMA embedding, and H&E staining provided excellent quality and resolution, as well as a single cell layer (0.5 to 1 μm). For ovarian tissues, the more routine fixation in Bouin's solution, paraffin embedding, and H&E staining (providing a section thickness of 3 to 5 μm) provides sufficient quality and resolution to allow histopathological appraisal and follicle-type quantification using light microscopy [5,6].

The procedures described above or similar for fixation, embedding, procedures and staining also appear to be appropriate for histopathological examination of gonadal tissues in zebrafish [7–10], medaka [11–16], and other small laboratory fish species. It is noteworthy, however, that the use of paraffin embedding for testicular appraisal often provides relatively thick sections (3 to 5 μm), thus multilayered testicular cells, and therefore is less than optimal for the quantitative assessment of testicular cell types.

7.3 PLANE OF GONAD SECTIONING FOR OPTIMAL ORGAN REPRESENTATION

To process (embed) and section the gonads appropriately such that a representative evaluation of the gonads can be achieved, it appears vital to understand the basic gonadal anatomy of the specific fish. Indeed, zebrafish and fathead minnow ovaries and testes are rather homogeneous with regard to the distribution of ovarian follicle stages and lobules containing cysts with different developmental stages of sperm cells, respectively, despite a cortex-to-medulla gradient with more juvenile germ cells (younger maturational stages) in the gonadal cortex. Conversely, an anterior-

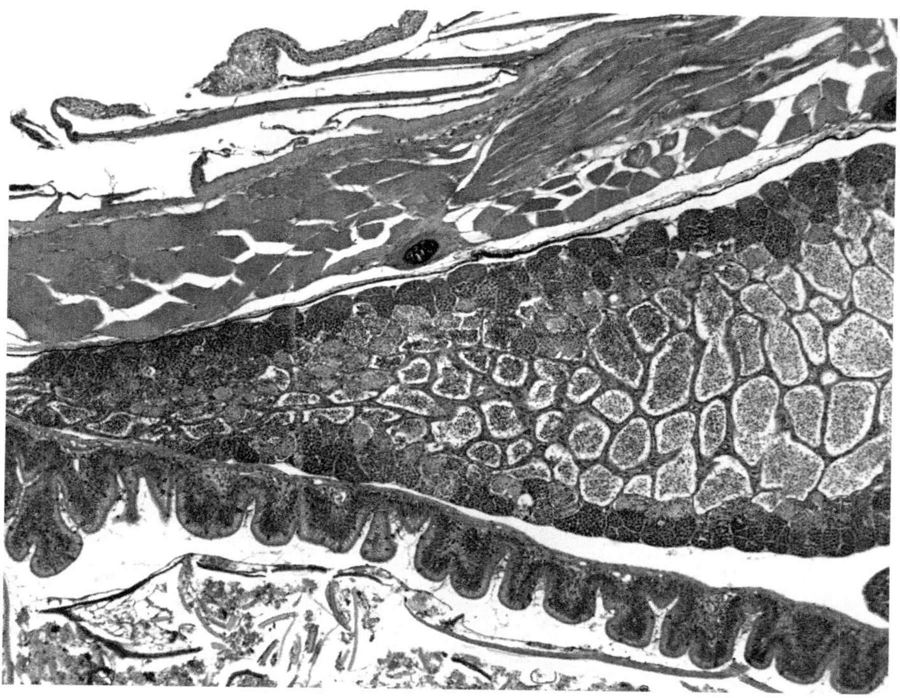

FIGURE 7.4 Transverse–oblique section through a medaka testis. (Courtesy of R. D. Johnson, U.S. EPA, Duluth, Minnesota.)

to-posterior as well as a cortex-to-medullary gradient of development is observed in the medaka, whereby the more juvenile germ cell stages are found in the anterior and cortical portion of the gonads. Consequently, transverse (i.e., cross-sectional) horizontal (frontal) or longitudinal (sagittal) (Figure 7.2A–C) sectioning of zebrafish or fathead minnow gonads can provide representative sections of the gonads whereby the choice of longitudinal versus transverse is largely dependent on the choice of gonad evaluation (see below). In medaka, a transverse oblique sectioning of the gonads (Figures 7.2D and 7.4) appears most favored, as this seems to provide the best representation of all germ cell stages present in the gonads [17], although a cautious approach is warranted, as little experience is presently available for an evaluation of this type of section.

Conversely, transverse or longitudinal sectioning of medaka gonads may provide sections with a bias toward one germ cell stage (maturational stage), largely dependent on the position of the section plane relative to the cortex or medulla and anterior, central, or posterior in the gonad.

REFERENCES

1. van der Ven, L.T.M., P.W. Wester, and J.G. Vos, Histopathology as a tool for the evaluation of endocrine disruption in zebrafish (*Danio rerio*). *Environmental Toxicology and Chemistry*, 2003, **22**(4):908–913.
2. Gimeno, S., Gonadal sexual differentiation as histological endpoint. *OECD Technical Workshop on Gonadal Histology of Small Laboratory Fish*, Bilthoven, The Netherlands, 2002.
3. Gimeno S., et al., Demasculinisation of sexually mature male common carp, *Cyprinus carpio*, exposed to 4-*tert*-pentylphenol during spermatogenesis. *Aquatic Toxicology*, 1998, **43**(2–3):93–109.
4. Gimeno S., et al., Feminisation of young males of the common carp, *Cyprinus carpio*, exposed to 4-*tert*-pentylphenol during sexual differentiation. *Aquatic Toxicology*, 1998, **43**(2–3):77–92.
5. Wolf J.C., et al., Qualitative and quantitative histomorphologic assessment of fathead minnow *Pimephales promelas* gonads as an endpoint for evaluating endocrine-active compounds: A pilot methodology study. *Toxicologic Pathology*, 2004, **32**(5):600–612.
6. Dietrich, D.R., et al., Optimization of tissue preparation for the quantification of germinal cells in histologic sections of fathead minnow gonads. *SETAC Europe 13th Annual Meeting*, Hamburg, Germany, 2003.
7. van der Ven L.T.M. and P.W. Wester, Normal histology and effects of hormonal active agents of reproductive organs of the zebrafish (*Danio rerio*). *OECD Technical Workshop on Gonadal Histology of Small Laboratory Fish as a Tool for Endocrine Disruption Screening and Testing*, Bilthoven, The Netherlands, 2002.
8. Hill L., Jr. and D.M. Janz, Developmental estrogenic exposure in zebrafish (*Danio rerio*): I. Effects on sex ratio and breeding success. *Aquatic Toxicology*, 2003, **63**(4):417–429.
9. Maack, G. and H. Segner, Morphological development of the gonads in the zebrafish (*Danio rerio*). *Journal of Fish Biology*, 2003, **62**(4):895–906.

10. Weber, L.P., R.L. Hill, Jr., and D.M. Janz, Developmental estrogenic exposure in zebrafish (*Danio rerio*): II. Histological evaluation of gametogenesis and organ toxicity. *Aquatic Toxicology*, 2003, **63**(4):431–446.

11. Hartley, W.R., et al., Gonadal development in Japanese medaka *(Oryzias latipes)* exposed to 17β-estradiol. *Marine Environmental Research*, 1998, **46**(1–5):145–148.

12. Kang, I.J., et al., Effect of 17β-estradiol on the reproduction of Japanese medaka (*Oryzias latipes*). *Chemosphere*, 2002, **47**(1):71–80.

13. Kawahara, T. and I. Yamashita, Sexual differentiation of gonadal cells in the medaka *Oryzias latipes*. *Zoological Science*, 1997, **14**(Suppl.):70.

14. Metcalfe T.L., et al., Gonadal development and endocrine responses in Japanese medaka (*Oryzias latipes*) exposed to o,p'-DDT in water or through maternal transfer. *Environmental Toxicology and Chemistry*, 2000, **19**(7):1893–1900.

15. Seki M., et al., Effect of ethinylestradiol on the reproduction and induction of vitellogenin and testis–ova in medaka (*Oryzias latipes*). *Environmental Toxicology and Chemistry*, 2002, **21**(8):1692–1698.

16. Wester, P.W. and J.H. Canton, Histopathological study of *Oryzias latipes* (medaka) after long-term β-hexachlorocyclohexane exposure. *Aquatic Toxicology*, 1986, **9**(1):21–45.

17. OECD, *Meeting of the Fish Drafting Group (FDG-15)*. OECD, Paris, 2003, p. 17.

CHAPTER 8

Evaluation of Effects in Fish Gonads

An important component of many fish reproductive biology studies, and consequently the parameter used to determine the impact of xenobiotics on fish reproduction, is the assessment of the gonadal development stage as well as the relative numbers of germ cell maturation stages present in each fish, notably when assessing seasonal breeders [1]. The determination of germ cell maturation stages concurrent with a histopathological assessment of the whole gonad (gonad staging) will thus provide information on the homogeneity or inhomogeneity of a (control) fish population with regard to germ cell development (reproductive state of the individual) and gonadal health but also additional information: for example, the impact of a chemical or other stressor on germ cell development (maturational inhibition, juvenilization, dedifferentiation) and germ and stromal cell histopathology (necrosis, apoptosis, atresia, fibrosis, inflammation), as discussed below. However, prior to the discussion of the pros and cons of qualitative (semiquantitative) versus quantitative evaluation, it is important to realize that either of the approaches is null and void if a proper experimental design was not carried out prior to the actual animal study.

8.1 QUALITATIVE (SEMIQUANTITATIVE) VERSUS QUANTITATIVE EVALUATION

Gonadal staging methods vary from highly detailed to cursory. Unfortunately, few reviews have been conducted regarding the reliability or usefulness of the various staging techniques. Generally, gonad staging involves the assessment of germ-cell-type (maturational-stage) proportions in order to identify the potential effects of exogenous chemicals or endogenously synthesized steroids and their precursors on gametogenesis. It can be performed quantitatively or qualitatively (semiquantitatively), with the latter being the more commonly reported procedure [2].

Because qualitative (semiquantitative) systems assess primarily the presence or absence and visual density of gametogenic precursors compared to mature gametocytes, shortcomings inherent in such systems can include problems with objectivity

Histological Analysis of Endocrine-Disruptive Effects in Small Laboratory Fish
By Daniel R. Dietrich and Heiko O. Krieger

and limitations in the amount of mechanistic information that can be obtained. The use of a quantitative staging approach, in which gonadal cell types (maturational stages) are identified and counted individually, has the potential to mitigate such shortcomings. There are several reports in which quantitative schemes were used for the staging of fish gonads [3–7]. In contrast to qualitative (semiquantitative) staging, a key requirement for quantitative staging is that each cell counted must be assigned to one of several well-defined categories according to predetermined histomorphological criteria. Consequently, to allow adequate visualization and differentiation of the various gonadal cell types (maturational stages), both the tissue collection procedures and preparation of the histological sections need to be optimized (see Chapter 7).

8.2 GONADAL STAGING IN THE TESTIS

Testicular staging of development in males is sometimes confusing, due to a lack of standardized classification of cell types and structures. This is due primarily to the fact that the literature contains data from investigative studies as well as of routine studies. Whereas in investigative studies, the individual cell can and will be inspected for correct classification under high magnification (including oil objectives, e.g., 100×), this is not normally done in the routine setting. Therefore, the use of different terms, many of them descriptive, for the same structures is common, especially when a clear classification of cell types (maturational stages) appears a priori impossible or the exposure to xenobiotics induced a morphological change of specific cell types such that the normal morphological criteria of a given cell type (maturational stage) are difficult to discern or no longer apply [3]. In view of the above, it becomes outwardly clear that comparison of histopathological effects in gonads reported in the literature [8,9], even of the same fish species, can prove difficult. As a further consequence, mistakedly confusing a certain cell type for another can have severe consequences with regard to the interpretation of the study and extrapolations thereof in a broader sense. Indeed, possible misinterpretations of spermatogonia for Sertoli cells in fathead minnows appeared to be present in the literature [8]. In consequence, the authors reported Sertoli cell hyperplasia in male fathead minnows as the result of 17β-estradiol exposure, whereas they probably should have reported an increase in spermatogonial cells or a juvenilization of the adult male fathead testes. Figure 8.1 illustrates the very similar morphological appearance of spermatogonia and Sertoli cells in fathead minnow. This stresses the strict requirement to illustrate the histopathological findings by relevant microphotographies of the control and affected organs.

8.2.1 Qualitative Staging in the Testis

Qualitative staging of fish testes can be based on criteria established in the U.S. Geological Survey Information and Technology Report, "Biomonitoring of Environmental Status and Trends (BEST) Program: Selected Methods for Monitoring Chemical Contaminants and Their Effects in Aquatic Ecosystems" [2], developed primarily for carp (*Cyprinus carpio*). This staging system was modified by Jeff Wolf (personal communication) to address the specific needs of fathead minnows (Figure 8.2) and thus

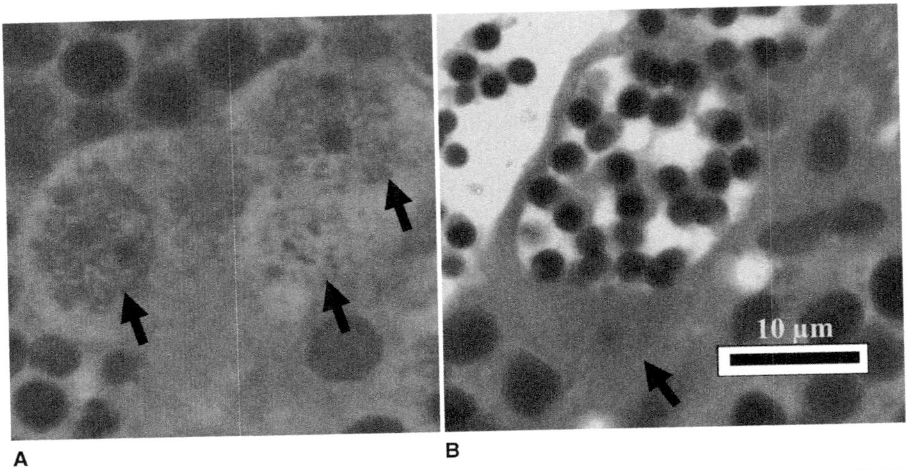

FIGURE 8.1 (A) Fathead minnow spermatogonia; (B) Sertoli cell. (Courtesy of J. Wolf, EPL Inc., Sterling, Virginia.) (See insert for color representation.)

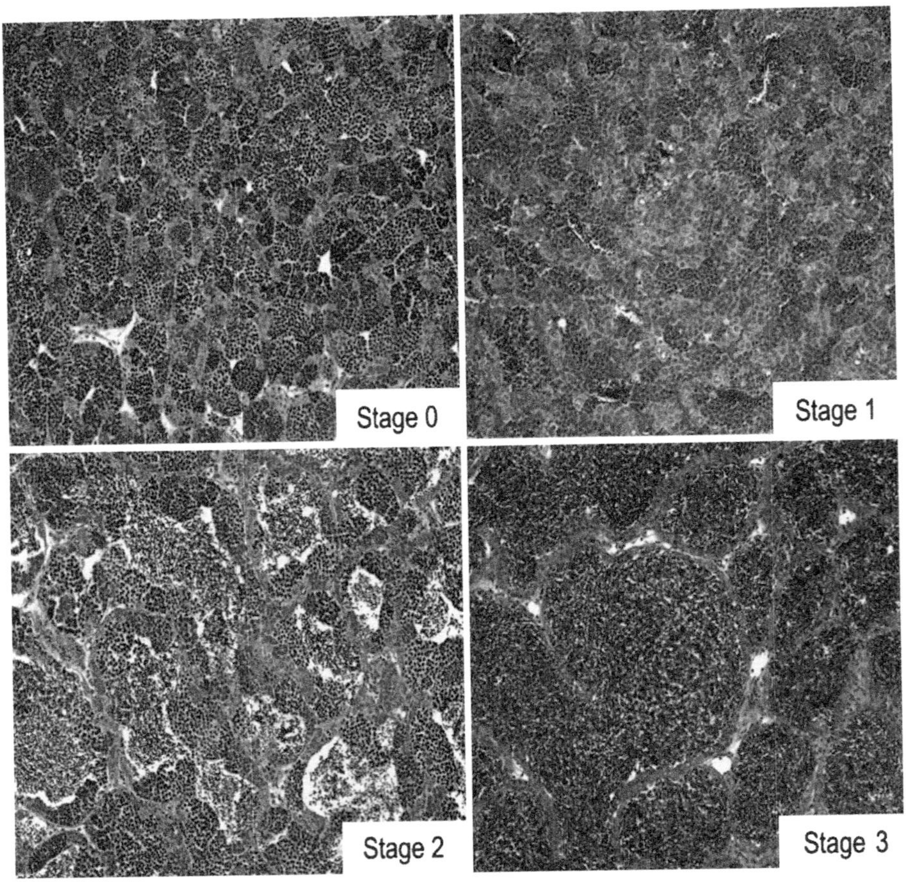

FIGURE 8.2 Gonadal stages in the testes of a fathead minnow. (Courtesy of J. Wolf, EPL Inc., Sterling, Virginia.) (See insert for color representation.)

modified may serve to accommodate the situation in zebrafish and medaka. It has to be noted here that this system describes development in a cycling testis, which does not apply to zebrafish and medaka; control adult males in these species always show stage 2 (see Chapters 3 and 5). Effects of toxicants may present differently in cycling and non cycling testes, because more processes of proliferation, differentiation, and maturation may be active simultaneously in the latter compared to the former, and changes in the noncycling testis may therefore be missed when applying a system that is based on the synchrony of processes.

The BEST system for testes [2], as modified by Wolf et al. [3], has the following stages:

Stage 0: undeveloped: exclusively immature phases (spermatogonia to spermatids with no spermatozoa)

Stage 1: early spermatogenic: immature phases predominate, but spermatozoa may also be observed

Stage 2: mid-spermatogenic: spermatocytes, spermatids, and spermatozoa are present in roughly equal proportions

Stage 3: late spermatogenic: all stages may be observed; however, mature sperm predominate

Stage 3a: mature sperm predominate; immature spermatogenic phases are still present consistently throughout the testis

Stage 3b: mature sperm predominate; immature spermatogenic phases are either completely absent or restricted to small scattered, nests of cells

Stage 4: spent: loose connective tissue with some remnant sperm

As is quite obvious, this type of an assessment can be performed rather rapidly and routinely, as it does not place great demands on the specific identification of individual cell types but rather, describes the gross impression. The quality (robustness and reproducibility) of this type of testicular assessment is, however, largely dependent on the experience of the reader of the sections. Consequently, qualitative staging may prove a useful tool to arrive at a general impression of whether or not a pathological effect is present. It must be emphasized, however, that only overt changes can be appraised with this system; subtle changes or shifts in the numbers of specific cell types and the associated underlying mechanisms cannot be detected and therefore are missed altogether.

8.2.2 Quantitative Staging in the Testis

Generally, two formats have been used in the past to quantify testicular germ cells: (a) counting (tagging) of individual cell types [4–6], or (b) determining the relative area of cysts filled with a given differential stage of germ cells [7]. As in qualitative (semiquantitative) staging, the quantification of testicular cell types (maturational stages) places emphasis on stringent descriptors of individual cell types (maturational stages) as well as on the experience of the evaluating scientist in distinguishing and categorizing individual cells consistently and reproducibly. Furthermore, the format by which the different cell types are counted can provide large differences in the final outcome of the evaluation. For example, Smith [6] and Sohoni et al. [5], using the direct assessment technique via microscopic quantification directly on histological slides, counted a maximum of 100 cells in three and eight individual locations per testis and fish, providing a total of 300 and 800 cells counted per fish, respectively. Both counted along an ocular grid in the light microscope and surprisingly, obtained comparable results (i.e., similar ratios of mature to immature sperm cells). The advantage of this method is that it is fairly simple to implement in the laboratory. The drawbacks of this form of evaluation are:

1. Proper categorization and counting of testicular cells is highly dependent on the quality (thinness, plane cut, even staining, lack of wrinkling) of the tissue sections.
2. The same or overlapping areas within the testis section may be counted and thus may provide for a bias.
3. There is no visual record of the cells identification and categorization.
4. The total number of cells counted is rather small and thus does not readily and adequately represent the situation in the testis.
5. Results are highly dependent on the experience and training of the pathologist categorizing the cell types.
6. Smaller germ cell types occurring with a high density (spermatids and spermatozoa) are clearly underrepresented, as they do not occupy the same space (area) as that occupied by larger germ cell types.

A similar but more technically advanced method was presented by Wolf et al. [3] and Dietrich et al. [4], whereby cell-type quantification was carried out on digital images and not under the light microscope using ImagePro-Plus (Media Cybernetics, Silver Spring, Maryland, version 4.1). These authors also laid a grid across the testicular image, allowing for 400 grid intersections. Cells touched by or underlying these grid intersections were categorized and counted (Figure 8.3). A total of four images were counted for each left and right testis, thus allowing for a maximum of 3200 cells that could be counted in total.

The advantages of these methods are:

1. The location of the images taken on the testis sections can be recorded such that no overlapping areas in the testis are being counted (only with the digital technique).
2. After categorization and counting of the cells on the image, this grid can be saved, providing a reviewable electronic record (only with the digital technique).
3. Categorization and counting off of images allows us to expand the number of cells counted until a statistically relevant number is reached (i.e., testicular representation is achieved). This can also be done in direct assessment by using a finer grid, additional fields of view, and so on.

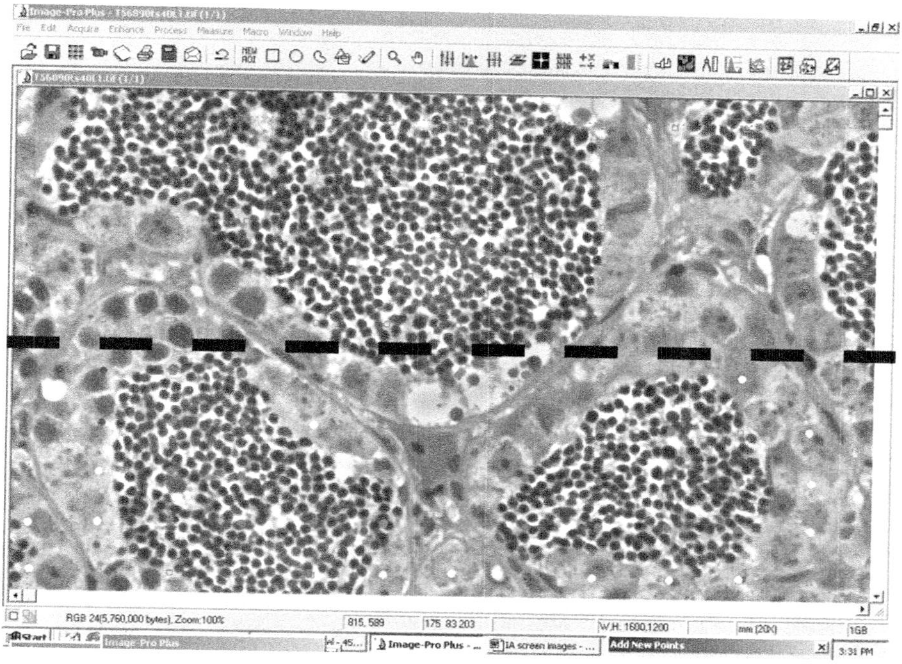

FIGURE 8.3 Manual tagging of digitized photomicrographic images is illustrated. A grid of green boxes (upper half) extends across the image, and cell-type differentiation is indicated by colored dots (lower half) that are individually applied to each box using the computer mouse. Boxes that remain untagged represent interstitial areas, areas devoid of tissue, or cells that cannot be reasonably classified due to suboptimal orientation or marked degeneration. Cell counts (>2000 cells are counted per male fish) are automatically exported to a spreadsheet for tabulation (not shown). (From [3], with permission.) (See insert for color representation.)

4. Empty spaces (cysts devoid or decreased numbers of germ cells) can be assessed numerically.
5. The total numbers of cell types counted can be transferred to a spreadsheet and statistically evaluated immediately, whereas in direct assessment this has to be done manually or via a computerized expert system.
6. Zones within the testes, as well as the left and right testes, can be compared with regard to homogeneity of cell-type distribution.
7. Subtle changes in cell-type ratios can be detected.

Apart from the general drawbacks mentioned for the direct light-microscopy method, the digital method additionally implies high initial cost in setting up the software and the imaging system and training the personnel. Indeed, cell type (maturational stage) representation within the testis is a major issue in the quantification of germ cell types within the testis. It must therefore be emphasized again here that the methods above either demand that the testis of the small fish species in question be

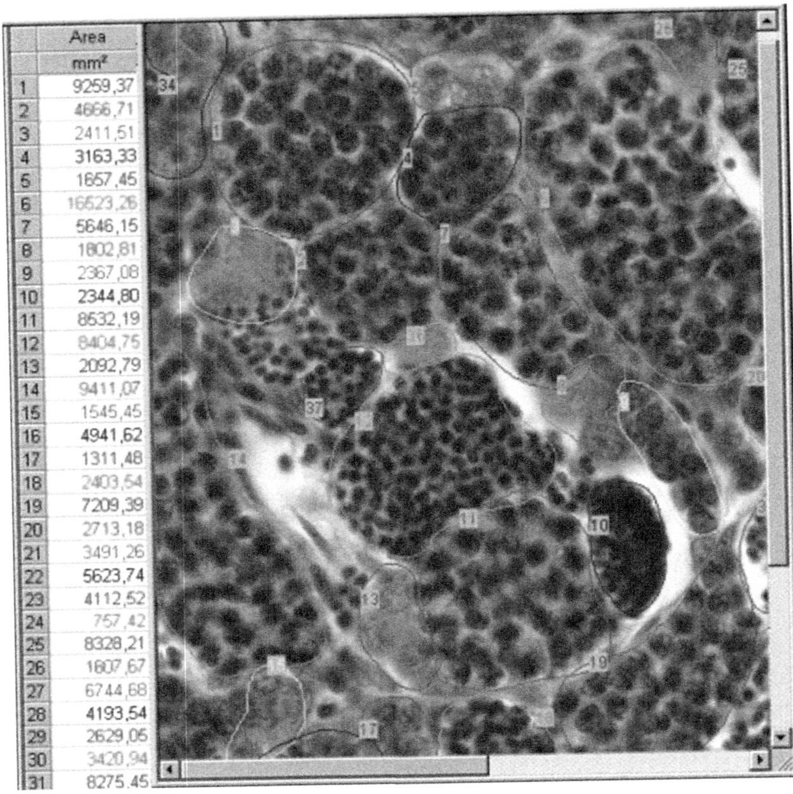

FIGURE 8.4 Part of a screenshot of the image analysis package (AnalySIS, Soft Imaging System, Münster, Germany) used for morphometry. Digital images were loaded into the package, magnification calibrated and adjusted, and testis structures outlined. The package generates a table with figures corresponding to the outlined areas (shown in the left column, $\times 10^7$). Each figure was identified with a histological classifier using three main stages of spermatogenesis: spermatogonium, spermatocyte, and spermatid. Thus, the substages spermatogonium A and B, and the substages leptotene, zygotene, and pachytene spermatocytes were pooled, as the transition from spermatogonium to spermatocyte (i.e., initiation of meiosis) and the transition from spermatocyte to spermatid, (i.e., initiation of final sperm maturation) were considered critical steps, sensitive to endocrine disruption. (Courtesy of L. van der Ven and P. Wester, RIVM, Bilthoven, The Netherlands.) (See insert for color representation.)

rather homogeneous (e.g., zebrafish, fathead minnow) in cell-type distribution or that the plane of tissue sectioning and number of sections evaluated per testis be adjusted for an expected inhomogeneity.

The possibility of better addressing germ-cell-type (maturational-stage) representation was developed by van der Ven et al. [7] for zebrafish. Digital microphotographs were processed in an image analysis package (AnalySIS, Soft Imaging System, Münster, Germany). All spermatogenic cysts in three fields of view (magnification 40×) of a routine section (27 to 55 cysts, average 42) were outlined and counted (Figure 8.4).

The spermatogenic cysts counted were those comprised of the three major classes: spermatogonia, spermatocytes, and spermatids. This procedure yielded the relative cyst size of the respective class, relative number of cysts of each class, and the relative area of cysts taken by each class. The ratio of cyst classes to one another could thus provide information (e.g., a shift from one class to another) as a consequence of the influence of a xenobiotic, for example. The main advantages of this method are:

1. The location of the images taken on the testis sections can be recorded such that no overlapping areas in the testis are being counted.
2. After categorization and counting of the cells on the image, this image can be saved, providing a reviewable electronic record (see [10]).
3. Categorization and counting off of images allows us to expand the number of cysts counted until a statistically relevant number is reached (i.e., testicular representation is achieved).
4. Empty spaces (cysts devoid of germ cells) can be assessed numerically.
5. The total numbers of cysts and areas of cysts determined can be transferred to a spreadsheet and statistically evaluated immediately.
6. Zones within the testes, as well as the left and right testes, can be compared with regard to homogeneity of germ cell class distribution.
7. Subtle changes in germ cell class ratios can be detected.

The drawbacks of this method are:

1. There is a high initial cost in setting up the software and imaging system and training personnel.
2. There is a high demand on the quality (thinness, plane cut, even staining, lack of wrinkling) of the tissue sections and consequently of the images taken.
3. Results are highly dependent on the experience and training of the pathologist categorizing the cell types.
4. Cysts must be well defined to be counted, and pathological changes (e.g., necrosis, inflammation, fibrosis) may compromise clear cyst distinction.
5. Morphological alteration in the appearance of the germ cell types may provide potential problems when categorization of the cysts in the major sperm cell classes is required.
6. Smaller germ cell types occurring with a high density (spermatids and spermatozoa) are clearly underrepresented, as the mere number of cysts and the area occupied by the cysts does not account for changes in density within the cysts.

Overall, both digital imaging techniques have great advantages but also inherent drawbacks. However, in comparison to rough qualitative staging or direct quantification under the light microscope, the digital imaging techniques appear far superior in providing an unbiased record and for the capability of detecting subtle changes in germ cell differentiation. It must be emphasized that despite the great advantages of the more technically advanced methods, the results derived thereof cannot stand by themselves but must be placed and interpreted in context with the overall histopathological appraisal of the whole testes to provide a scientifically sound interpretation of the data.

8.3 GONADAL STAGING IN THE OVARY

Similar to the situation in the testes, the reproductive state of an ovary can be staged based on the degree of germ cell (follicle) differentiation. Obviously, in small fish that are continuous spawners, follicles comprising all developmental (maturation) stages are present at all times, whereas in seasonal spawners (e.g., wild fathead minnows or medaka) the relative numbers of different developmental stages of oocytes depend on the time in the season when the fish are sampled. As already discussed above for the testes, staging involves two major formats: Qualitative (semiquantitative) and quantitative staging. Both formats depend entirely on the stringent description and nomenclature of the different follicular stages present during oogenesis. Consequently, the quality and dependability of the staging results hinge largely on the training and experience of the evaluating scientist as well as on the consistency and frequency of ovarian section readings by the particular scientist.

8.3.1 Qualitative Staging in the Ovary

In many studies, ovaries are classified by the most advanced type of follicle present, regardless of how numerous they are. However, some authors base their staging schemes on relative numbers of different types of follicles [1]. Staging of fish gonads can be based on the criteria established in the U.S.Geological Survey Information and Technology Report, "Biomonitoring of Environmental Status and Trends (BEST) Program: Selected Methods for Monitoring Chemical Contaminants and Their Effects in Aquatic Ecosystems" [2], developed primarily for carp. This staging system was modified by Jeff Wolf (personal communication) to address the specific needs of fathead minnows (Figure 8.5) and thus may serve, although modified to accommodate the situations in the respective fish, to zebrafish and medaka. It has to be noted here that this system describes development in cycling ovaries and so does not apply to zebrafish and medaka (see Section 4.2.2.1).

The BEST system for ovary [2], as modified by Wolf et al. [3], has the following stages:

Stage 0: undeveloped: perinucleolar pre-vitellogenic follicles exclusively; no cortical alveoli

Stage 1: early development: $>90\%$ pre-vitellogenic, predominantly perinucleolar through cortical alveolar

Stage 2: mid-development: at least half of observed follicles are early and mid-vitellogenic

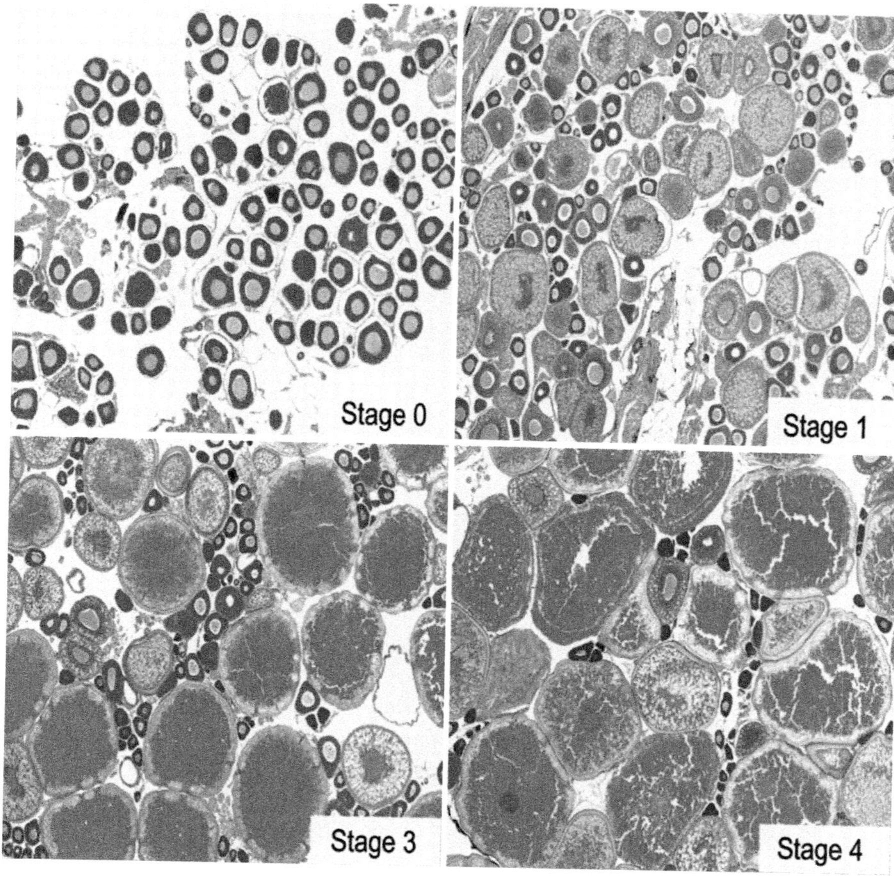

FIGURE 8.5 Gonadal stages in the ovaries of a fathead minnow. (Courtesy of J. Wolf, EPL Inc., Sterling, Virginia.) (See insert for color representation.)

Stage 3: late development: the majority of developing follicles are late vitellogenic

Stage 4: late development/hydrated: the majority is late vitellogenic; follicles are much larger

Stage 5: post-ovulatory: spent follicles, remnants of theca externa and granulosa

Similar to this scheme, Ankley et al. [11] used a staging technique conforming to the suggestions put forth by Leino and McCormick [12] and Selman and Wallace [13] and adapted to ovaries of the fathead minnow: stage 1, primary growth; stage 2, cortical alveolus; stage 3, early vitellogenic; stage 4, late vitellogenic; stage 5, mature/spawning oocyte. As an addition to the pure "ranking" of the ovaries into one of these stages, Ankley et al. [11] suggested that the number of pre-ovulatory atretic follicles per section should be counted. The mere comparison between the qualitative (semiquantitative) proposed by BEST [2] and Ankley et al. [11] demonstrates how crucial an issue it is to maintain conformity and stringency in the description of stages. Indeed, while BEST [2] describes stage 3 as a late development (late vitellogenic), this would correspond more closely to stage 4 in the Ankley et al. [11] staging system.

Yet another staging system, adapted largely from the staging systems and nomenclature used for mammalian gonad histopathology, which is a hybrid between the strict qualitative and a quantitative assessment, was employed by Nichols et al. [9] and Miles-Richardson et al. [14]. These authors defined follicular stages, as described below, then counted 50 follicles per ovary, and finally, calculated the relative proportion of each follicular stage in the ovary:

- *Primary follicular stage*: large nucleus, abundant basophilic cytoplasm, and no yolk vesicles
- *Secondary follicular stage*: presence of numerous eosinophilic yolk vesicles
- *Post-ovulatory* (referred to in the publication as Graafian): large follicle bordered by an amphiphilic egg membrane with a central core of eosinophilic yolk protein
- *Atretic*: degenerative follicles with shrunken, irregular borders

As in the cases of testes staging above, the variety of methods proposed does not ease the comparison of effects observed. Furthermore, it may sometimes even be difficult to distinguish between an atretic follicle that failed to ovulate and a developing oocyte that has become atretic before reaching its full size [15]. This should be kept in mind especially with regard to gonadal staging in adult female fish. Furthermore, the distinction of a post-ovulatory follicle and an atretic follicle is not clearcut and is probably likely largely dependent on the point within degeneration that the follicle is inspected. Thus, inclusion of post-ovulatory follicles or atretic follicles without a stringent morphological descriptor and corresponding microphotographic illustration may provide for more confusion than enlightment in the evaluation of histopathological changes in the ovaries. In addition, the distinction of a post-ovulatory follicle and an atretic follicle may be worsened (diluted) following the influence of stressors (xenobiotics or biophysical factors) [16–18]. One major drawback in all the systems described above is that all of them are conducted under light-microscopic conditions *in situ* only and rarely on digital images of the ovarian sections. This provides the lack of a reviewable record and thus makes a comparison of results among scientists even more difficult.

8.3.2 Quantitative Staging in the Ovary

Recently, quantitative methods of analyzing histological sections of gonads based on point-count techniques have been introduced. These methods provide estimates of the volume fractions of the various components and may also be used to count follicles in different stages of development. Indeed, histomorphometric quantification of oocyte size (surface area, diameter) may be helpful to analyze xenobiotic effects [7,19].

Van der Ven et al. [7] counted and measured all previtellogenic and vitellogenic oocytes present (75–170) in a digital microphotograph of a single section of an ovary in a zebrafish that was processed in an image analysis package (AnalySIS, Soft Imaging System, Münster, Germany). The oocyte (maturational stage) classes were defined a priori by the diameter range: previtellogenic oocytes (0.0015 to 0.05 mm^2) and vitellogenic oocytes (0.05 to 0.2 mm^2). This procedure yielded the relative oocyte size of the respective class, the relative number of oocytes of each class, and the relative area of oocytes taken by each class. The ratio of oocyte classes to one another could thus provide information (e.g., a shift from one class to another as a consequence of the influence of a xenobiotic, for example). However, inclusion of two oocyte classes could only potentially leave changes in the smallest oocytes (oogonia) or post ovulatory/atretic follicles undetected.

A similar approach to that of van der Ven et al. [7], or by Wolf et al. [3], and Dietrich et al. [4] was taken for the quantitative evaluation of fathead minnow ovaries. Two digital photomicrographic images (2.2×2.9 mm subject area) were obtained from each of the left and right ovary sections (i.e., four images total per fish) using a $4\times$ microscope objective. To ensure the acquisition of non-overlapping ovary areas, an ultrafine permanent marker was used to draw a line on the underside of the glass slide that bisected the ovary section approximately perpendicular to its long axis. Two tissue areas for digital photography, from each half of the ovary, were selected arbitrarily. For each digital image, the ImagePro-Plus (IPP; Media Cybernetics, Silver Spring, Maryland, version 4.1) was used by a pathologist to manually tag individual ovary cells as to cell type. The contrast of each image was globally enhanced using IPP's "Best Fit" equalization algorithm. Via the computer mouse, the pathologist then sequentially tagged each ovarian follicle with one of six different solid-colored squares that each represented one of the following six follicle cell types (maturational stages): perinucleolar, cortical alveolar, early vitellogenic, late vitellogenic, mature/spawning, and atretic (Figure 8.6). A minimum of 100 follicles was counted per female fathead minnow. Counts per image of the different follicle types were exported to a spreadsheet.

This procedure yielded percentages of the totals for each follicle type per ovary (left and right) and for each follicle type per female fathead minnow. Consequently, this method has the potential to detect a shift of follicle cell types in a given direction (juvenilization, abrogated development, early atresia). Furthermore, as the size of the digital images is defined, this method can also define the number of follicles and specific follicle types per area inspected and thus give a relative impression of the total potential fecundity.

8.4 QUALITATIVE ASSESSMENT OF HISTOPATHOLOGICAL CHANGES

To achieve a scientifically sound interpretation of gonadal changes in conjunction with environmental stressors, including endocrine disrupters, mere quantification of

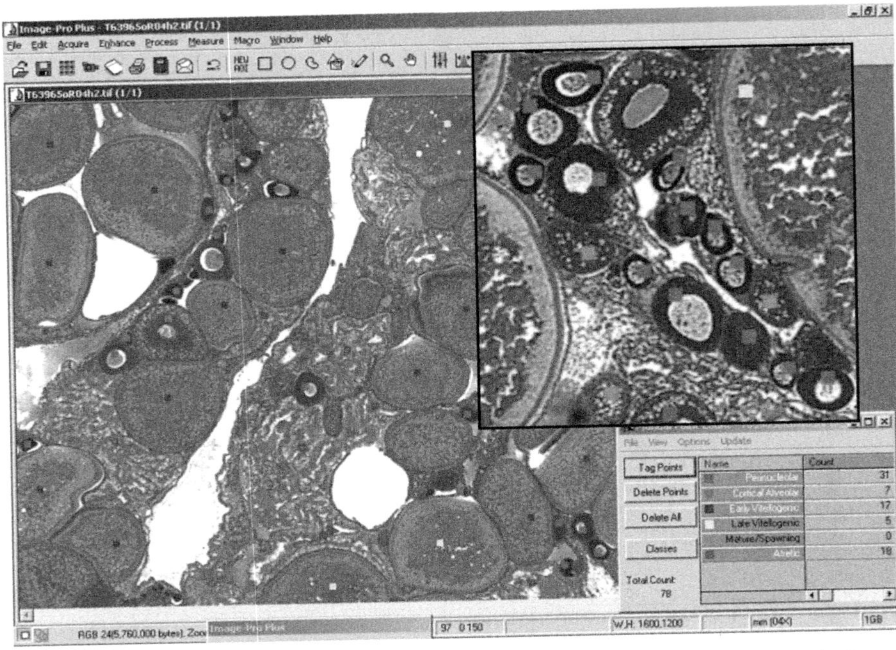

FIGURE 8.6 Manual tagging of digitized photomicrographic images is illustrated. Due to the small number of cells in the ovary relative to the testis, a grid overlay is not utilized, and all germinal cells are counted. A different-colored square (see the inset) represents a single stage of follicular development. (From [3], with permission.) (See insert for color representation.)

germ cells or a qualitative staging will not suffice. Indeed, changes in germ cell numbers and cell types (maturational stage of the germ cells) must be seen within the context of the entire reproductive organ, and thus an overall histopathological appraisal of the entire gonad is a prerequisite. Under light microscopy, morphologic abnormalities and potential treatment-related changes can be graded by the level of severity. The severity of any degenerative, infectious, inflammatory, or proliferative changes (i.e., apoptosis, necrosis, mineralization, infiltration of inflammatory cells, fibrosis, hypertrophy, interstitial accumulation of proteinaceous fluids, intracellular proteinaceous droplet formation) can be graded or ranked on a scale ranging from minimal, to mild, moderate, moderately severe, and severe. The combination of overall histopathological assessment and germ cell staging or quantification can then provide a basis that allows mechanistic deductions in the case of single-compound investigations [7,14,18,20–35] or assessment of the impact of mixtures (e.g., sewage treatment effluents) on the reproductive organ *in toto* [34,36].

Finally, it should be noted that pathological alterations in a gonad could equally be induced by intercurrent, non-test compound (in this context, endocrine disturbance) related etiology, so the pathologist needs to be aware of and conversant with these possibilities.

REFERENCES

1. West, G., Methods of assessing ovarian development in fishes: A review. *Australian Journal of Marine and Freshwater Research*, 1990, **41**:199–222.
2. McDonald, K.K., et al., Reproductive indicators, in *Biomonitoring of Environmental Status and Trends (BEST) Program: Selected Methods for Monitoring Chemical Contaminants and their Effects in Aquatic Ecosystems.* U.S. Department of the Interior, U.S. Geological Survey, Biological Resources Division, Washington, DC, 2000, pp. 37–42.
3. Wolf, J.C., et al., Qualitative and quantitative histomorphologic assessment of fathead minnow *Pimephales promelas* gonads as an endpoint for evaluating endocrine-active compounds: a pilot methodology study. *Toxicologic Pathology*, 2004, **32**(5):600–612.
4. Dietrich, D.R., et al., Optimization of tissue preparation for the quantification of germinal cells in histologic sections of fathead minnow gonads. *SETAC Europe 13th Annual Meeting*, Hamburg, Germany, 2003.
5. Sohoni, P., et al., Reproductive effects of long-term exposure to bisphenol A in the fathead minnow (*Pimephales promelas*). *Environmental Science and Technology*, 2001, **35**(14):2917–2925.
6. Smith, R.J.F., Seasonal changes in the histology of the gonads and dorsal skin of the fathead minnow *Pimephales promelas. Canadian Journal of Zoology*, 1978, **56**:2103–2109.
7. van der Ven, L.T.M., P.W. Wester, and J.G. Vos, Histopathology as a tool for the evaluation of endocrine disruption in zebrafish (*Danio rerio*). *Environmental Toxicology and Chemistry*, 2003, **22**(4):908–913.
8. Miles-Richardson, S.R., et al., Effects of waterborne exposure of 17 β-estradiol on secondary sex characteristics and gonads of fathead minnows (*Pimephales promelas*). *Aquatic Toxicology*, 1999, **47**(2):129–145.
9. Nichols, K.M., et al., Effects of exposure to municipal wastewater in situ on the reproductive physiology of the fathead minnow (*Pimephales promelas*). *Environmental Toxicology and Chemistry*, 1999, **18**(9):2001–2012.
10. van der Ven, L.T.M. and P.W. Wester, Toxicological pathology atlas of small laboratory fish with special reference to normal histology and effects of endocrine disruption in the zebrafish (*Danio rerio*). http://www.rivm.nl/fishtoxpat, 2003.
11. Ankley, G.T., et al., Evaluation of the aromatase inhibitor fadrozole in a short-term reproduction assay with the fathead minnow (*Pimephales promelas*). *Toxicological Sciences*, 2002, **67**:121–130.
12. Leino, R.L. and J.H. McCormick, Reproductive charactersitics of the ruffe, *Gymnocephalus cernuus*, in the St Louis River estuary on western Lake Superior: A histological examination of the ovaries over one annual cycle. *Canadian Journal of Fisheries and Aquatic Sciences*, 1997, **54**:256–263.
13. Selman, K. and R.A. Wallace, Gametogeneis in *Fundulus heteroclitus. American Zoologist*, 1986, **26**:173–192.
14. Miles-Richardson, S.R., et al., Effects of waterborne exposure to 4-nonylphenol and nonylphenol ethoxylate on secondary sex characteristics and gonads of fathead minnows (*Pimephales promelas*). *Environmental Research, Section A*, 1999, **80**(2):S122–S137.
15. Tyler, C.R. and J.P. Sumpter, Oocyte growth and development in teleosts. *Reviews in Fish Biology and Fisheries*, 1996, **6**:287–318.

16. Janz, D.M. and G.J. van Der Kraak, Suppression of apoptosis by gonadotropin, 17β-estradiol, and epidermal growth factor in rainbow trout preovulatory ovarian follicles. *General and Comparative Endocrinology*, 1997, **105**:186–193.
17. Janz, D.M., et al., Recovery of ovary size, follicle cell apoptosis, and HSP70 expression in fish exposed to bleached pulp mill effluent. *Canadian Journal of Fisheries and Aquatic Sciences*, 2001, **58**:620–625.
18. Janz, D.M., et al., Elevated ovarian follicular apoptosis and heat shock protein-70 expression in white sucker exposed to bleachad kraft pulp mill effluent. *Toxicology and Applied Pharmacology*, 1997, **147**(2):391–398.
19. van der Ven, L.T.M. and P.W. Wester, Normal histology and effects of hormonal active agents of reproductive organs of the zebrafish (*Danio rerio*). *OECD Technical Workshop on Gonadal Histology of Small Laboratory Fish as a Tool for Endocrine Disruption Screening and Testing*, Bilthoven, The Netherlands, 2002.
20. Wester, P.W. and J.G. Vos, Toxicological pathology in laboratory fish: An evaluation with two species and various environmental contaminants. *Ecotoxicology*, 1994, **2**:21–44.
21. Wester, P.W. and J.H. Canton, Histopathological study of *Oryzias latipes* (medaka) after long-term β-hexachlorocyclohexane exposure. *Aquatic Toxicology*, 1986, **9**(1): 21–45.
22. Weber, L.P., et al., Increased cellular apoptosis after chronic aqueous exposure to nonylphenol and quercetin in adult medaka (*Oryzias latipes*). *Comparative Biochemistry and Physiology, Part C: Toxicology and Pharmacology*, 2002, **131**(1):51–59.
23. Seki, M., et al., Effect of ethinylestradiol on the reproduction and induction of vitellogenin and testis–ova in medaka (*Oryzias latipes*). *Environmental Toxicology and Chemistry*, 2002, **21**(8):1692–1698.
24. Kinnberg, K., B. Korsgaard, and P. Bjerregaard, Concentration-dependent effects of nonylphenol on testis structure in adult platyfish *Xiphophorus maculates*. *Marine Enviromental Research*, 2000, **50**:169–173.
25. Kinnberg, K., et al., Effects of nonylphenol and 17β-estradiol on vitellogenin synthesis and testis morphology in male platyfish *Xiphophorus maculates*. *Journal of Experimental Biology*, 2000, **203**:171–181.
26. Kinnberg, K. and G. Toft, Effects of estrogenic and antiestrogenic compounds on the testis structure of the adult guppy (*Poecilia reticulata*). *Ecotoxicology and Environmental Safety*, 2003, **54**:16–24.
27. Kiparissis, Y., et al., Effects of the antiandrogens, vinclozolin and cyproterone acetate on gonadal development in the Japanese medaka (*Oryzias latipes*). *Aquatic Toxicology*, 2003, **63**(4):391–403.
28. Palace, V.P., et al., Induction of vitellogenin and histologic effects in wild fathead minnows from a lake experimentally treated with the synthetic estrogen, ethynylestradiol. *Water Quality Research Journal of Canada*, 2002, **37**:637–650.
29. van den Belt, K., et al., Effects of ethynylestradiol on the reproductive physiology in zebrafish (*Danio rerio*): time dependency and reversibility. *Environmental Toxicology and Chemistry*, 2002, **21**(4):767–775.
30. Gray, M.A., A.J. Niimi, and C.D. Metcalfe, Factors affecting the development of testis–ova in medaka, *Oryzias latipes*, exposed to octylphenol. *Environmental Toxicology and Chemistry*, 1999, **18**(8):1835–1842.

31. Islinger, M., et al., Effects of 17α-ethinylestradiol on the expression of three estrogen-responsive genes and cellular ultrastructure of liver and testes in male zebrafish. *Aquatic Toxicology*, 2003, **62**:85–103.

32. Weber, L.P., R.L. Jr., Hill, and D.M. Janz, Developmental estrogenic exposure in zebrafish (*Danio rerio*): II. Histological evaluation of gametogenesis and organ toxicity. *Aquatic Toxicology*, 2003, **63**:417–429.

33. Wester, P.W. and H.H. Canton, Histopathological effects in *Poecilia reticulata* (guppy) exposed to methyl mercury chloride. *Toxicological Pathology*, 1992, **20**(1):81–92.

34. Wester, P.W., et al., Identification of endocrine disruptive effects in the aquatic environment: a partial life cycle assay in zebrafish. Rijksinstituut voor Volksgzondheid en Milieu, Bilthoven, The Netherlands, 2003, p. 112.

35. Laenge, R., et al., Effects of the synthetic estrogen 17α-ethynylestradiol on the life-cycle of the fathead minnow (*Pimephales promelas*). *Environmental Toxicology and Chemistry*, 2001, **20**(6):1216–1227.

36. Jobling, S., et al., Altered sexual maturation and gamete production in wild roach (*Rutilus rutilus*) living in rivers that receive treated sewage effluents. *Biology of Reproduction*, 2002, **66**(2):272–281.

CHAPTER 9

Experimental Design and Statistics

9.1 BASIC CONSIDERATIONS IN EXPERIMENTAL DESIGN

To arrive at meaningful experimental data, it is important at the outset of an experiment to stringently clarify what one wants to determine (effect), at what level of difference (∂) the effect is considered of biological importance, and with which level of confidence (type I error) this difference in effect should be detected. In most cases exposure groups will be compared to a control group. Consequently, it is important to know the variability of a given effect within the control and, if possible, within the historical control (control groups from several different laboratories or control data from several different field investigations) [1]. Based on this knowledge, the necessary number of animals per replicate and number of replicates in each of the control and exposure groups can be estimated using power analyses, whereby the assumptions are that the probability of a type I error (erroneously defining a difference between control and exposure groups) is less than 5% ($\alpha < 0.05$) and that the type II error (erroneously negating a difference between control and exposure groups) is less than 10 to 20% ($\beta < 0.10$ to 0.20) [2,3]. Consequently, power analyses can determine prior to the actual experiment whether or not an endpoint chosen [e.g., ranked qualitative (semiquantitative) or quantitative gonadal cell-type staging] can be determined with the necessary precision and power to allow distinction between control and exposure groups with the appropriate statistical test and thus reliably detect an effect.

Thus, power analysis in the experimental design is essential in helping to choose the most appropriate endpoints or endpoint methodologies. However, albeit all of the above is true, it is critical to understand that the statistical analysis of one single variable is not a stand-alone issue. Indeed, in most experiments, multiple variables will be analyzed, and therefore an overall type I error level of $\alpha \leq 0.05$ will have to be applied to the entire experiment (i.e., the error level will have to be split over the number of variables analyzed; see Section 9.4.1 to 9.4.2). The consequence of this is that the higher the number of variables that are being tested, the higher the number of replicates required to allow the detection of differences with the necessary precision

Histological Analysis of Endocrine-Disruptive Effects in Small Laboratory Fish
By Daniel R. Dietrich and Heiko O. Krieger

and power, as discussed in more detail below. Thus, it is also of utmost importance to understand what variables should and can be measured and why (relationship of the variable with the experimental objective and the ensuing biological interpretation at the outset of experimental planning).

9.1.1 Defining the Objectives

Despite the temptation to add variables that can be measured easily, one must keep in mind that "Not everything that can be counted counts, and not everything that counts can be counted" (Albert Einstein). Therefore, it is crucial to consider and clearly define the objective of the experiment beforehand:

1. Does the compound have an adverse effect on life-history variables of a species?
2. Can general toxicity, including endocrine-disrupting activity, be detected and attributed to the compound treatment?
3. Can an adverse effect on gonadal germinal cells (i.e., germinal cell type frequency in male and female of a species) be detected and attributed to the compound treatment?

The consequence of reducing the objective to more or less precise questions [3] is a well-defined number of variables to be measured and therefore a higher power for detection significant changes despite a technically limited number of replicates (see Sections 9.3.6 to 9.3.5).

9.1.2 Asking Precise Questions

Precisely defining the objective will then allow answering some principal statistical questions, generally using historical data (prior experience) or data from preliminary experiments, which are prerequisite for a proper experimental design:

- Which variables (parameters) are being employed, and what are their statistical characteristics?
- How does the number of variables determined and thus the number of statistical comparisons affect the significance level and thus the likelihood of being able to detect compound-induced effects?
- How does replication and number of variables affect the minimum observable effect (MOE) when planning the experiment, and the minimum significant difference (MSD) after completion of the experiment?
- How does replication affect the detection of compound-induced effects (the power to detect an effect)?
- How does the experimental unit (e.g., aquaria) and pseudoreplication affect statistical evaluation and thus detection of compound-induced effects?

9.1.3 The Principal Question: Is There an Effect or Is There Not?

Most experimental designs are set up with the aim of detecting an effect of a given environmental stressor (e.g., temperature, light, food, population density) or compound exposure. Simplistically speaking, any experiment employing the most extreme conditions will always provide some sort of measurable effect (e.g., reduced survival), whether or not the exposure regimen and/or the effect determined is biologically or ecologically reasonable or relevant. Contrary to the latter and irrespective of whether the experiments carried out are geared toward the determination of a NOEC (no-observed-effect concentration) as a toxicological summary statistic (see Section 9.1.4), it must be understood that a lack of significant changes observed following treatment is no proof of the absence of an effect [3–5], but rather, may be the consequence of the inability to observe an effect due to the experimental design. Simply, it is impossible to prove that a treatment has no effect. However, it is possible to prove that the effect of a treatment is less than some specified value (experimental threshold value), affects only a small subset of the population tested (e.g., EC_5; see Section 9.1.4), or that the endeavor to produce a significant effect may become technically and biologically impossible (see Sections 9.3.6 and 9.3.4). However, as highlighted recently by EFSA [6], defining an experimental threshold value demands intimate knowledge of the variable measured as well as an accompanying biological interpretation that allows determination of an acceptable effect size (change from normal).

9.1.4 Basic Mechanistic Research, Risk Assessment, and the NOEC, LOEC, and EC_x Question

Although in principle the same fundamental statistical considerations apply to basic mechanistic research as to experiments geared toward risk assessment, the two differ dramatically with regard to the experimental design. Indeed, whereas the objectives of basic mechanistic research involve primarily the elucidation of a very specific effect (e.g., the effect of 17β-estradiol on spermatogenesis of male zebrafish), the objectives of an experiment geared toward risk assessment are manifold: to determine the (adverse) effect of, for example, 17β-estradiol on zebrafish to derive a NOEC (no-observed-effect level), LOEC (lowest-observed-effect level), or EC_x (effect-concentration affecting $x\%$ of the population tested) to develop sufficiently broad and robust data sets that would allow extrapolation of the experimental data to a chronic or lifetime exposure setting and to whole zebrafish populations or to endemic populations of fish species in general.

Obviously, experiments of basic mechanistic research will require only a limited number of variables and concentrations. Primarily, the researcher wants to answer the question: Does it or does it not have an effect? Consequently, employing sufficiently high numbers of true replicates (not pseudoreplicates, see Section 9.3.5) in the experimental design will ensure sufficient precision and power to allow detection of the hypothesized changes at a high level of confidence (very low α). The latter not only pertains to the experimental design but also to the optimal methods employed in the analysis of the variables. Indeed, errors are cumulative, so reducing an error in one part of the experiment results in a lower aggregate error for the entire experiment.

Conversely, in the risk assessment setting, a rather high number of a priori unrelated variables (e.g., mortality, wet weight, vitellogenin levels, histopathological lesions, clinical chemistry) are determined. Moreover, as prior to the experiment the direction (increase or decrease) of the change in a given variable is not known (see Section 9.3.2), most variables will have to be tested in a two-tailed fashion. The latter will decrease the level of significance (α) per comparison (see Sections 9.4.1 and 9.4.2). In addition, a high number of variables and compound concentrations will demand splitting the significance level across the variables and comparisons (see Sections 9.4.1 and 9.4.2), the consequence being that a higher number of replicates will be required to allow determination of a demonstrable effect (MOE) with the necessary power (Sections 9.3.5 and 9.3.6). Many experiments used for risk assessment estimate an NOEC, LOEC, and EC_x via a concentration–response model employing geometrically spaced multiple concentrations (often >5). NOEC and LOEC are test concentrations and thus depend entirely on the sample size (number of concentrations employed) employed for statistical evaluation. The larger the sample size, the lower the LOEC and NOEC.

Both summary statistics depend largely on the compound, the species, the observed variables, and the level of significance (generally 5%) used. Moreover, if the experimental test is carried out accurately, thus providing a smaller sample error, NOEC and LOEC will be smaller [7]. However, it must be remembered that the NOEC is treated routinely as if it were a no-effect concentration, albeit it must be clear that it is impossible to demonstrate the absence of an effect (see Section 9.1.3). Consequently, whenever a NOEC is given, this value should be accompanied by the minimal significant difference (MSD) value (see Section 9.3.6) of the specific experiment reported.

In contrast to the NOEC and LOEC, estimation of an EC_x from a dose–response curve is not very difficult. However, it must be kept in mind that depending largely on the x to be determined, the actual concentrations tested should be spaced closely around the EC_x expected, concurrently with a high degree of replication in the exposure groups and $k^{1/2}$ times as many replicates in the control (k being the number of concentrations tested), such that the EC_x can be estimated with small confidence intervals. However, more often than not, EC_x are estimated from experimental designs employing geometrically spaced concentrations, therefore not catering primarily to the "as best as possible" estimation of an EC_x [8]. Consequently, the EC_x has to be estimated from a given concentration-response curve, which, inherently, provides for fairly broad confidence intervals at both ends of the curve (i.e., in the effect range 0–10% and 90–100%). The consequence of the latter is that the scientist will have to decide which level of effect is just enough to have the potential for generating ecological consequences for the species tested. In many opinions, this level is set at 5 or 10% (i.e., an EC_5 or an EC_{10} is estimated) [5,7]. The question is, then, how this EC_x value will be employed:

- As an alternative to an NOEC, meaning that any environmental concentration below the estimated EC_x is assumed to have no effect on the population?
- As an alternative to a LOEC, whereby the assumption is made that the estimation of the EC_x is sufficiently robust that environmental concentration above the estimated EC_x will provide for a detrimental effect on the population?

9.2 VARIABLES TO BE DETERMINED AND THEIR INHERENT BIOLOGICAL AND MATHEMATICAL CHARACTERISTICS

Of the variables that are considered quintessential for the experimental objective, it is not only important to understand their causal relationship with the objective (e.g., the measure of an effect on a fish population or the measure of an effect on reproduction). Indeed, it is vital to understand how a variable behaves biologically (biological characteristics, see Section 9.2.1): for example, seasonal variability in reproduction (recrudescence of reproductive tissue, vitellogenin levels, etc.) under normal circumstances [i.e., without the undue influence of a compound (xenobiotic) or stressor (temperature, light, water quality parameters, etc.)]. The higher the variability of a variable in a given population, the more important it will be to randomize the experimental population properly when assigning individuals to control or exposure groups. Moreover, the higher the variability of the variable, the more important it will be to increase the number of replicates in the control groups to allow for detection of significant changes with sufficient power (see Sections 9.3.5 and 9.3.6).

It is also important to understand the inherent mathematical characteristics of the respective variables (see Section 9.2.2; e.g., independence of the variable, interdependence of the variable) as well as the methodological limitations in measuring the respective variable [e.g., limit of detection (LOD) and limit of quantification (LOQ) of a given method], in order to allow their proper statistical testing.

9.2.1 Biological Characteristics

The following is an incomplete subset of the numerous variables determined in fish experiments, which will serve as examples for the presentation of principle. The questions asked for each variable may, however, serve as a means to enhance a scientist's aptitude to extrapolate the same principle to variables not discussed here. But whichever variable(s) are determined in an experiment, it is critical to understand the normal (background) variability of the respective variable in a given population. The latter understanding will not only allow estimation of the required replication within an exposure group to allow for a given MOE when planning an experiment, but will also allow identification of outliers within the control group after the experiment.

- *Survival (mortality).* Survival is rather easy to determine; however, as an apical variable it has little value for the mechanistic understanding of compound exposure or stress. Any significant change in mortality in the experimental control group from the normal (background) incidence may signify that the population used for experimentation may not be healthy, stressed, or otherwise compromised.

- *Life-history variables*
  - *Length*: increase or decrease in length, but neither correlating directly with increased or decreased fitness of the individual or population. For some fish species (e.g., salmonids), however, body size (including length) is correlated positively with sexual dominance and thus capability to reproduce.
  - *Wet weight* (body weight): reduced body weight may be an indication of decreased fitness, for example, in conjunction with a reduced or impaired food uptake and metabolism. Increased body weight may be an indication for decreased fitness when associated with abnormal metabolism (e.g., lipid metabolism), but also an indication for increased fitness resulting from optimal use of energy resources. For some fish species (e.g. salmonids), body size (weight and length) [9–13] is correlated positively with sexual dominance and thus capability to reproduce [11,12,14–20].
  - *Organ weight* (e.g., gonad weight): whether increased or decreased, may reflect a histological change such as a change in cell type (e.g., scarring of tissue due to necrosis with ensuing fibrosis), cell type number (e.g., hyperplasia: an increased cell proliferation of a specific cell type), or cell physiology/morphology (e.g., cell hypertrophy due to increased glycogen or lipid storage, or proliferation of cell organelles).
  - *Hepatosomatic index* (HSI): the ratio of liver weight to whole-body weight. Thus, if the liver is heavier than normal yet the body weight remains within the normal range, the HSI would reflect this situation. As the HSI is normalized via the respective body weight, the HSI may provide for a more sensitive detection of liver weight change than the liver weight per se, as the liver weight may be subject to enormous biological variability. A decrease in HSI may reflect a pathological change (e.g., cell loss due to necrosis or apoptosis and ensuing fibrosis, anabolic or catabolic inhibition) or mere feeding stress (hunger) as could also be encountered during normal reproductive activity. An increased HSI may reflect changes due to excessive feeding (e.g., lipid or glycogen storage) but also a pathological change (e.g., hyperplasia, hypertrophy).
  - *Gonadosomatic index* (GSI): calculated similar to the HSI. However, when a GSI is tabulated it must be clarified whether or not the sum of gonad weights (left and right gonad in case of unfused organs) or individual gonad to body weight indices are tabulated. Whenever the GSI of single gonads are calculated, it must be kept in mind that the weights of left and right gonads in a single fish may differ substantially, thus providing a much higher variability of this parameter. In contrast, the GSI, calculated with the sum of gonad weights provides a sensitive measure of gonad weight changes due to normalization via the respective body weight. Increased GSI may reflect

increased reproductive activity (especially in seasonal fish) but could also indicate the presence of pathological changes (e.g., abrogated gamete differentiation and thus accumulation of certain germinal cell stages). Similarly, a low GSI could be encountered under normal circumstances (e.g., at the end of the reproductive period in seasonal fish) but just as well as the result of overt gonad toxicity and thus cell loss (somatic as well as germinal cells).

- *Vitellogenin* (VTG). Synthesis of the hepatically synthesized egg yolk protein vitellogenin [21–30] is under estrogenic control in the female fish. Correspondingly, male fish synthesize little VTG [31,32]. Consequently, methodological assessment of VTG levels will have to accommodate the different plasma or serum VTG levels expected (i.e., determination of VTG in control males will demand having a much lower LOD and LOQ [21] than corresponding assessments in females). Expression of VTG will vary with reproductive activity, especially in seasonal fish (e.g., salmonids or fathead minnows) as well as with exposure to environmental estrogens and xenobiotics [33–37].
- *Fecundity*. The reproductive capacity of a fish can be assessed via fecundity: the number of eggs or sperm spawned. Depending largely on the fish species, assessment of fecundity can prove to be rather difficult unless fish are kept individually, which itself may influence spawning. When fish spawn in batches (e.g., fathead minnows), fecundity can be expressed as the number of eggs per batch, number of batches or eggs per female in a given time period. In the case of fish, where females spawn freely over a graveled surface (e.g., zebrafish), assessment of fecundity (eggs per spawn) may be more difficult. Similar difficulties will be encountered in the assessment of male fecundity. Consequently, only robust and highly reproducible methodologies for fecundity assessment will allow determination of changes in fecundity. Increases or decreases in fecundity may depend on seasonal reproductive patterns, which can still be present even in laboratory-held species, feeding, temperature, and light regimes, appropriate mating partners (*dislike is not an exclusively human trait!*), environmental conditions (sufficient spawning areas or surfaces, water composition), stress (predation or disturbance), or xenobiotic influences [38].
- *Histopathological lesions*. Pathological changes in a fish are assessed on an organ/tissue-specific basis. For each organ or tissue, specific lesions are noted, and for each lesion the severity of the change is determined. An increase in number and severity of pathologic al lesions may be associated with the concurrent exposure to a xenobiotic; however, other sources (e.g., parasitism, bacterial or viral infections, congenital malformations due to inbreeding and lack of proper randomization, species- and strain-specific diseases, malnutrition) must be kept in mind or countered or minimized prior to the experiment. Some of the lesions (e.g., hepatic necrosis, fibrosis) may be causally related and therefore dependent, whereas other lesions (e.g., glomerulonephritis, interstitial fibrosis) may not and therefore are considered independent. Moreover, the

occurrence of lesions in an organ may or may not be causally related to the presence of lesions in another organ or tissue. In principle, the severity of lesions is graded according to a severity scale (ranked/scalar data; see Section 9.2.2). It is important to note that current discussions among pathologists and statisticians differ with regard to blind vs. nonblind histopathological assessment of tissue sections [3]. Pathologists prefer nonblind reading, as they would like to be able to detect subtle changes in exposed fish, and this can be done only if the histological section can be compared directly with the concurrent control. The latter approach may, however, provide an unintentional bias in grading of lesions within the experiment. In contrast, "blind" reading of randomized tissue sections will minimize the occurrence of a pathologist-induced bias but may be less sensitive for detecting subtle changes. Bias in nonblind reading can be minimized if multiple readings of the control group are carried out intermittently throughout the reading of a whole study.

- *Quantitative histology*
  - *Germinal cell type counts:* quantification of germinal cell types [39,40] as a measure of gametogenic normality or abnormality. Decreased germinal cell-type counts can be due to normal fluctuations during the seasonal reproductive pattern of a given species, to an individual's compromised health (infections, parasitism), to social hierarchy [10,11,17,41,42], to nutritional status [9,12,16,42–48], to stress [9,11–14,17,18,20,41,44–57], to intoxication with compounds specifically affecting reproductive organ tissue in general (including gametogenic, Theca, Leydig, or Sertoli cells), or as a result of endocrine-active compounds [39,40]. Whenever germinal cell types are counted, it must be kept in mind that germinal cells are interdependent, whereas somatic cells (e.g., Leydig or Sertoli cells) are not. Consequently if one observes a major increase in the number of spermatogonia, for example, one would expect to see a corresponding decrease in the number of the ensuing germinal differentiation stages (e.g., spermatocytes, spermatids, and spermatozoa). Detection of changes in gametogenic cell types is highly dependent on the methods employed for counting [40] and the number of replicates used in the control and exposure groups (see Sections 9.2.2 and 9.3.3).
  - *Germinal cell type area determination:* can be measured as to their relative area (see Chapter 8), similar to germinal cell type counting, cysts of spermatogonia, spermatocytes, spermatids, or spermatozoa as well as oogonial cell types [58–61], the assumption being that the larger the area, the higher the number of germinal cell types within the respective cyst. This holds true, however, only if the germinal cell type density is comparable throughout the experimental population. As is the case for germinal cell type counting, changes in the germinal cyst area may result from a number of influences (see above). Moreover, the image analysis technique employed needs to be evaluated carefully with regard to the system-inherent LOD and LOQ.

9.2.2 Mathematical Characteristics

The distribution types of the individual variables are listed in Table 9.1 with the corresponding statistical tests that can be employed. The other variable-inherent characteristics are discussed below.

- *Survival (mortality).* As survival is a binary variable, introduced bias is limited to experimental or procedural problems (e.g., lack of immediate removal of dead animal from replicate tanks or a compromised experimental population at the start of the experiment (see Section 9.2.1).
- *Life-history variables*
  - Length, wet weight (body weight), and organ weight (e.g., gonad weight) are continuous variables with inherently high biological variability. Consequently, a high degree of replication in the control group is a prerequisite for the detection of small changes (see Sections 9.3.5 and 9.3.6). It must be noted that fish held within the same aquarium are not independent and thus may influence each other with regard to growth due to social hierarchy [10,11,16,17,41,42].
  - The hepatosomatic index (HSI) and gonadosomatic index (GSI) are continuous variables with primarily log-normal distribution [62]. As HSI and GSI are ratios and therefore the result of a mathematical data transformation, they are subject to the same influences as the other life-history variables.
- *Vitellogenin.* The distribution of VTG levels in plasma or tissue can be heavily skewed, therefore not following a normal distribution. VTG levels can be log-transformed and then tested again for normalcy of distribution. Whenever near-normal distribution cannot be achieved, even following log transformation, or when a priori a high number of replicates per concentration are employed (see Section 9.4.4), nonparametric testing should be used.
- *Fecundity.* Fecundity (e.g., number of eggs per female) usually follows a normal distribution. Whenever females are not housed individually but rather, are housed in numbers ≥ 2 per aquarium, they may influence one another [53]. The measure of fecundity would thus be the "mean" number of eggs per female per replicate aquarium, whereby the means of the replicate aquaria are used to generate a mean $\pm$ standard deviation of the concentration and control group, respectively, for the ensuing Student's-*t*-test or Dunnett's test.
- *Histopathological lesions.* Histopathological tissue classification employs ordered scales. As this scale is used for statistical testing, the tissue classification is summarized as the median of the classification in the individual female and male fish per aquarium. The latter stands in contrast to other variables where either the mean or the variance of a variable is being tested.

- *Quantitative histology*
  - *Germinal cell type counts:* as germinal cell types are interdependent, it is assumed that the frequencies of the germinal (reproductive) cell types sum to 1 (100%). They are therefore highly correlated with one another and hence cannot be tested independently. Thus, the germinal cell types should be ordered from most to least mature, and then the frequency of each germinal cell type in the group of equally and less mature cell types calculated. These relative frequencies can all take values between 0 and 1, independent and irrespective of the value observed for any of the other relative frequencies. Note that for the least mature cell type, this frequency is by definition 1, and therefore irrelevant. To allow the use of tests based on normal distribution, it is important to transform the frequencies (values between 0 and 1) to variables that can take any value between $-\infty$ and ∞. Two transformations are commonly used: the arcsine–square root transformation and the logit transformation. Arcsine–square root transformation is appropriate only in cases where each frequency is a frequency within the same number of observed cells. However, as the latter is rarely the case, the logit transformation should be employed.
  - *Germinal cell type area determination:* depending on the format of cell type the determination of the statistical approach will vary.

    When, for example, a total of five cysts of each germinal cell type (spermatogonia) are measured, a mean area of the spermatogonia per male fish per replicate aquaria can be calculated. As the area of a cyst is a continuous variable, the data will follow a normal distribution. Hence, the mean area $\pm$ standard deviation of spermatogonial cysts in exposed fish can be compared with the concurrent control group using a Student-*t*-test or Dunnett's test.

    When all spermatogenic cysts in three fields of view (magnification 40×) of a routine section are evaluated as proposed for the zebrafish by van der Ven et al. [61], a statistical approach similar to that of counting germinal cell types appears to be the proper solution. Consequently, it is assumed that due to the fact that germinal cell types are interdependent, germinal cysts containing the different stages of germinal cell types are interdependent as well. Hence, it must be assumed that the mean relative areas of each germinal cyst type sum to 1 (100%). They are therefore highly correlated with one another and hence cannot be tested independently. Thus, the germinal cyst types should be ordered from most to least mature, and the relative mean area of each germinal cyst type in the group of equally and less mature cyst types is calculated. These relative mean areas can all take values between 0 and 1, independent and irrespective of the value observed for any of the other relative mean areas. Note that for the least mature germinal cyst type, this relative mean area is by definition 1, and therefore irrelevant. To allow the use of tests based on normal distribution, it is important to logit-transform the relative mean areas (values between 0 and 1) to variables that can take any value between $-\infty$ and ∞.

TABLE 9.1 Tests That Can Be Employed in a Comparison of the Values of a Single Variable Observed in Several Treatments with Concurrent Control Value

Data Type	Continuous Data	Score, Ordered, or Non-normal Distributed Continuous Data	Counts with No Known Upper Limit	Counts or Proportions (Quantal Data)
Distribution	Normal or log normal	Non-normal, often unknown	Poisson or negative binomial; normality can often be achieved by transformation	Binomial
Variables observed in a fish study	Length Wet weight Gonad weight Gonadosomatic index (GSI) Hepatosomatic index (HSI) Fecundity Logit of the (rel.) germinal or somatic cell frequencies Area of spermatogenic cysts in male fish	Pathological lesion scores Gonadal staging Vitellogenin levels Fraction of surviving fish per aquarium	Number of offspring of individual fish pairs	Survival (mortality) Sex Color

Pairwise comparison[a]	Student *t*-test	Wilcoxon test	If *successfully* transformed to normal distribution: parametric tests (e.g., Student *t*-test) If transformed *unsuccessfully*: non-parametric tests (e.g., Wilcoxon test)	Fisher 2 x 2 exact test x^2-test (only if the *n* is very large and the number of positive findings per unit is >5)
(Multiple) comparison with a control	ANOVA[b]; Dunnett's test	Steel test	If transformed *successfully* to normal distribution: parametric tests (e.g., Dunnett's test, ANOVA[b]) If transformed *unsuccessfully*: non-parametric tests, (e.g., Steel test)	
Trend test	William's test	Jonckheere–Terpstra test	If transformed *successfully* to normal distribution: William's test If transformed *unsuccessfully*: Jonckheere–Terpstra test	Cochran–Armitage test

[a] If a pairwise comparison is used, the significance level must be split over the comparisons by variations of the Bonferroni method (see Section 9.4.1).
[b] If analysis of variance is used, homogeneity of variances should be confirmed using Bartlett's test.
Source: As suggested by Chapman et al. and van der Hoeven [3,5,63].

9.3 PREREQUISITE STATISTICAL CONCEPTS

9.3.1 Sample Independence: The Experimental Unit

A statistical significance test compares the variance between independent aquaria exposed to the same treatment (control) with the variance between the mean values for the different treatments (compound concentrations). If the variance between treatments is large compared to the variance between the independent aquaria of the control, it is concluded that the compound treatment has an effect, the key point being that the aquaria within each treatment (compound or control) should be independent. However, it is also important to recognize what "independence" constitutes within the setting of an experiment. Therefore, this will be exemplified briefly via a virtual experiment: In this experiment, each experimental unit will be influenced by many factors, one of which is the compound treatment. Other influencing factors may be temperature, light regime, water quality, or genetics.

In order that all experimental units within the compound treatment experiment are independent, each of the influencing factors must either be identical for all experimental units at all treatment levels (e.g., all units are kept in the same climate room), or the influencing factor should be assigned at random to each experimental unit (e.g., each experimental unit should be placed in a random location within the set of all available locations). If, however, the influencing factor is inherent to the experimental unit (e.g., the genetic composition of a fish or the water quality in a aquarium), the experimental units should be distributed randomly over the treatment levels. If the experimental units are not independent, the differences observed between treatment levels might not be due to the treatment but rather due to one of the other influencing factors differing between the experimental units. For example, if the treatment was the 17β-estradiol concentration in the water and several fish are kept per aquarium, fish housed within the same aquarium are dependent; that is, they have a common factor influencing all fish within this aquarium but not influencing the fish housed in other aquaria (see Section 9.2.2). The common influencing factor is thus the aquarium itself. If the ensuing statistical tests are then based on individual fish within the same aquaria (pseudoreplicates; see Hurlbert [64] and Section 9.3.5), these tests use the estimates of the variance between fish to decide whether the difference between two treatments (e.g., control and compound treatment) is significant. In this case, observed differences might just as well be due to differences between aquaria than to treatment. This is especially true when only one aquarium is available per treatment (such as proposed in the OECD 21-day fish test [65,66]). Unfortunately, design flaws, prohibiting statistical unequivocal identification of compound treatment–related effects, apply to many of the recent publications (e.g., on bisphenol A–mediated effects in carp [67], the role of sex ratio on spawning performance in common dentex [68], and the impacts of 17β-estradiol in zebrafish [69]). These examples demonstrate that statistically robust experimental designs are key to providing meaningful and biologically interpretable experimental results.

9.3.2 One- or Two-Sided Testing

Each of the tests in Table 9.1 can be used either one- or two-sided. If it is reasonable to assume a priori that the effect of the test chemical is in a specific direction: for instance, a decrease in survival, a decrease in wet weight, or a decrease in the frequency of the more mature germinal cell type, a one-sided test is preferable. Note that this needs to be decided *prior* to the experiment and thus to testing the data. If before the actual experiment it was decided, for example, that only a decrease in wet weight should be noted as a toxic effect, an observed increase in wet weight, however large, can never be significant.

9.3.3 Observable Effect Size: Minimum Observable Effect

The *minimum significant difference* (MSD), described in Section 9.3.6, is based on the values observed in the experiment. But the observed value, for instance the observed mean wet weight in the control, is only an estimate of the "real" average wet weight. A new experiment with another sample of all possible fish would lead to another mean value, therefore to another estimate of the "real" average wet weight. Our interest is not primarily whether the values observed in a control and an exposure group differ, but whether the real average values for these two groups differ. The difference Δ^* between the two real averages can be observed with a certain probability [i.e., the power $P = 1 - \beta$ (see Section 9.3.4)]. This power depends on the number of replicates (n), the real standard deviation of the variable observed in both control and treatment, and the significance level α chosen. If both the power and the significance level have been chosen and at least a reasonable estimate of the standard deviation is known, the *minimum observable effect* (MOE) can be calculated based on the number of replicates in the experiment. Or, based on a predefined MOE, the necessary number of replicates in the experiment can be calculated (see Section 9.3.4) before the experiment is carried out [5,7]. So, for example, as shown in Figure 9.1, if a difference Δ of 20% were chosen as the MOE and the coefficient of variation (CV) were 15%, the relative difference $D = \Delta/\text{CV}$ is 1.33. Thus, this MOE can be observed with a power of 80%, 90%, or 95% only if a minimum of 9, 11, or 13 replicates, respectively, are used per concentration. Note that in the latter hypothetical example, $k = 5$ compound concentrations are compared with a concurrent control using a one-sided Dunnett's test. According to van der Hoeven [8], the control group should consist of $k^{1/2}$ as many replicates as the compound concentration groups. In this case, this would mean using 2.23 times as many replicates in the control group as in the concentrations groups (i.e., 20, 25, and 29 replicates in the controls to allow detection of the MOE of 20% with a power of 80, 90, and 95%, respectively). The latter example clearly demonstrates that the lower the MOE, the higher the number of replicates required to allow detection of the MOE with a high power. Moreover, the higher the power chosen (the probability of observing the effect in an experiment), the higher the number of replicates required in the exposure groups, and even more so in the control groups.

* Note that Δ represents the "real" difference between real averages (e.g., average wet weight in the control and exposure groups.)

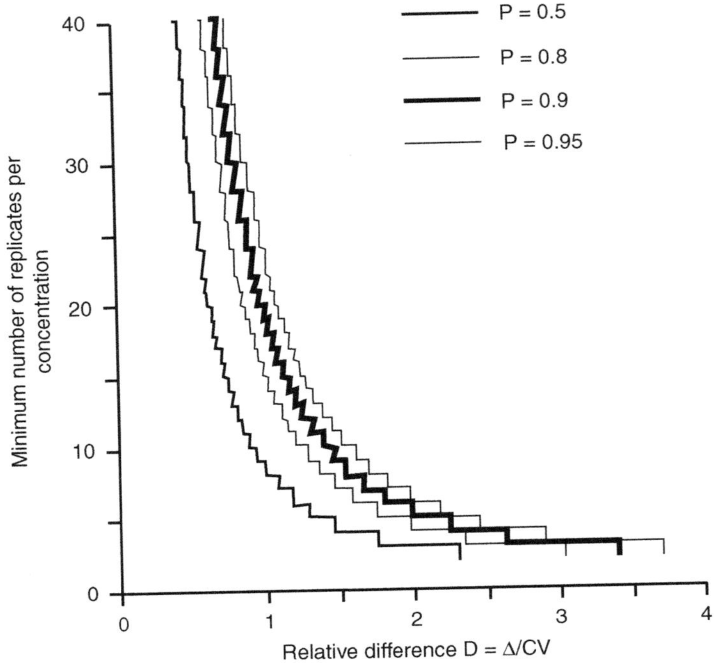

FIGURE 9.1 Minimum number of replicates needed per concentration tested to observe a relative difference $D = \Delta/\mathrm{CV}$ with a power P. A one-sided Dunnett's test ($\alpha = 5\%$) is used and five compound concentrations are compared with a concurrent control. From left to right: $P = 0.5$, 0.8, 0.9, and 0.95. (Courtesy of Nelly van der Hoeven EcoΣtat, Leiden, The Netherlands.)

Conversely, employing the same experimental design as described in Section 9.3.6 (i.e., 10 replicates per group), $\alpha = 0.05$ (one-tailed unpaired t-test), a power $1-\beta$ 0.8, and a standard deviation of 1.02 g, the MOE for wet weight would be 1.18 g. Thus if the control group had a real average wet weight of 2.52 g, the real average wet weight in the exposure group would have to be more than 3.70 g to be observed in 80% of the experiments performed according to this protocol. If, indeed, as in the former example, five groups are compared to the control and thus the significance level $\alpha = 5\%$ (type I error probability) had to be split across the five concentrations, control comparisons would result in a per comparison significance level of $\alpha' = 1.02\%$. Concurrently, the MOE for wet weight would be 1.57 g; that is, with 10 replicates the experiment has a power of 80% to detect a real difference between wet weight means of 1.57 g with a significance level α of 0.01 (one-tailed).

9.3.4 Power ($1 - \beta$) to Detect an Effect

Keeping in mind that each variable (e.g., wet weight) is assumed to have a real or true average value, the value observed is only an estimate of that real value. For example, the mean wet weight of control male fathead minnows exposed for 43 days to a specific feeding regime is 2.4 g [70]. If another small number of fish, stemming from the same original population, were also exposed for 43 days to this regime, the mean wet weight will always differ from the value observed in the first group, and both values will probably differ from the real average value. Thus when as in the case of the BPA experiment by Sohoni et al. [70], two or more groups of fish are compared, albeit originating from the same stock and differing only in that they were exposed to different concentrations of bisphenol A (BPA) and exposure to BPA had no effect, theoretically the average wet-weight value of male fathead minnows in all of the BPA group will be the same, say 2.4315 g. However, the average wet-weight values observed between two groups of male fathead minnows will always differ, whether or not they were exposed to BPA.

Thus, statistical testing is designed such that the erroneous determination (type I error) of a significant difference in real average wet-weight of the control and one of the BPA-exposed groups occurs maximally with an error probability of α also known as significance level α. If the difference (Δ) between the real wet-weight averages of the two groups (control and one of the BPA-exposure groups) is unequal to zero, the probability to observe a significant effect is known as the *power* $1-\beta$ to detect this real difference Δ. Consequently, $1-\beta=\alpha$ if Δ is zero (i.e., $1-\beta$ is small if Δ is small). If the power $1-\beta$ to observe a difference Δ is small (e.g., only 50%), the difference can be considered unobservable because only a fraction $(1-\beta)$ (50%) of the experiments will lead to the conclusion that the (presumably) observed effect really exists, and the effect will be missed in β (50%) of the experiments. For example, after 43 days of exposure to BPA [70] mean wet-weight values of 2.2 and 1.8 g were observed in male fathead minnows exposed to 640 and 1280 μg BPA/L, respectively. However, the fact that the significance level $\alpha = 5\%$ (type I error probability) had to be split across the five BPA concentration–control comparisons resulted in a per comparison significance level of $\alpha' = 1.02\%$ (see Section 9.3.6). In view of the enormous standard deviations within the groups (SD = 1.02 g) and the low α', the power to observe the difference Δ of 0.6 g (between the average wet-weight value of the control and the 1280-μg BPA/L group) would be between 40 and 50% (power calculations carried out using Graphpad Statmate, $\alpha = 0.01$ and a one-tailed unpaired *t*-test), despite use of 30 individual fish per group. In the study in question the 30 fish were actually split into 15 fish per aquaria, thus providing two independent replicates only, which, by definition, would not allow detection of differences in wet weight between the control and BPA-exposed groups (see Section 9.3.1 and 9.3.5).

Conversely, using an a priori choice of the MOE, a defined significance level α, and power to observe the MOE as well as a reasonable estimate of the standard deviation of a given variable allows calculation of the number of replicates required for the experiment and therefore would have allowed appropriate designing of the experiment. Indeed, using the same example as that taken from [70], with α' 1.02%, SD = 1.00 g, the difference Δ of 0.6 g being the MOE, and a required power of 0.8, would have required using 60 independent replicates. The term *independent replicates* (see also Sections 9.3.1 and 9.3.5) would have meant using 60 different aquaria to allow the observation of a real decrease of 0.6 g in wet weight in the BPA-exposed fish (i.e., a decrease of 25%, assuming that the real control wet weight is

2.4 g). As the latter calculations were carried out using a one-tailed unpaired t-test, only a weight decrease could have been observed. Had the preexperiment decision been that it is unknown whether a BPA-induced weight increase or a decrease could occur, a two-tailed unpaired t-test would have to be employed, thus demanding 70 independent replicates to see a change of $\Delta \geq 0.6$ g in the real average wet weight.

9.3.5 Replication and Pseudoreplication

Individual fish housed together in an aquarium are not true replications, but rather, represent pseudoreplicates [64]. Using pseudoreplicates in an experimental design is a priori undesirable, as this design can give the impression that the ensuing statistical test is much more precise and powerful than it really is. This is of greatest importance in experiments where low numbers of replicates (few replicate aquaria within the same concentration group) but high numbers of pseudoreplicates (fish within the same aquarium) are being used. Indeed, Figure 9.2 clearly demonstrates the effect of pseudoreplication with one or more variables determined. In this hypothetical example, use of four pseudoreplicates confined per aquarium and using two true replicates ($n = 2$), yet analyzing only one variable, will provide a 123% observable effect, when in reality, with two true replicates, only 100% should be observed. This phenomenon is exacerbated when a higher number of variables (10 or 50) are being determined (i.e., in

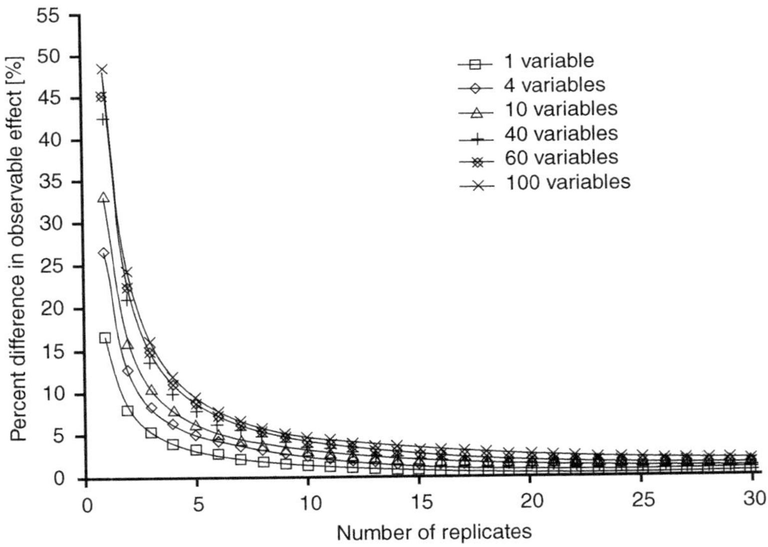

FIGURE 9.2 The percent increase in the size of the observable effect if pseudoreplicates (the fish) are used incorrectly for the statistical test instead of true replicates (the aquaria), whereby the effect size obtained with true replicates is set at 100%. Based on a one-sided Dunnett test with four treatment groups, a control group, groups consisting of >2 true replicates with four pseudoreplicates per true replicate, and a significance level of $\alpha = 5\%$. (Courtesy of Nelly van der Hoeven, EcoΣtat, Leiden, The Netherlands.)

the case of 50 variables the test design suggests 170% more precision and power than is really available). Thus, using pseudoreplicates in exchange for proper replication may provide statistically unsound results (i.e., improper significant differences where no differences are observable). In contrast, increasing the number of true replicates from 5 to 15 in an experimental design demonstrates that the "dubious advantage" of pseudoreplication is nearly lost even when a large number of variables are determined, emphasizing that sufficient true replication will provide for robust statistical analyses in complex experimental designs, whereas this may not necessarily be the case for experiments using pseudoreplication.

9.3.6 Minimum Significant Difference

In most experiments the mean value observed in the control and in a treatment group will differ. If the difference observed, δ^*, is larger than some critical value, known as the minimum significant difference (MSD), this difference is assumed to be significant. The MSD is dependent on the number of observations in the control and treatment groups (n_c and n_t, respectively), the variance observed, and the significance level α of the statistical test. If several treatment concentrations are compared with the same control, the MSD also depends on the number of treatments (k). For Dunnett's test, the MSD can be calculated as

$$\text{MSD} = d_{\alpha,k,\nu} s \sqrt{\frac{1}{n_c} + \frac{1}{n_t}}$$

where $d_{\alpha,k,\nu}$ is the critical value of Dunnett's test (depending on the significance level α, the number of treatments compared with the same control k, the degrees of freedom ν, and s, the residual standard deviation). If the MSD is given as a percentage of the control, the MSD also depends on the control value. From the equation above it follows that:

- The lower the variability of the variable determined, the lower the MSD.
- Increasing the number of replicates in the control will lower the MSD for the comparisons of all treatment groups with the control.
- Increasing the number of replicates in only one of the treatment groups will lower the MSD only for the comparison of this particular treatment group and the control but not for comparisons of other treatment groups with the control.

For example, given an experiment where 10 and 20 replicate fish are used in each of five exposure groups and in the control group, respectively, the only variable of interest is wet weight, the average wet weight observed in the control is 2.4 g, and the residual standard deviation of all groups together is 1.02 g. The significance level α is set at 5%.

[*] Note that δ represents the "observed" difference between measured (observed) mean values (e.g., control and exposure groups.)

According to the equation and with $n_c = 10$ and $n_t = 10$, the critical value for of the Dunnett test $d_{\alpha=0.05,k=5,\nu=54} = 2.212$ and $s = 1.02$, the resulting MSD = 1.01; conversely, with a control group n_c of 20, $d_{\alpha=0.05,k=5,\nu=64} = 2.245$ and the resulting MSD = 0.89.

9.4 STATISTICAL TESTS AND TESTING SITUATIONS ENCOUNTERED ROUTINELY

9.4.1 Significance Level of Multiple Comparisons: Variable-wise Significance

In an experiment where a number of different concentrations are tested, the variable significance level per comparison must be calculated by splitting the variable significance level over the number of concentrations that were tested. This is necessary to keep the simultaneous probability of wrongly concluding that an effect had occurred (type I error) at a certain concentration below a chosen significance level α ($\alpha = 5\%$, the simultaneous significance level). Splitting the significance level per concentration comparison over the number of concentrations tested can be done by variations of the Bonferroni method, of which the Dunn–Šidák method [71,72] is the least conservative. Using this method, the significance level per comparison is $\alpha = 1 - (1 - \alpha)^{1/k}$ if the simultaneous significance level is α and k different concentration groups are employed.

9.4.2 Significance Level for Many Different Comparisons: Simultaneous Significance

In cases where a high number of different variables are tested, the significance level per variable comparison must be split over the variables that were tested. This is necessary to keep the simultaneous probability of wrongly concluding that an effect had occurred (type I error) at a certain concentration below a chosen significance level α ($\alpha = 5\%$, the simultaneous significance level). Splitting the significance level per-variable comparison over the variables can be done by variations of the Bonferroni method, of which the Dunn–Šidák method [71,72] is the least conservative. Using this method, the significance level per comparison is $\alpha' = 1-(1 - \alpha)^{1/m}$ if the simultaneous significance level is α and m different comparisons are made. For each variable, each concentration is compared with the control. Many significance tests (e.g., Dunnett, Williams, and nonparametric trend tests) use the simultaneous significance level per variable instead of the concentration-wise significance level for the comparison of one concentration with the control of that variable. However, when a pairwise comparison is used [e.g., when using the Student t-test or the Wilcoxon test (Table 9.1)], the variable-wise significance level should be split over all pairwise comparisons. A quick alternative is first to test the highest concentration at the variable-wise significance level. Only if the effect at the highest concentration is significant is the next-highest concentration tested at the variable-wise significance level. This is continued with subsequent concentrations until the first nonsignificant concentration is reached [73].

In cases where one group of variables becomes predominant (e.g., histopathological lesions in multiple organs and tissues), it is preferable not to lower the significance level for variables in one group (e.g., life-history variables) by increasing the number of variables in another group (e.g., histopathological lesions). The latter can be avoided by grouping the variables into m groups, followed by splitting the simultaneous significance level α over these m groups. Within each group the significance level is then equally divided over the m_n variables within that group (e.g., six different lesion types in the liver).

9.4.3 Multivariate Tests

An alternative to splitting the significance level over all variables is to perform a multivariate test. The use of redundancy analysis (RDA), as described by ter Braak [74], allows finding the linear combination of the variables observed which differs most between the treatment levels; this method can be used to test whether treatment levels have different effects and allows the determination of those variables most related to the treatment effect. However, the method does not allow determining which of the variables are affected significantly. Combining variables in multivariate testing can best be done for variables measured on a comparable scale (e.g., the severity scale of histopathological lesions), thus emphasizing another advantage of grouping the variables: Within a group, the variables are more related and often measured on a similar scale.

9.4.4 Parametric Versus Nonparametric Testing

In cases where variables are not normally distributed and cannot be transformed so as to fit a normal distribution, nonparametric testing (e.g., Kruskal–Wallis test) can be employed. Similarly, the use of nonparametric tests may be preferable over parametric tests for multiple-variable testing. This is especially valid in cases where the experiment significance level is distributed across the number of variables tested, resulting in a very small significance level per variable. Indeed, parametric testing of variables with a very small significance level per variable may be problematic, as the underlying assumption of normal by distributed data will at best be only approximately correct.

As nonparametric tests have a power comparable to that of parametric tests at high replicate numbers but a considerable smaller power for a very small number of replicates, the biological (variability) and mathematical (data type) understanding of the variables to be tested and thus sufficient replicates in the experimental design appear of utmost importance for any biological experiment aiming at reliable interpretation of treatment-related effects.

REFERENCES

1. Underwood, A.J., Techniques of analysis of variance in experimental marine biology and ecology. *Oceanography and Marine Biology Annual Reviews*, 1981, **19**:513–605.
2. Clarke, K.R. and R.H. Green, Statistical design and analysis for a "biological effects" study. *Marine Ecology*, 1988, **46**:213–226.

3. Chapman, P.F., et al., Improving the quality of statistics in regulatory ecotoxicity tests. *Ecotoxicology*, 1996, **5**(3):169–186.
4. Chapman, P.F., et al., *Asking the Right Questions: Ecotoxicology and Statistics*. SETAC, Royal Holloway University of London, Egaham, Surrey, UK, 1995.
5. van der Hoeven, N., F. Noppert, and A. Leopold, How to measure no effect: Part I. Towards a new measure of chronic toxicity in ecotoxicology. Introduction and workshop results. *Environmetrics*, 1997, **8**(3):241–248.
6. EFSA, Opinion of the scientific panel on plant protection products and their residues on a request from the commission related to the revision of Annexes II and III to Council Directive 91/414/EEC concerning the placing of plant protection products on the market: Ecotoxicological studies. *EFSA Journal*, 2007, **461**:1–44.
7. van der Hoeven, N., How to measure no effect: Part III. Statistical aspects of NOEC, ECx and NEC estimates. *Environmetrics*, 1997, **8**(3):255–261.
8. van der Hoeven N., Power analysis for the NOEC: What is the probability of detecting small toxic effects on three different species using the appropriate standardized test protocols? *Ecotoxicology*, 1998, **7**(6):355.
9. Bernier, N.J., The corticotropin-releasing factor system as a mediator of the appetite-suppressing effects of stress in fish. *General and Comparative Endocrinology*, 2006, **146** (1):45–55.
10. Hoglund, E., P.H. Balm, and S. Winberg, Skin darkening, a potential social signal in subordinate arctic charr (*Salvelinus alpinus*): the regulatory role of brain monoamines and pro-opiomelanocortin-derived peptides. *Journal of Experimental Biology*, 2000, **203** (Pt. 11):1711–1721.
11. Overli, O., C.A. Harris, and S. Winberg, Short-term effects of fights for social dominance and the establishment of dominant-subordinate relationships on brain monoamines and cortisol in rainbow trout. *Brain, Behavior and Evolution*, 1999, **54**(5): 263–275.
12. Overli, O., C. Sorensen, and G.E. Nilsson, Behavioral indicators of stress-coping style in rainbow trout: Do males and females react differently to novelty? *Physiology and Behavior*, 2006, **87**(3):506–512.
13. Sloman, K.A., et al., Plasma cortisol concentrations before and after social stress in rainbow trout and brown trout. *Physiological and Biochemical Zoology*, 2001, **74**(3): 383–389.
14. Bender, N., et al., The relationship between social status, behaviour, growth and steroids in male helpers and breeders of a cooperatively breeding cichlid. *Hormones and Behavior*, 2006, **50**(2):173–182.
15. DiBattista, J.D., et al., The effects of cortisol administration on social status and brain monoaminergic activity in rainbow trout *Oncorhynchus mykiss*. *The Journal of Experimental Biology*, 2005, **208**(Pt. 14):2707–2718.
16. DiBattista, J.D., et al., Growth depression in socially subordinate rainbow trout *Oncorhynchus mykiss*: more than a fasting effect. *Physiological and Biochemical Zoology*, 2006, **79**(4):675–687.
17. Overli, O., et al., Stress coping style predicts aggression and social dominance in rainbow trout. *Hormones and Behavior*, 2004, **45**(4):235–241.
18. White, S.A., T. Nguyen, and R.D. Fernald, Social regulation of gonadotropin-releasing hormone. *Journal of Experimental Biology*, 2002, **205**(Pt. 17):2567–2581.

19. Winberg, S. and O. Lepage, Elevation of brain 5-HT activity, POMC expression, and plasma cortisol in socially subordinate rainbow trout. *American Journal of Physiology*, 1998, **274**(3 Pt. 2):R645–R654.

20. Wingfield, J.C. and R.M. Sapolsky, Reproduction and resistance to stress: when and how. *Journal of Neuroendocrinology*, 2003, **15**(8):711–724.

21. Biosense, L., Carp vitellogenin EIA kit. Bergen, Norway, 2002.

22. Gronen, S., et al., Serum vitellogenin levels and reproductive impairment of male Japanese medaka (*Oryzias latipes*) exposed to 4-*tert*-octylphenol. *Environmental Health Perspectives*, 1999, **107**(5):385–390.

23. Mellanen, P., et al., Expression of the vitellogenin gene in the liver of juvenile whitefish (*Coregonus lavaretus* L. *s.l.*) exposed to effluents from pulp and paper mills. *Ecotoxicology and Environmental Safety*, 1999, **43**:133–137.

24. Monteverdi, G.H. and R.T. Di Giulio, Vitellogenin-associated maternal transfer of exogenous and endogenous ligands in the estuarine fish, *Fundulus heteroclitus. Marine Environmental Research*, 2000, **50**(1–5):191–199.

25. Palace, V.P., et al., Induction of vitellogenin and histologic effects in wild fathead minnows from a lake experimentally treated with the synthetic estrogen, ethynylestradiol. *Water Quality Research Journal of Canada*, 2002, **37**(3):637–650.

26. Korte, J.J., et al., Fathead minnow vitellogenin: Complementary DNA sequence and messenger RNA and protein expression after 17β-estradiol treatment. *Environmental Toxicology and Chemistry*, 2000, **18**(4):972–981.

27. Hahlbeck, E., et al., The juvenile three-spines stickleback (*Gasterostues aculeatus* L.) as a model organism for endocrine disruption: II. Kidney hypertrophy, vitellogenin and spiggin induction. *Aquatic Toxicology*, 2004, **70**(4):311–326.

28. Hemmer, M.J., et al., Vitellogenin mRNA regulation and plasma clearance in male sheepshead minnows, (*Cyprinodon variegatus*) after cessation of exposure to 17 beta-estradiol and p-nonylphenol. *Aquatic Toxicology*, 2002, **58**(1–2):99–112.

29. Van der Ven, L.T.M., et al., Vitellogenin expression in zebrafish *Danio rerio*: Evaluation by histochemistry, immunohistochemistry, and in situ mRNA hybridisation. *Aquatic Toxicology*, 2003, **65**(1):1–11.

30. Berg, A.H., L. Westerlund, and P.E. Olsson, Regulation of Arctic char (*Salvelinus alpinus*) egg shell proteins and vitellogenin during reproduction and in response to 17β-estradiol and cortisol. *General and Comparative Endocrinology*, 2004, **135** (3):276.

31. Schmid, T., et al., Determination of vitellogenin kinetics in male fathead minnows (*Pimephales promelas*). *Toxicology Letters*, 2002, **131**:65–74.

32. Carlson, D.B. and D.E. Williams, Sex-specific vitellogenin production in immature rainbow trout. *Environmental Toxicology and Chemistry*, 1999, **18**(10):2361–2363.

33. Monteverdi, G.H. and R.T. Di Giulio, Vitellogenin association and oocytic accumulation of thyroxine and 3,5,3′-triiodothyronine in gravid *Fundulus heteroclitus. General and Comparative Endocrinology*, 2000, **120**(2):198–211.

34. Folmar, L.C., et al., Vitellogenin-induced pathology in male summer flounder (*Paralichthys dentatus*). *Aquatic Toxicology*, 2001, **51**(4):431–441.

35. Foran, C.M., et al., Influence of exposure history on vitellogenin induction in medaka: a case of 'estrogen memory'? *Marine Environmental Research* 2000, **50** (1–5):196.

36. Mills, L.J. and C. Chichester, Review of evidence: Are endocrine-disrupting chemicals in the aquatic environment impacting fish populations? *Science of the Total Environment*, 2005, **343**(1–3):1–34.
37. Mills, L.J., et al., Relationship between reproductive success and male plasma vitellogenin concetrations in cunner, *Tautogolabrus aspersus*. *Environmental Health Perspectives*, 2003, **111**(1):93–99.
38. Nash, J.P., et al., Long-term exposure to environmental concentrations of the pharmaceutical ethynylestradiol causes reproductive failure in fish. *Environmental Health Perspectives*, 2004, **112**(17):1725–1733.
39. Sohoni, P., P.A. Lefevre, J. Ashby, and J.P. Sumpter, Possible androgenic/anti-androgenic activity of the insecticide fenitrothion. *Journal of Applied Toxicology*, 2001, **21**(3):173–178.
40. Wolf, J.C., et al., Qualitative and quantitative histomorphologic assessment of fathead minnow *Pimephales promelas* gonads as an endpoint for evaluating endocrine-active compounds: A pilot methodology study. *Toxicology Pathologic*, 2004, **32**(5):600–612.
41. Elofsson, U.O., et al., Intermale competition in sexually mature arctic charr: Effects on brain monoamines, endocrine stress responses, sex hormone levels, and behavior. *General and Comparative Endocrinology*, 2000, **118**(3):450–460.
42. Vera Cruz, E.M. and C.L. Brown, The influence of social status on the rate of growth, eye color pattern and insulin-like growth factor-I gene expression in Nile tilapia, *Oreochromis niloticus*. *Hormones and Behavior*, 2007, **51**(5):611–619.
43. Chang, C.-F., et al., The concentrations of plasma sex steroids and gonadal aromatase during controlled sex differentiation in grey mullet, *Mugil cephalus*. *Aquaculture*, 1999, **177**(1–4):37–45.
44. Doyon, C., et al., Corticotropin-releasing factor and neuropeptide Y mRNA levels are modified by glucocorticoids in rainbow trout, *Oncorhynchus mykiss*. *General and Comparative Endocrinology*, 2006, **146**(2):126–135.
45. Summers, C.H., et al., Does serotonin influence aggression? Comparing regional activity before and during social interaction. *Physiology and Biochemistry of Zoology*, 2005, **78**(5):679–694.
46. Lepage, O., et al., Serotonin, but not melatonin, plays a role in shaping dominant-subordinate relationships and aggression in rainbow trout. *Hormones and Behavior*, 2005, **48**(2):233–242.
47. Larson, E.T., et al., Social stress affects circulating melatonin levels in rainbow trout. *General and Comparative Endocrinology*, 2004, **136**(3):322–327.
48. Overli, O., et al., Differences in behaviour between rainbow trout selected for high- and low-stress responsiveness. *Journal of Experimental Biology*, 2002, **205**(Pt. 3):391–395.
49. Hoeger, B., et al., Sex and low-level sampling stress modify the impacts of sewage effluent on the rainbow trout (*Oncorhynchus mykiss*) immune system. *Aquatic Toxicology*, 2005, **73**(1):79–90.
50. Campbell, P.M., T.G. Pottinger, and J.P. Sumpter, Preliminary evidence that chronic confinement stress reduces the quality of gametes produced by brown and rainbow trout. *Aquaculture*, 1994, **120**(1–2):151–169.
51. Pankhurst, N.W., G. van der Kraak, and R.E. Peter, Evidence that the inhibitory effects of stress on reproduction in teleost fish are not mediated by the action of cortistol on ovarian steroidogenesis. *General and Comparative Endocrinology*, 1995, **99**(3):249–257.

52. Hallare, A.V., H.R. Kohler, and R. Triebskorn, Developmental toxicity and stress protein responses in zebrafish embryos after exposure to diclofenac and its solvent, DMSO. *Chemosphere*, 2004, **56**(7):659.

53. McCormick, M.I., Mothers matter: Crowding leads to stressed mothers and smaller offspring in marine fish. *Ecology*, 2006, **87**(5):1104–1109.

54. Hoglund, E., N. Kolm, and S. Winberg, Stress-induced changes in brain serotonergic activity, plasma cortisol and aggressive behavior in Arctic charr (*Salvelinus alpinus*) is counteracted by L-DOPA. *Physiology and Behavior*, 2001, **74**(3):381–389.

55. Overli, O., et al., Brain monoaminergic activity in rainbow trout selected for high and low stress responsiveness. *Brain, Behavior and Evolution*, 2001, **57**(4):214–224.

56. Clement, T.S., et al., Behavioral coping strategies in a cichlid fish: The role of social status and acute stress response in direct and displaced aggression. *Hormones and Behavior*, 2005, **47**(3):336–342.

57. Doyon, C., V.L. Trudeau, and T.W. Moon, Stress elevates corticotropin-releasing factor (CRF) and CRF-binding protein mRNA levels in rainbow trout (*Oncorhynchus mykiss*). *Journal of Endocrinology*, 2005, **186**(1):123–130.

58. van der Ven, L.T.M. et al., Comparative histopathology of zebrafish gonads after disruption of the sex hormone system. *SETAC Europe 12th Annual Meeting*, Vienna, Austria; 2002.

59. van der Ven, L.T.M. and P.W. Wester, Normal histology and effects of hormonal active agents of reproductive organs of the zebrafish (*Danio rerio*). *OECD Technical Workshop on Gonadal Histology of Small Laboratory Fish as a Tool for Endocrine Disruption Screening and Testing*, Bilthoven, The Netherlands, 2002.

60. van der Ven, L.T.M. and P.W. Wester, Toxicological pathology atlas of small laboratory fish with special reference to normal histology and effects of endocrine disruption in the zebrafish (*Danio rerio*). http://www.rivm.nl/fishtoxpat, 2003.

61. van der Ven, L.T.M., P.W. Wester, and J.G. Vos, Histopathology as a tool for the evaluation of endocrine disruption in zebrafish (*Danio rerio*). *Environmental Toxicology and Chemistry*, 2003, **22**(4):908–913.

62. Rees, B.B., F.A. Sudradjat, and J.W. Love, Acclimation to hypoxia increases survival time of zebrafish, *Danio rerio*, during lethal hypoxia. *Journal of Experimental Zoology*, 2001, **289**(4):266–272.

63. van der Hoeven, N., Current issues in statistics and models for ecotoxicological risk assessment. *Acta Biotheoretica*, 2004, **52**(3):201.

64. Hurlbert, S.H., Pseudoreplication and the design of ecological field experiments. *Ecological Monographs*, 1984, **54**(2):187–211.

65. OECD, OECD Final Draft Report of the Initial Work Towards the Validation of the Fish Screening Assay for the Detection of Endocrine Active Substances: Phase 1A. OECD Environmental Health and Safety Publications, Paris, 2006.

66. OECD, OECD Final Draft Report of Phase 1B of the Validation of the 21-Day Fish Assay for the Detection of Endocrine Active Substances. OECD Environmental Health and Safety Publications, Paris, 2006.

67. Mandich, A., et al., In vivo exposure of carp to graded concentrations of bisphenol A. *General and Comparative Endocrinology*, 2007, **153**(1–3):15–24.

68. Pavlidis, M., L. Greenwood, and A.P. Scott, The role of sex ratio on spawning performance and on the free and conjugated sex steroids released into the water by common

dentex (*Dentex dentex*) broodstock. *General and Comparative Endocrinology*, 2004, **138**(3):255–262.

69. Brion, F., et al., Impacts of 17beta-estradiol, including environmentally relevant concentrations, on reproduction after exposure during embryo-larval-, juvenile-, and adult-life stages in zebrafish (*Danio rerio*). *Aquatic Toxicology*, 2004, **68**(3):193–217.
70. Sohoni, P., et al., Reproductive effects of long-term exposure to bisphenol A in the fathead minnow (*Pimephales promelas*). *Environmental Science and Technology*, 2001, **35** (14):2917–2925.
71. Sokal, R.R. and F.J. Rohlf, *Introduction to Biostatistics*, 2nd ed. W.H. Freeman, New York, 1996.
72. Ury, H.K., A comparison of four procedures for multiple comparisons among means (pairwise contrasts) for arbirtary sample sizes. *Technometrics*, 1976, **18**:89–97.
73. Tamhane, A.C., et al., Multiple test procedures for identifying the maximum safe dose. *Journal of the American Statistical Association*, 2001, **96**(455):835–843.
74. Ter Braak, C.J.F., Canonical community ordination: Part I. Basic theory and linear methods. *Ecoscience*, 1994, **1**:127–140.

CHAPTER 10

Conclusions

The presence and continuous contamination of surface waters with hormones and hormonally active compounds called endocrine disrupters (EDCs), primarily via effluents from sewage treatment plants, is an indisputable fact [1–4]. Consequently, it should not be too surprising that in close proximity to the point of sewage treatment plant effluent inflow into the surface waters, effects on the indigenous fish populations are reported [5–19]. A question remains, however, as to whether or not the phenomenon of endocrine disruption is as widespread as is currently being reported [12], or conversely, whether or not this phenomenon is restricted to the site of sewage treatment effluent inflow into the receiving surface waters and thus corresponds to a chronic moderate-to-high dose exposure of local fish populations. Although the term *moderate-to-high* dose may be disputable, especially when one is referring to 17β-estradiol effluents concentrations ranging between <0.1 and 20 ng/L [3,20], a comparison of the effluent concentrations with the plasma levels of estrogens present in developing male fish demonstrates a difference of several orders of magnitude.

In light of this and in conjunction with the vast breadth of literature describing moderate to high dose effects in small fish (e.g., fathead minnows, zebrafish and medaka), one would expect to observe nearly 100% male-to-female conversion in wild fish populations. Although a high percentage of intersex males (with testis–ova) were reported in roach (*Rutilus rutilus*) and gudgeon (*Gobio gobio*) in UK streams receiving sewage treatment effluents [5,10], a good proportion of these intersex males were able to reproduce. Indeed, had 100% sexual reversal or sexual dysfunction taken place, no roach or gudgeon populations could have been maintained and thus found at the locations of highest effluent inflow. Although the latter observation does not question the impact of endocrine disrupters on fish, a comparison of the type and severity of pathological alterations in wild fish with those reported in small laboratory fish (e.g., fathead minnows, zebrafish, and medaka) does demonstrate substantial differences in susceptibility to endocrine disrupters and type of lesions observed, and therefore questions the feasibility of extrapolating findings from laboratory experiments for predicting potential effects in the wild. Hopefully, this book has demonstrated that despite a vast breadth of literature on the reproductive and developmental biology, physiology, endocrinology, and effects of endocrine disrupters in the best studied small fish (medaka, fathead minnows, zebrafish, and other fish species), current knowledge and understanding are still too fraught with weaknesses to allow a thorough understanding of the underlying mechanisms of action of many of the endocrine disrupters.

In this context, the book highlights the high degree of care and planning to which scientific studies must be subjected to allow a mechanistic interpretation of the findings [21]. Indeed, the present literature clearly demonstrates that gonadal histopathology should not be assessed without concurrent pathological analysis of the whole fish or at least of the most important organs [22–27]. Furthermore, the often large biological variability among a cohort of fish stringently dictates a thorough understanding of the biology of the fish species in question, including seasonal [28], geomorphological influences and parasitism-mediated effects [29], the specific histopathological and physiological parameters addressed, and consequently, a sufficient number of control and exposed animals to allow a proper assessment of an endocrine effect in the laboratory or the field [30–39]. In addition, an analysis of the methodology employed to allow histopathological assessment of endocrine disrupters clearly demonstrated that there is still a great need for more concordance in the approaches taken. It appears critical that some of the "typical" endocrine effects for some compounds and in some species, (e.g., the zebrafish) have been reported to occur at higher concentrations than overt cell toxicity [23], thus undermining the classification of the given compound as an endocrine disrupter and hampering environmental risk extrapolation. The latter example strongly demonstrates that the reduction of histopathological assessment to only a few parameters and organs may result in inadequacies in the study results and, more problematic, in seriously flawed interpretation of the data [21].

In conclusion, the present review of the literature currently available demonstrates that despite a fairly solid understanding of fish reproductive and developmental biology and anatomy/histology, there is still insufficient knowledge to allow a thorough understanding of the impact of endocrine disrupters at the mechanistic level. Carefully planned and carried out within a holistic appraisal of the whole fish and its environment, histopathology can be an excellent tool to further our mechanistic understanding of endocrine disruption and thus allow for a better assessment of the "risks" to the environment (i.e., fish populations).

Finally, the currently available studies on EDC-induced effects on gonadal development and on gametogenic and support cells suggest an interaction of EDCs at several levels within endocrine physiology of fish:

- Direct interference via positive or negative feedback signaling in steroid-producing cells (testis and ovary)
- Direct effects on the viability and functionality of steroid-producing cells in the testis or ovary
- Direct interference with expression levels or function of enzymes involved in steroid synthesis (e.g., Δ^5-3β-hydroxysteroid dehydrogenase, aromatase) or steroid product processing (vitellogenin processing with the follicle)

- Direct interference with expression levels of vitellogenin (liver)
- Direct effects on gonadal duct cells (reproductive tract)
- Direct and indirect interference within the positive and negative feedback signaling of the hypothalamic–pituitary–gonadal axis, with subsequent effects on gonad development and gametogenesis

All of the above are observed primarily at fairly high concentrations of the respective EDCs and often employ only a limited number of endpoints (parameters). As evaluation of only a single endpoint or a limited number of endpoints can easily produce under- or overestimation of the actual hazard, most presently available laboratory and field studies on the effects of EDCs in fish do not lend themselves readily to the determination of risk for a specific fish population or general environmental risk assessment. However, given sufficient knowledge and care in the selection of fish species and endpoints to be assessed, concomitant confirmation of actual exposure concentrations by adequate analytical methods, and a profound understanding of the statistical prerequisites of the experimental design, it is hoped that more robust data sets will become available in the near future.

REFERENCES

1. Kolpin, D.W., et al., Pharmaceuticals, hormones, and other organic wastewater contaminants in U.S. streams, 1999–2000: A national reconnaissance. *Environmental Science and Technology*, 2002, **36**(6):1202–1211.
2. Desbrow, C., et al., Identification of estrogenic chemicals in STW effluent: 1. Chemical fractionation and in vitro biological screening. *Environmental Science and Technology*, 1998, **32**(11):1549–1558.
3. Larsson, D.G.J., et al., Ethinyloestradiol: An undesired fish contraceptive? *Aquatic Toxicology*, 1999, **45**(2–3):91–97.
4. O'Brien, E. and D.R. Dietrich, Hindsight rather than foresight: Reality versus the EU draft guideline on pharmaceuticals in the environment. *Trends in Biotechnology*, 2004, **22**(7):326–330.
5. Jobling, S., et al., Altered sexual maturation and gamete production in wild roach (*Rutilus rutilus*) living in rivers that receive treated sewage effluents. *Biology of Reproduction*, 2002, **66**:272–281.
6. Janz, D.M., et al., Recovery of ovary size, follicle cell apoptosis, and HSP70 expression in fish exposed to bleached pulp mill effluent. *Canadian Journal of Fisheries and Aquatic Sciences*, 2001, **58**(3):620–625.
7. Rodgers-Gray, T.P., et al., Exposure of juvenile roach (*Rutilus rutilus*) to treated sewage effluent induces dose-dependent and persistent disruption in gonadal duct development. *Environmental Science and Technology*, 2001, **35**(3):462–470.
8. Sumpter, J.P. and S. Jobling, Vitellogenesis as a biomarker for oestrogenic contaminants in the aquatic environment. *Environmental Health Perspectives*, 1995, **103**(Suppl. 7): 173–178.

9. Taylor, M.R., et al., A research strategy for investigating the ecological significance of endocrine disruption: Report of a UK workshop. *Science of the Total Environment*, 1999, **233**(1–3):181–191.
10. van Aerle, R., et al., Sexual disruption in a second species of wild cyprinid fish (the gudgeon, *Gobio gobio*) in United Kingdom freshwaters. *Environmental Toxicology and Chemistry*, 2001, **20**(12):2841–2847.
11. Vos, J.G., et al., Health effects of endocrine-disrupting chemicals on wildlife, with special reference to the European situation. *Critical Reviews in Toxicology*, 2000, **30**(1):71–133.
12. Jobling, S., et al., Widespread sexual disruption in wild fish. *Environmental Science and Technology*, 1998, **32**:2498–2506.
13. Janz, D.M., et al., Elevated ovarian follicular apoptosis and heat shock protein-70 expression in white sucker exposed to bleachad kraft pulp mill effluent. *Toxicology and Applied Pharmacology*, 1997, **147**(2):391–398.
14. Larsson, D.G.J., H. Hällman, and L. Förlin, More male embryos near a pulp mill. *Environmental Toxicology and Chemistry*, 2000, **19**(12):2911–2917.
15. Ellis, R.J., et al., In vivo and in vitro assessment of the androgenic-potential of a pulp and paper mill effluent. *Environmental Toxicology and Chemistry*, 2003, **22**(7):786–798.
16. Denton, T.E., et al., Masculinisation of female mosquitofish by exposure to plant sterols and *Mycobacterium smegmatis*. *Bulletin for Environmental Contamination and Toxicology*, 1985, **35**(5):627–632.
17. Damstra, T., et al., eds., *Global Assessment of the State-of-the-Science of Endocrine Disruptors*. WHO/PCS/EDC/02.2. World Health Organisation, Geneva, 2002.
18. Cooper, R.L. and R.J. Kavlock, Endocrine disruptors and reproductive development: A weight-of-evidence overview. *Journal of Endocrinology*, 1997, **152**(2):159–166.
19. Zhou, T., et al., Endocrine disruption: Thyroid dysfunction in mummichogs (*Fundulus heteroclitus*) from a polluted habitat. *Marine Environmental Research*, 2000, **50**(1–5): 393–397.
20. Robinson, C.D., et al., Effects of sewage effluent and ethynyl oestradiol upon molecular markers of oestrogenic exposure, maturation and reproductive success in the sand goby (*Pomatoschistus minutus*, Pallas). *Aquatic Toxicology*, 2003, **62**(2):119–134.
21. van der Ven, L.T., et al., Effects of the estrogen agonist 17beta-estradiol and antagonist tamoxifen in a partial life-cycle assay with zebrafish (*Danio rerio*). *Environmental Toxicology and Chemistry*, 2007, **26**(1):92–99.
22. van der Ven, L.T.M. P.W. Wester, and J.G. Vos, Histopathology as a tool for the evaluation of endocrine disruption in zebrafish (*Danio rerio*). *Environmental Toxicology and Chemistry*, 2003, **22**(4):908–913.
23. Weber, L.P., R.L. Hill, Jr., and D.M. Janz, Developmental estrogenic exposure in zebrafish (*Danio rerio*): II. Histological evaluation of gametogenesis and organ toxicity. *Aquatic Toxicology*, 2003, **63**(4):431–446.
24. Wester, P.W. and J.H. Canton, Histopathological effects in *Poecilia reticulata* (guppy) exposed to methyl mercury chloride. *Toxicologic Pathology*, 1992, **20**(1):81–92.
25. Wester, P.W. and J.H. Canton, The usefulness of histopathology in aquatic toxicity studies. *Comparative Biochemistry and Physiology, Part C: Toxicology and Pharmacology*, 1991, **100**(1–2):115–117.

26. Wester, P.W. and J.G. Vos, Toxicological pathology in laboratory fish: An evaluation with two species and various environmental contaminants. *Ecotoxicology*, 1994, **2**(1):21–44.
27. Wester, P.W., et al., Aquatic toxicology: Opportunities for enhancement through histopathology. *Environmental Toxicology and Pharmacology*, 2002, **11**(3–4):289–295.
28. Koger, C.S., S.J. Teh, and D.E. Hinton, Variations of light and temperature regimes and resulting effects on reproductive parameters in medaka (*Oryzias latipes*). *Biology of Reproduction*, 1999, **61**:1287–1293.
29. Arme, C., *Ligula intestinalis*, a tape worm contraceptive. *Biologist*, 2002, **49**(6):265–269.
30. Fent, K., B.R. Woodin, and J.J. Stegeman, Effects of triphenyltin and other organotins on hepatic monooxygenase system in fish. *Comparative Biochemistry and Physiology, Part C: Toxicology and Pharmacology*, 1998, **121**(1–2):277–288.
31. McAllister, B.G. and D.E. Kime, Early life exposure to environmental levels of the aromatase inhibitor tributyltin causes masculinisation and irreversible sperm damage in zebrafish (*Danio rerio*). *Aquatic Toxicology*, 2003, **65**(3):309–316.
32. Czech, P., K. Weber, and D.R. Dietrich, Effects of endocrine modulating substances on reproduction in the hermaphroditic snail *Lymnaea stagnalis L. Aquatic Toxicology*, 2001, **53**(2):103–114.
33. Schulte-Oehlmann, U., et al., Effects of endocrine disruptors on prosobranch snails (Mollusca: Gastropoda) in the laboratory: Part II. Triphenyltin as a xenoandrogen. *Ecotoxicology*, 2000, **9**(6):399–412.
34. Kuiper, R., et al., Toxicity of tetrabromobisphenol A (TBBPA) in zebrafish (*Danio rerio*) in a partial life-cycle test. *Archives of Toxicology*, 2007, **81**(1):1.

Appendix: Fish Preparation and Microdissection of Organs

A.1 FISH PREPARATION

- After collection and euthanization [in at least 0.4% tricaine methanesulfonate (MS-222)], fish are weighted and measured (distance in millimeters from the snout to the base of the caudal fin), and gender (based on externally visible secondary sex characteristics) is recorded.
- All unusual external features (e.g., lordosis, kyphosis, scoliosis), neoplastic changes, infectious disease, and other abnormalities are recorded.
- After examination of the caudal and the posterior region of a fish for obvious lesions, the tail is removed by an angular cut beginning near the anal vent and proceeding anteriorly to the dorsal side of the fish (adjust the length of the carcass to fit into the tissue cassette).
- Using irridectomy scissors, a midventral slit is made from the anal vent to the operculum.
- To ensure similar sectioning of each fish of a study, the plane of sectioning must be determined before gross trimming.
- Transverse or longitudinal sections of the whole fish allow the assessment of nearly all organs and tissues *in situ*.
- Exact positioning of the fish is very important to get representative areas of gonads, this pertains even more to midbody transverse sections of fathead minnows.
- Qualitative staging of gonads is possible, but quantification of cell types is not, unless numerous sections and multiple levels of tissue sections or blocks are employed.
- If whole fish are microtomed transversely instead of longitudinally, a far greater number of sections would have to be generated and examined to ensure adequate spatial representation of the gonads.

Histological Analysis of Endocrine-Disruptive Effects in Small Laboratory Fish
By Daniel R. Dietrich and Heiko O. Krieger

- Longitudinal (sagittal and parasagittal) or transverse whole-fish sectioning might be more appropriate for gonad cell quantification in smaller fishes such as Japanese medaka and zebrafish. For larger fish (e.g., fathead minnow), whole-fish dissection is considered less ideal.
- Even if the head and the tail are removed, standard plastic tissue cassettes are not roomy enough to accommodate the trunk segments of large adult fathead minnow males if positioned longitudinally.
- The fathead minnow gonads are elongated tubular structures that are oriented parallel to the axis of the fish; therefore, the entire length of a gonad cannot be recovered consistently on one slide using longitudinal whole-body sectioning.
- It is difficult to identify individual gonads accurately as right or left when longitudinal whole-body sectioning is performed.

A.2 MICRODISSECTION OF ORGANS

- The fish (e.g., fathead minnow) is placed in right lateral recumbency on the stage of a dissecting microscope.
- Fine dissecting instruments are used to remove the left body wall and to remove the gonads by severing their mesenteric attachments.
- Grossly evident external or internal lesions should be noted (e.g., on an individual fish necropsy form).
- Gonad excision avoids microtoming artifacts that can be caused by the presence of firmer tissues (such as bone and cartilage) in the whole-body sections.
- Additionally, gonad excision allows these organs to be visualized macroscopically *in situ*, and it is possible to obtain gonad weights if desired (e.g., for calculation of the gonadosomsatic index).
- Using the gonadal microdissection technique, later steps of fixation and preservation of gonads can be optimized.
- Problem: other organs must be preserved separately at the same time to enable an overall pathological assessment of the fish.

The gonads or the whole fishes are then placed into compartments of a plastic screen pre-stamped tissue cassette prior to immersion with fixation solution.

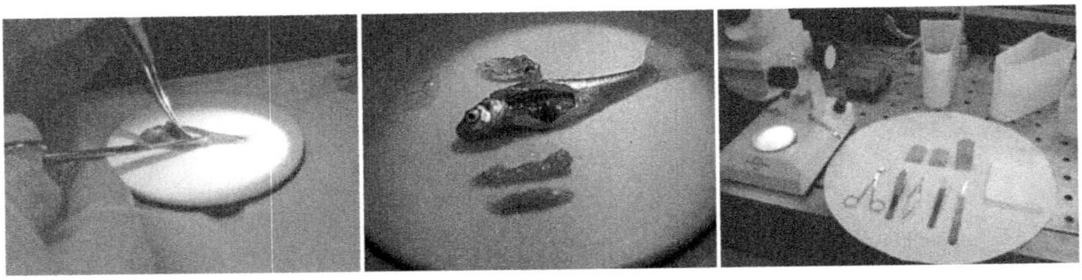

A.3 TISSUE FIXATION

More in-depth information can be found in reviews by Hopwood [1] and Latendresse et al. [2] and books by Humason [3], Gabe [4], and Pearse [5].

- *Note*: "No one fixative is ideal for all situations" [1].
- *However*: Any fixative should:
  - Penetrate rapidly to prevent postmortem changes.
  - Coagulate cell contents into insoluble substances.
  - Protect tissue against shrinkage and distortion during dehydration, embedding, and sectioning.
  - Allow cell parts, selectively, to be made clearly visible.
- A single chemical seldom has all the qualities of a good fixative; thus a reliable fixing solution is composed of a combination of chemicals: one or more coagulant chemicals in combination with one or more noncoagulant chemicals.
- Coagulants, used alone, may form too coarse a network for the best cytological detail or may induce the formation of artefacts.
- Noncoagulants produce fewer artifacts, but if used alone they give the tissue a poor consistency for embedding.
- Rapid fixation is essential to prevent postmortem changes and non-exposure-related artifacts.
- Commonly used fixation solutions: neutral buffered formalin, Bouin's solution, Davidson's solution, or glutaraldehyde as described by Latendresse and co-workers [2].
- If samples are designated to be shipped, it is recommended that glutaraldehyde-fixed organs be refrigerated at 4°C, or samples fixed with Bouin's or Davidson's fixative be placed in 70% ethanol.
- First consideration in the choice of a fixative is the purpose of the study and the future use of the tissue (routine all-purpose fixative adequate, or preservation of special parts of a cell?).
- An extremely dense tissue needs a fixative that penetrates easily and quickly.
- The pieces to be fixed should be as small as possible: the smaller the bulk, the better the fixation.
- Use a large volume of fixing fluid, at least 10 times the volume of the tissue.
- If there is any doubt concerning the future needs for a tissue, fix it in formalin; this can be followed by post-fixation treatments.
- Prolonged action in this case is considered to be rarely harmful, due to the progressive action of formalin.

A.3.1 Neutral Buffered Formalin

- Composition as described by Gabe [4] and Pearse [5]:

formaldehyde (37–40%)	100 mL
distilled water	900 mL
sodium phosphate monobasic ($NaH_2PO_4 \cdot H_2O$)	4.0 g
sodium phosphate dibasic (anhydrous) (Na_2HPO_4)	6.5 g

- Sample may remain in fixative indefinitely, due to its progressive action; wash in water to remove excess formalin.
- Formalin, when reacting with proteins, appears to form links between adjacent protein chains.
- The quantity of irreversibly bound protein drops as the pH of the solution rises above pH 10; the most efficient reaction takes place at pH 7.5 to 8.0.
- Formalin has a moderate speed of penetration, but its action is low.
- Formalin is a good fixative for lipids: it does not dissolve lipids or fats.
- Formalin does not fix soluble carbohydrates, and it does dissolve some glycogen and urea.
- Formalin fixation alone is considered to be "one of the poorest preparations for paraffin-embedding" [4].
- Formalin fixation was reported to be incomplete even after 7 days of fixation process [5].
- Formalin fixation can induce cellular shrinking and membrane wrinkling and thus is not recommended for gonadal tissues of small fish.

A.3.2 Bouin's Fixative

- Composition:

picric acid, saturated aqueous	750 mL
formaldehyde (37–40%)	250 mL
glacial acetic acid	50 mL

- Mode of action:
  - Formaldehyde fixes the cytoplasm, but retards paraffin penetration. It fixes chromatin poorly and makes cytoplasm basophilic.
  - Picric acid coagulates cytoplasm so that it admits paraffin, leaves the tissue soft, fixes chromatin, and makes the cytoplasm acidophilic. It shrinks tissue and makes chromatin acidophilic.
  - Acetic acid compensates for the defects of both formaldehyde and picric acid.
- Bouin's fixative provides a minimum of artifacts and is capable of decalcification of bony structures (whole-fish fixation).
- Fix for 4 to 48 hours (less than 24 hours is optimal); change to 70% ethanol after fixation. Longer fixation makes tissues brittle and difficult to section. (*Note*: Long-term storage in 70% ethanol may lead to tissue shrinkage.)

- Several weeks causes no damage, but long periods (months) result in poor nuclear staining.
- Due to the acetic acid content, do not use Bouin's for cytoplasmatic inclusions.
- Large vacuoles often form in tissues.
- The yellow color must disappear before sections are stained; the color is usually removed in the alcohol series, but if not, treat slides in 70% alcohol plus a few drops of saturated lithium carbonate until the yellow disappears.
- It is felt, that tissue fixed with Bouin's and washed with water loses some of its soluble picrates and therefore should be transferred from the fixative directly in 50% alcohol.

A.3.3 Lillie's Fixative

- Lillie's fixative is closely related to Bouin's fixative, with acetic acid being replaced by formic acid.
- Lillie's provides better decalcification properties than Davidson's.
- Composition:

picric acid, 1–2% saturated aqueous	850 mL
formaldehyde (37–40%)	100 mL
formic acid, 90–95% aqueous	50 mL

- The fixative is used for 1 to 2 days.
- To extract some of the yellow color from the picric acid, place in 70 to 80% alcohol for 2 to 3 days.

A.3.4 Davidson's Fixative

- Although Bouin's fixative greatly improves the preservation of nuclear and cellular details, picric acid presents a health and safety hazard for technicians and an environmental problem regarding waste disposal. Davidson's fluid lacks picric acid and also provides good preservation of cellular details.
- The ratio of fixative to tissue volume should be at least 10 : 1 to ensure good fixation.
- Fix for 24 hours; store in 70% ethanol.
- Composition:

formaldehyde (37–40%)	200 mL
glycerol	100 mL
glacial acetic acid	100 mL
absolute ethanol	300 mL

- Davidson's solution is an excellent choice for preserving the structure of the tissues. In addition, tissue sections fixed in Davidson's solution can later be stained by different histochemical methods, as well as *in situ* hybridization with DNA probes.

- Composition (according to the Animal and Plant Health Inspection Service):

95% ethyl alcohol	330 mL
100% formalin	220 mL
(a saturated 37–39% aqueous solution of formaldehyde gas)	
glacial acetic acid*	115 mL
tap water (for marine animals, seawater may be substituted)	335 mL

A.3.5 Dietrich's Fixative

- Dietrich's fixative provides for good cellular detail with little shrinkage or distortion of tissue.
- Specimens can be stored in Dietrich's for two to three months without hardening the tissue.
- Tissues stored in Dietrich's for a longer period must be washed in running water for 24 hours prevent the formation of black formalin precipitates.
- Composition:

distilled water	750 mL
95% ethanol	375 mL
formaldehyde (37–40%)	125 mL
glacial acetic acid	25 mL

A.3.6 Glutaraldehyde

- Composition (2%–6% solution):

50% solution of glutaraldehyde	4–12 mL
0.1–0.2 M phosphate buffer	100 mL

- The pH of the solution should be 7.4.
- If shrinkage occurs, use the 2% solution.
- 1% sucrose can be added to the glutaraldehyde solution to improve tissue preservation; store the solution in the refrigerator.
- Glutaraldehyde-fixed tissue is frequently post-fixed in 1% osmic acid buffered with 0.1 M phosphate buffer and glucose or veronal–acetate–sucrose after cacodylate-buffered glutaraldehyde.
- Cacodylate buffer (pH 7.2) composition:

1 M cacodylic acid (137.99 g/1000 mL)	54.6 mL
1 N NaOH (40 g/1000 mL)	50.0 mL
distilled water	895.4 mL

*Do not substitute other acids, such as HCl, for acetic acid. Histological sections prepared from HCl–Davidson's solution are not suitable for routine hematoxylin and eosin histological staining.

- Physiologic buffer:

solution A: 2.26% $NaH_2PO_4{\cdot}H_2O$	41.5 mL
solution B: 2.52% NaOH	8.5 mL
solution C: 5.4% glucose	
solution D: isotonic disodium phosphate	

Adjust to pH 7.4 - 7.6 with solution B.

- Phosphate-buffered fixative composition:

solution D	45.0 mL
osmic acid	0.5 g
solution C	5.0 ml

- This solution is stable for several weeks at 4°C.
- Wash tissue in buffer several hours or overnight.
- Maximal stability of glutaraldehyde is at concentrations between 2 and 10%.
- Low temperatures and storage in the dark are advisable.

A.4 EMBEDDING

- Tissues fixed in aqueous solutions maintain a high water content; therefore (except in special cases, i.e., freezing, water-soluble waxes, special cell contents), the tissue must be dehydrated before further processing.
- During fixing and washing, tissues lack the ideal consistency for sectioning (i.e., they may be soft, contain hollow spaces and can be deformed easily by sectioning, or the fluid content of cells could be released while cutting, causing them to collapse.
- To preclude these problems, the fluids must be replaced by a medium that hardens and thus provides for a firm, easily sectionable material.
- The cells are filled intracellularly (impregnation or infiltration) and enclosed extracellularly (embedding) with the medium, thereby being protected during subsequent handling or storage.

A.4.1 Paraffin

- Paraffin is suitable for most histological purposes when thin sections (1 to 15 μm) are desired, serial sections are necessary, and sections need to be prepared in a short time.
- Before embedding can take place, all water has to be removed from the tissue.
- Dehydration is normally achieved using a series of solutions of ethanol in water, with gradually increasing percentages of alcohol (30%, 50%, 70%, 80%, 95%, and absolute ethanol).
- Shrinkage of paraffin embedded tissue can amount to as much as 20%.

- Time required for each step depends on the size (volume) of the tissue: 30 minutes to 2 hours (up to 3 hours for large, dense tissue types); a second change of 100% alcohol should be included to ensure complete removal of water.
- Prolonged dehydration with 95% or 100% alcohol tends to harden the material, making it brittle and difficult to section.
- Because alcohol does not mix with paraffin, an important intermediate step in the paraffin-embedding procedure is the clearing step.
- Benzene, toluene, or xylene is commonly used for clearing, with toluene considered to be the safest clearing chemical.
- Exposure time for the toluene step is about 1/2 to 1 hour; repeat this procedure once.
- After clearing, the tissue is normally transferred directly into the embedding medium; after 30 minutes to 1 hour the tissue is placed in fresh embedding medium for the same time period.
- During the procedure, keep the oven temperature just high enough to maintain the paraffin in a melted phase, not higher. This reduces the danger of overheating the tissue, leading to hardening and shrinkage of tissue.
- For soft tissues and relatively thick sections, use "soft paraffin" (i.e., with a melting point of 50 to 55°C) for hard tissues and thinner sections, use "hard paraffin" (i.e., with a melting point of 56 to 68°C); the choice of the suitable paraffin wax may also depend on the environmental temperature in the lab; hot weather will demand the use of a harder paraffin.

  Note: Melted paraffin kept warm in a warm oven for several days or weeks is better for infiltrating and embedding than freshly melted paraffin.
- When infiltrating friable tissues, the following precautionary step must be added: After the tissue is well cleared, begin to saturate the solution with fine shavings of paraffin until some of the paraffin remains undissolved. Leave the tissue in the saturated clearer for 4 to 6 hours (or overnight for large pieces); then use a warm spatula to transfer the tissue to the melted paraffin.
- The use of a vacuum oven for infiltrating will remove air from some tissues (e.g., lungs) and eliminate holes in the final paraffin block.
- As soon as the tissue is thoroughly infiltrated with paraffin, it is ready to be embedded (blocked); the paraffin is allowed to solidify around and within the tissue.
- The tissue is placed in a small container or paper box filled with melted paraffin, which is rapidly cooled in water; transfer the tissue with warmed instruments to prevent congealing of paraffin on metal surfaces.
- Handle the tissue and paraffin as rapidly as possible to prevent the paraffin from solidifying before the tissue is oriented in it.

  Note: Orientation is important!
- If the tissue is placed in a known position and marked carefully (e.g., with a slip of paper in the hardening paraffin), it is easy to determine the proper surface for sectioning.

- While using a paper or foil box, the orienting mark may be made directly on one of its flaps.
- Small amounts of paraffin ready to be solidified can be cooled immediately in water (preferably at a temperature of 10 to 15°C).
- Water colder than 10°C causes the block to contract too strongly, and it may crack.
- Additionally, normal crystals may form in the center, but abnormal crystals may form in the periphery of the block.
- In a perfect block the paraffin crystals are contiguous and the paraffin block appears clear and homogeneous.
- Since paraffin may contain 7 to 15% air dissolved in it (clear if evenly distributed), pockets of air produce milky spots, a condition called *crystallization.*
- Slow hardening in the air or too rapid cooling may cause this effect, particularly in large blocks; quick hardening of the outer surface will trap the air.

Paraffin Timing Schedule

1. Fix overnight or 24 hours (Bouin's, formalin).
2. Wash in water, running if possible: 6 to 8 hours or overnight (Bouin's fixed tissue can be transferred directly to 50 or 70% alcohol without washing).
3. Transfer to 50% ethanol for 1 hour, to 70% ethanol for 1 hour, and to 95% ethanol for 1 to $1\frac{1}{2}$ hour.
4. Transfer to absolute ethanol no. 1 for $\frac{1}{2}$ to 1 hour, and to absolute ethanol no. 2 for $\frac{1}{2}$ to 1 hour (alternatively, use isopropyl alcohol).
5. Transfer to toluene no. 1 for $\frac{1}{2}$ to 1 hour, and to toluene no. 2 for $\frac{1}{2}$ to 1 hour.
6. Transfer to melted paraffin no. 1 for $\frac{1}{2}$ to 1 hour, and to melted paraffin no. 2 for $\frac{1}{2}$ to 1 hour.
7. Embed.

A.4.2 Glycolmethacrylate (GMA)

- In comparison to GMA, one major drawback of paraffin embedding is shrinkage: a trout liver preparation subjected to routine paraffin embedding can result in up to 25% shrinkage compared to 1 to 2% shrinkage when embedding in GMA.
- Tissues to be embedded in GMA may be dehydrated through a graded series of aqueous to nonaqueous methacrylate monomer solutions (preserves enzyme activity normally destroyed by alcohol dehydration), or through a graded series of ethanol.
- A problem with GMA used to be the exothermic nature of the polymerization reaction: introduction of several cyclic diketo carbon acids has resolved this.

- GMA embedding allows microtoming at a thickness ranging between 0.5 and 4 μm.

GMA Example Timing Schedule Use glutaraldehyde-fixed tissue.

1. Place tissue in 70% ethanol for 45 minutes, then transfer sample to 80% ethanol for 45 minutes (all steps are at room temperature).
2. Transfer to 95% ethanol for 45 minutes; repeat this step once.
3. Transfer to absolute ethanol for 1 hour; repeat this step once.
4. Put the sample in GMA infiltration solution (prepared according to the manufacturer's instruction) for 2 to 4 hours.
5. Transfer to fresh GMA for 15 to 20 hours.
6. Transfer to fresh GMA for 20 to 24 hours; repeat this step once.
7. Embed.

The schedule may be shortened (depending on the tissue's size and/or consistency) by replacing steps 3 to 7 as follows:

3. Transfer to absolute ethanol for 1 hour.
4. Transfer to GMA for 1 hour; repeat this step once in fresh GMA.
5. Transfer to fresh GMA for 20 to 22 hours.
6. Embed.

For embedding, fill embedding molds just below the rim with GMA embedding solution, place the tissue accurately oriented in a mold, and allow the GMA to set overnight. The GMA should now be hard with the tissue set in it. In conditions of high humidity, store GMA blocks in a desiccator to prevent water uptake.

A.5 TISSUE SECTIONING

- Cells and tissues can be adequately viewed only in thin sections of biological materials, so the next step in the preparation of histological material involves microtomy.
- Steel knives, tungsten carbide knives, glass knives, or specially adapted razor blades, and a rotary microtome are usually used.
- The pre-trimmed (two edges parallel) paraffin (or GMA) block, in its holder, is clamped firmly inside the block carrier of the microtome.
- Before sectioning, the block should be given a final trimming, permitting serial sections through the area of interest without much space being committed to excess paraffin (or GMA); less deformation and wrinkling of sections occur when paraffin (or GMA) margins are small.

- The microtome knife must be held in the clamps at a proper angle for optimal sectioning, producing a minimal amount of compression and allowing the sections to adhere to each other.
- When placed into the microtome, the knife must be tilted just enough so that the cutting facet next to the block clears the surface of the block.
- If the tilt is not sufficient, the surface of the block is pressed down with the wedging effect of the facet, and no section results; the next stroke of the knife will then result in a thick section.
- Too great a tilt of the knife makes it scrape through like a chisel rather than cut through the tissue.
- Scratches on the sections can be remedied by cleaning the knife from fragments of paraffin (using a camel's hair brush or fingertips).

  Note: discard the first section after cleaning the knife—it may be a thick one.
- During cutting, a ribbon of sections will form, but only if the parallel edges of the block are cut clean and parallel.
- As the ribbon forms, hold it away from the knife with a camel's hair brush and ease it forward so that the sections do not remain on the knife.
- The ribbons can be mounted directly on slides, floated on a water bath, or laid in order in a box (especially when the slides cannot be made immediately, or when serial sections have to be cut).
- In dry weather, static is frequently a problem during sectioning: the sections stick to the knife or to parts of the microtome, breaking apart or staying bunched up instead of lying flat.
- To relieve this situation, wear cotton clothes rather than woollen or nylon clothes (especially when using plastic-covered chairs), use a static eliminator (e.g., an alpha-radiating static eliminator strip) or antistatic spray.
- For routine paraffin work, section thicknesses of 4 to 6 μm are common.
- Using embedding media such as glycolmethacrylate, sections as thin as 0.5 to 1 μm can be obtained.
- When adequately stained, the latter provide better resolution than thicker sections cut through paraffin.

A.6 SAMPLE MOUNTING

- After sectioning, samples have to be mounted on (glass) slides.
- Commonly used adhesives for slides are egg albumen, gelatine, and L-lysine, or direct mounting on commercially available sialinated slides.
- To mount sections (held in a water bath) on a slide, dip an albumenized slide under them and with a needle hold them against the slide while removing it from the bath.

- After the excess water drains off (facilitated by touching a piece of cleansing tissue or filter paper to the edge of the slide and around the sections), permit slides to dry overnight on a warm plate or in an oven.

A.7 TISSUE SLIDE STAINING

For the processed tissue slides there are a number of staining methods. Two common procedures are described in detail on the following slides:

- Hematoxylin and eosin staining
- Masson's trichrome staining

There are numerous other stains for special purposes in fish (e.g., staining blood cells, gills, liver, or cell organelles):

- (Perl's) Prussian blue (haemosiderin, lipofuscin, melanin)
- Azan, Mallory Heidenhain, Kornhauser's quad (connective tissue, proteins)
- Giemsa (differential leukocyte counts)
- Lillie's cold Schiff reagent (liver glycogen)
- Oil red O technique (lipids)

A.7.1 Hematoxylin and Eosin

H&E staining is one of the most commonly used histochemical staining procedures:

- Hematoxylin lakes are basic and therefore form salts with acids such as DNA and RNA.
- Because other tissue components are alkaline and will not form salts with a basic dye, a counterstain may be used.
- Eosin is the counterstain for hematoxylin, but strictly speaking, it should be termed a secondary stain because it stains cytoplasmic inclusions specifically.
- Eosin stains cytoplasm pink.

Schedule for Mayer's H&E Staining Procedure

1. Use partly deparaffinized slides (in an oven overnight, 56 to 60°C).
2. Complete deparaffinization and rehydrate the slides:
   a. Place slides in xylene for 2 to 3 minutes.
   b. Transfer slides to fresh xylene for 1 to 2 minutes.
   c. Transfer slides to absolute ethanol for 1 to 2 minutes.

d. Sequentially transfer slides to 95% ethanol, 70% ethanol, and 50% ethanol for each.
e. Place slides in distilled water.
3. Prepare and apply stains:
a. Prepare Mayer's hematoxylin.
b. Prepare eosin solution.
c. Place slides in Mayer's hematoxylin for 15 minutes.
d. Wash slides in running tap water for 20 minutes.
e. Dip slides in the eosin solution for 15 seconds to 2 minutes according to the depth of counterstain desired.
4. Differentiate slides by placing them in two 2-minute changes of 95% ethanol until the excess eosin is removed; verify the latter under the microscope (sections should have a clear background and slightly pink to orange cytoplasm).
5. Transfer slides at once to plain 95% ethanol, then transfer them to two 2-minute changes of absolute alcohol. Clear slides in two 2-minute changes of xylene and apply mounting medium (e.g., Permount or Histoclad); finally, place a coverslip on each.
6. In sections properly stained with Mayer's H&E, nuclei are blue, and regions of cytoplasm with extensive rough endoplasmatic reticulum show a wispy blue to purple hue. The remainder of cytoplasm is various shade of pink.

Alternative Schedule for the H&E Staining Procedure

1. Wash slides in running tap water.
2. Rinse well in distilled water.
3. Put slides in Mayer's hematoxylin for 30 seconds to 2 minutes.
4. Rinse in running tap water to remove excess stain and place in lukewarm water to "blue" sections (if placed too long in water, the chlorine will bleach the hematoxylin).
5. Rinse in distilled water.
6. Put slides in 50% ethanol, than in 70% ethanol for 1 to 2 minutes, respectively.
7. Dip slides five times (30 seconds) in 0.5% alcoholic Eosin or Eosin and Phloxine B, slightly acidified with glacial acetic acid.
8. Put slides in 95% ethanol, three changes for 30 seconds each. Agitate slides during the procedure to remove excess eosin; otherwise, the overlay of eosin gives the sections a muddy appearance.
9. Dehydrate slides in absolute ethanol for 1 to 2 minutes; repeat in fresh ethanol.
10. Agitate slides in ethanol–xylene for 30 seconds. If sections look milky, dehydration is not complete; take back to clean absolute ethanol and return to fresh ethanol–xylene.

11. Put slides in xylene through several changes, until sections are clear.
12. Mount.

COMPOSITION OF MAYER'S HEMATOXYLIN

hematotoxylin crystals	1.0 g
distilled water	1.0 L
sodium iodate	0.2 g
citric acid	1.0 g
chloral hydrate	50.0 g

COMPOSITION OF EOSIN AND PHLOXINE B

1% eosin Y	100 mL
1% phloxine B	10 mL
95% ethanol	780 mL
acetic acid	4 mL

COMPOSITION OF ALCOHOLIC EOSIN

Stock solution:

eosin Y	1.0 g
distilled water	20 mL
95% EtOH	80 mL

Preparation of the working solution (1%): Add 1 part stock solution to 3 parts 80% ethanol. Just before use, add 0.5 mL of glacial acetic acid per 100 mL stain and stir.

A.7.2 Masson's Trichrome Stain

Masson's trichrome stain technique may be generally applied to tissues. It is used primarily to demonstrate fibrin or young collagen fibers, and to differentiate muscle from connective tissue. Bouin's is the preferred tissue fixative, but tissues fixed in 10% neutral buffered formalin may also be used.

1. Slide preparation:
   a. Deparaffinize and hydrate sections (see Staining–Hematoxylin & Eosin)
   b. Prepare Bouin's (see Bouin's fixative)
   c. Place slides in Bouin's, cover, and leave overnight at room temperature.
   d. Wash slides in running tap water until sections are colorless.
2. Prepare and apply stains:
   a. Prepare an iron hematoxylin solution (see below).
   b. Prepare a ponceau–acid fuchsin solution (see below).

c. Stain nuclei by placing slides in iron hematoxylin for 10 minutes. Wash slides in running tap water for 10 minutes and rinse in distilled water for 1 to 2 minutes.

d. Place slides in Ponceau–acid fuchsin solution for 2 to 3 minutes.

e. Wash slides in water and differentiate sections in a 1% aqueous phosphomolybdic acid solution for 5 to 15minutes at room temperature.

f. Wash well in water.

3. Counterstain sections with a working light green solution for 1minute (stock: 2% solution of light green in acetic acid; working solution is a 1 : 10 dilution of the stock with distilled water).

4. Wash slides in distilled water, then dehydrate, clear, and mount the sections. In well-stained slides, nuclei are blue-black. Muscle, fibrin, and red blood cells are red. Connective tissue is green.

COMPOSITION OF THE IRON HEMATOXYLIN SOLUTION

Solution A	Solution B
hematotoxylin crystals 1.0 g	aluminum chloride 10 g
95% ethanol 100 mL	ferrous sulfate 10 g
	distilled water 100 mL

COMPOSITION OF PONCEAU—ACID FUCHSIN SOLUTION

Mix equal volumes of 0.5% solution of Ponceau 2R in 1% acetic acid and a 0.5% solution of acid fuchsin in 1% acetic acid.

A.8 FINAL PROCESSING

- Final processing of paraffin sections on slides to make permanent preparations for storage and subsequent examination involves removal of all water and alcohol.
- Additionally, a medium must be applied that will:
  - Maintain tissues in a clear and transparent condition.
  - Not alter color or intensity of stains.
  - Hold the cover glass (if used) in place.
- Water is removed bypassing the slides through successively increasing concentrations of ethanol until absolute ethanol is reached (as done for preparation of slides for paraffin infiltration/embedding).
- The final reagent in the dehydration procedure is xylene (or a similar solvent) that removes alcohol and renders the sections transparent.
- Finally, a mounting medium is applied and coverslips (if used) are lowered into place, covering the sections completely.
- The solvent for most mounting media is xylene or toluene.
- Of the resinous mounting media, synthetic resins are superior to natural resins in most respects; their composition can be controlled, they are stable and inert, and they dissolve readily in xylene or toluene.

Note: The refractive index of the dried mounting media must be approximately that of dried protein (i.e., 1.53 to 1.54) for maximum transparency of stained tissues.

A.9 EXAMPLES

Hematoxylin and Eosin Staining Example: Female Fathead Minnow

Fathead minnow ovaries with varying fixation and embedding combinations all stained with hematoxylin and eosin stain (approx. 40×; bar size = 25 μm).

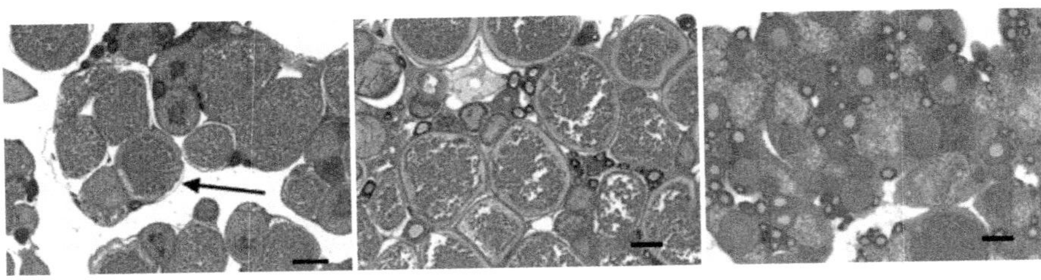

Ovaries fixed with **formalin**, embedded in **paraffin**

Ovaries fixed with **Bouin's**, embedded in **paraffin**

Ovaries fixed with **glutaraldehyde**, embedded in **GMA**

Hematoxylin and Eosin Staining Example: Male Fathead Minnow

Fathead minnow testes with varying fixation and embedding combinations all stained with hematoxylin and eosin stain (approx. 40×; bar size = 25 μm).

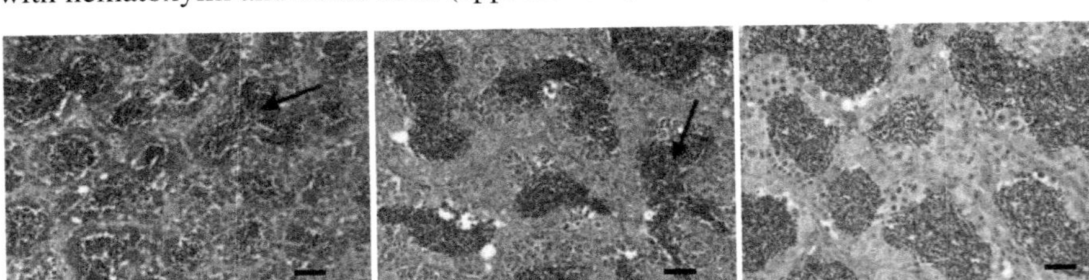

Testes fixed with **formalin**, embedded in **paraffin**. Arrow: relatively "thick" section (3 to 5 μm); note the poor demarcation of the individual cells, due to formalin fixation, enhancing the cellular shrinkage.

Testes fixed with **Bouin's**, embedded in **paraffin**. Arrow: relatively "thick" section (3 to 5 μm); demarcation of cells is improved, yet a lack of marked differential staining of individual cell types still does not provide for optimal cell type recognition.

Testes fixed with **glutaraldehyde**, embedded in **glycolmethacrylate** (**GMA**); note the optimal resolution due to lack of overlapping cell layers of the section due to thin sectioning (0.5 to 2 μm) and the excellent demarcation and staining of the individual cells.

All pictures courtesy of J. Wolf, EPL Inc., Sterling, Virginia.

Masson's Trichrome Staining Example: Female Fathead Minnow

Fathead minnow ovaries with varying fixation and embedding combinations all stained with Masson's trichrome stain (approx. 40×; bar size = 25 μm).

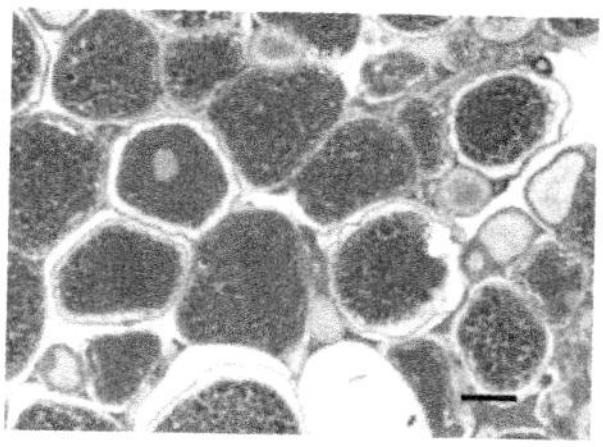

Ovaries fixed with **formalin**, embedded in **paraffin**

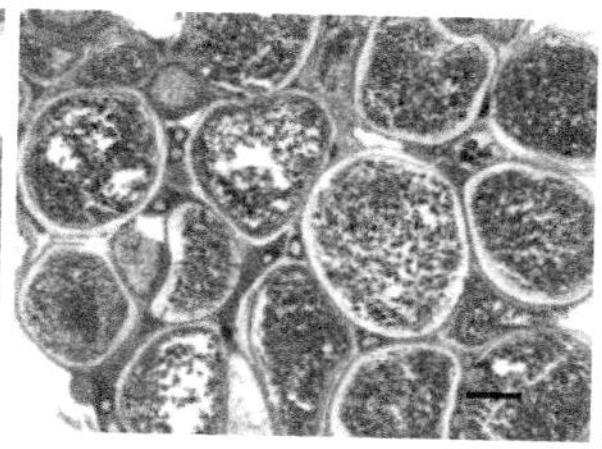

Ovaries fixed with **Bouin's**, embedded in **paraffin**

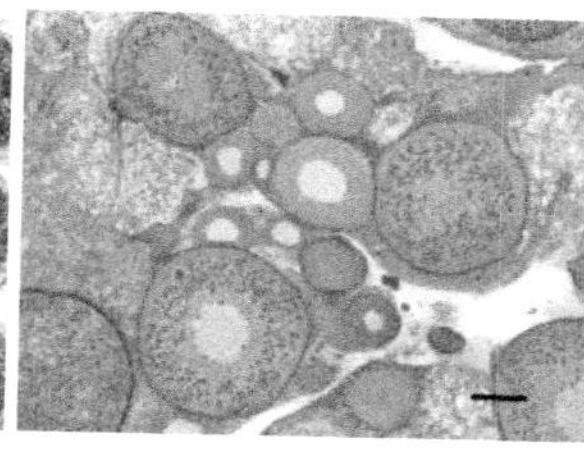

Ovaries fixed with **glutaraldehyde,** embedded in **GMA**

Masson's Trichrome Staining Example: Male Fathead Minnow

Fathead minnow testes with varying fixation and embedding combinations all stained with Masson's trichrome stain (approx. 40×; bar size = 25 μm)

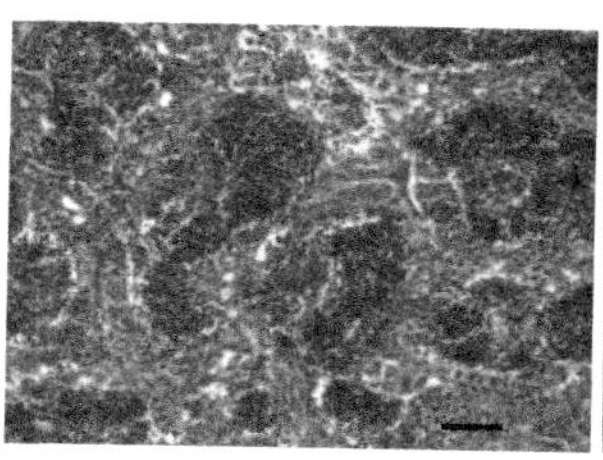

Testes fixed with **formalin**, embedded in **paraffin**

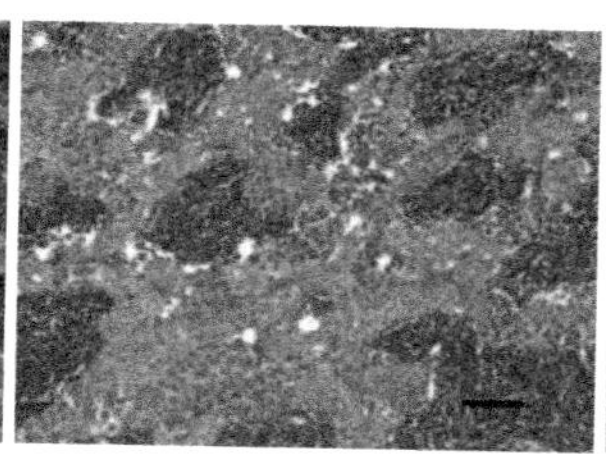

Testes fixed with **Bouin's**, embedded in **paraffin**

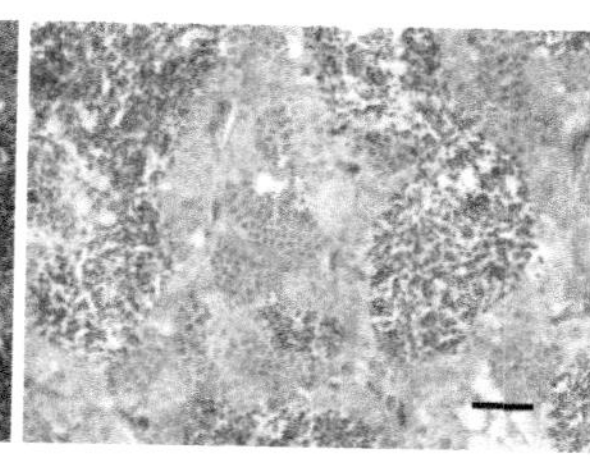

Testes fixed with **glutaraldehyde**, embedded in **GMA**

All pictures courtesy of J. Wolf, EPL Inc., Sterling, Virginia.

A.10 SUMMARY

- Optimal fixation and preservation of gonads can be achieved using the gonadal microdissection technique.
- Glutaraldehyde fixation provides excellent sample quality regarding the fixation of male gonads.
- Fixation of female gonads using Bouin's fixative (or Davidson's) results in optimal sample quality compared to the other fixation techniques and solutions.
- For qualitative staging, 3- to 5-μm sections provide sufficient information, but at this thickness, additional cell layers may lead to distortion of cell morphology.
- Accurate cell type differentiation of testes is only possible with very thin sections (i.e., 0.5 to 2 μm, reaching a monolayer of cells).

REFERENCES

1. Hopwood, D., Fixatives and fixation: A review. *The Histochemical Journal*, 1969, **1**(4): 323–360.
2. Latendresse, J.R., et al., Fixation of testes and eyes using a modified Davidson's fluid: Comparison with Bouin's fluid and conventional Davidson's fluid. *Toxicologic Pathology*, 2002, **30**(4):524–533.
3. Humason, G.L., *Animal Tissue Techniques*, 3rd ed. W.H. Freeman. San Francisco, CA, 1972.
4. Gabe, M., *Histological Techniques*. Masson, Paris, 1976, pp. 177–191.
5. Pearse, A.G., *Histochemistry: Theoretical and Applied*, 4th ed., Vol. 1. Churchill Livingstone, Edinburgh, UK, 1980, p. 101.

INDEX

Histological Analysis of Endocrine-Disruptive Effects in Small Laboratory Fish
By Daniel R. Dietrich and Heiko O. Krieger

Printed and bound by CPI Group (UK) Ltd, Croydon, CR0 4YY
16/09/2026
14963444-0001